Tributes
Volume 38

Logic, Intelligence, and Artifices
Tributes to Tarcísio H. C. Pequeno

Volume 28
Conceptual Clarifications. Tributes to Patrick Suppes (1922-2014)
Jean-Yves Béziau, Décio Krause and Jonas R. Becker Arenhart, eds.

Volume 29
Computational Models of Rationality. Essays Dedicated to Gabriele Kern-Isberner on the Occasion of her 60th Birthday
Christoph Beierle, Gerhard Brewka and Matthias Thimm, eds.

Volume 30
Liber Amicorum Alberti. A Tribute to Albert Visser
Jan van Eijck, Rosalie Iemhoff and Joost J. Joosten, eds.

Volume 31
"Shut up," he explained. Essays in Honour of Peter K. Schotch
Gillman Payette, ed.

Volume 32
From Semantics to Dialectometry. Festschrift in Honour of John Nerbonne.
Martijn Wieling, Martin Kroon, Gertjan van Noord, and Gosse Bouma eds.

Volume 33
Logic and Computation. Essays in Honour of Amílcar Sernadas
Carlos Caleiro, Fransciso Dionísio, Paula Gouveia, Paulo Mateus and João Rasga, eds.

Volume 34
Models: Concepts, Theory, Logic, Reasoning, and Semantics. Essays Dedicated to Klaus-Dieter Schewe on the Occasion of his 60th Birthday
Atif Mashkoor, Qing Wang and Bernhrd Thalheim, eds.

Volume 35
Language, Evolution and Mind. Essays in Honour of Anne Reboul
Pierre Saint-Germier, ed.

Volume 36
Logic, Philosophy of Mathematics and their History.
Essays in Honor of W. W. Tait
Erich H. Reck, ed.

Volume 37
Argumentation-based Proofs of Endearment. Essays in Honor of Guillermo R. Simari on the Occasion of his 70th Birthday
Carlos I. Chesñevar, Marcelo A. Falappa, Eduardo Fermé, Alejandro J. García, Ana G. Maguitman, Diego C. Martínez, Maria Vanina Martinez, Ricardo O. Rodríguez, and Gerardo I. Simari, eds.

Volume 38
Logic, Intelligence and Artifices. Tributes to Tarcísio H. C. Pequeno
Jean-Yves Béziau, Francicleber Ferreira, Ana Teresa Martins and Marcelino Pequeno, eds.

Tributes Series Editor
Dov Gabbay   dov.gabbay@kcl.ac.uk

Logic, Intelligence, and Artifices
Tributes to Tarcísio H. C. Pequeno

edited by

Jean-Yves Béziau

Francicleber Ferreira

Ana Tereas Martins

Marcelino Pequeno

© Individual authors and College Publications 2018. All rights reserved.

ISBN 978-1-84890-297-8

College Publications
Scientific Director: Dov Gabbay
Managing Director: Jane Spurr

http://www.collegepublications.co.uk

Cover design by Laraine Welch
Cover photograph by Márcia Farias

All rights reserved. No part of this publication may be reproduced, stored in a retrieval system or transmitted in any form, or by any means, electronic, mechanical, photocopying, recording or otherwise without prior permission, in writing, from the publisher.

CONTENTS

HUGO D. MACEDO AND EDWARD H. HAEUSLER
Yoneda's Embedding and Post-Completeness   1

SHEILA R. M. VELOSO, PAULO A. S. VELOSO, MARIO R. F. BENEVIDES, AND ISAQUE M. S. LIMA
On General Graphical Model-Checkers   13

ANA TERESA MARTINS AND FRANCICLEBER FERREIRA
A Proof Theory for the Inconsistent Default Logic   33

RICARDO S. SILVESTRE
Modality, Paraconsistency and the Concept of Inductive Plausibility   55

ITALA M. LOFFREDO D'OTTAVIANO AND ROGÉRIO J. DE R. DA SILVA JÚNIOR
Paraconsistent Modalities as a Possible Way of Treating Epistemic-Doxastic Paradoxes   71

ODINALDO RODRIGUES
An Investigation into Reduction and Direct Approaches to the Computation of Argumentation Semantics   97

VLÁDIA PINHEIRO, TARCÍSIO PEQUENO, VASCO FURTADO
A Semantic-Inferentialist Framework for Natural Language Understanding   121

ABEL LASSALLE CASANAVE AND LUIZ CARLOS PEREIRA
A Short Note on the Formalism of Johann von Neumann   141

ANDRÉ LECLERC
Context and Computation   155

MARCOS SILVA
How to Understand the Normativity of Logic in the Context of Logical Pluralism: a Pragmatist Proposal   165

CARLOS EDUARDO FISCH DE BRITO
Language, Tools and Machines   183

CANDIDA DE SOUSA MELO AND DANIEL VANDERVEKEN
On the Limits of Language, Thought and Experience   197

DANIEL VANDERVEKEN
Principia Ethica Illocutionary Acts in Deontic Logic   217

GUIDO IMAGUIRE
How and Why to be a Minimalist   247

DESIDÉRIO MURCHO
How to Get Rid of Logic and Be Happy   261

CÍCERO ANTÔNIO CAVALCANTE BARROSO
Necessity and Logic   275

MANFREDO ARAÚJO DE OLIVEIRA
Contextualism, Universal Pragmatics and Metaphysics   289

OSWALDO CHATEAUBRIAND
Sense, Reference, and Connotation   317

MATTHIAS SCHIRN
Reflections on Frege's Platonism   323

JOÃO BRANQUINHO
The Need for Indexical *Sinn*   337

DAVID MILLER
A Hoard of Hidden Assumptions   355

VERA VIDAL
Nous Sommes des Prisonniers du Discours: Theorie- Langage- Processus Cognitif dans L'Epistémologie de Quine   363

JEAN-YVES BÉZIAU
Dice: a hazardous symbol for chance?   379

List of Contributors

Cícero Antônio Cavalcante Barroso — Federal University of Ceará.
cicero@lia.ufc.br

Mario R. F. Benevides — Computing and System Engineering Program, COPPE, Federal University of Rio de Janeiro (UFRJ).
mariorfb@gmail.com

Jean-Yves Béziau — Federal University of Rio de Janeiro.
jyb@ufrj.br

João Branquinho — Faculdade de Letras. Universidade de Lisboa.
jbranquinho@campus.ul.pt

Carlos Eduardo Fisch de Brito — Federal University of Ceará.
carlos@lia.ufc.br

Abel Lassalle Casanave — Federal University of Bahia and CNPq.
abel.lassalle@gmail.com

Oswaldo Chateaubriand — Pontifical Catholic University of Rio de Janeiro and CNPq.
ochateaubriand@gmail.com

Itala M. Loffredo D'Ottaviano — Centre for Logic, Epistemology and the History of Science and Philosophy Department University of Campinas
itala@cle.unicamp.br

Francicleber M. Ferreira — Federal University of Ceará.
fran@lia.ufc.br

Vasco Furtado — Programa de Pós-Graduação em Informática Aplicada, Universidade de Fortaleza
furtado.vasco@gmail.com

Edward H. Haeusler — Departamento de Informática, PUC-Rio.
hermann@inf.puc-rio.br

Guido Imaguire — Federal University of Rio de Janeiro and CNPq.
guido_imaguire@yahoo.com

André Leclerc — Universidade de Brasília and CNPq.
andre.leclerc55@gmail.com

Isaque M. S. Lima — Computing and System Enginneering Program, COPPE , Federal University of Rio de Janeiro (UFRJ).
lima.isaque@gmail.com

Hugo D. Macedo — Department of Engineering, Aarhus University.
hdm@eng.au.dk

Ana Teresa Martins — Federal University of Ceará.
ana@lia.ufc.br

David Miller — University of Warwick.
d.w.miller@warwick.ac.uk

Desidério Murcho — Federal University of Ouro Preto.
desiderio.murcho@gmail.com

Manfredo Araújo de Oliveira — Federal University of Ceará.
manfredo.oliveira2012@gmail.com

Luiz Carlos Pereira — Pontifical Catholic University of Rio de Janeiro, State University of Rio de Janeiro and CNPq.
luiz@inf.puc-rio.br

Vládia Pinheiro — Programa de Pós-Graduação em Informática Aplicada, Universidade de Fortaleza
furtado.vasco@gmail.com

Odinaldo Rodrigues — Department of Informatics, King's College London.
odinaldo.rodrigues@kcl.ac.uk

Matthias Schirn — Munich Center for Mathematical Philosophy, University of Munich.
matthias.schirn@lrz.uni-muenchen.de

Marcos Silva — Federal University of Alagoas.
marcossilvarj@gmail.com

Rogério J. de R. da Silva Júnior — Centre for Logic, Epistemology and the History of Science and Philosophy Department University of Campinas.
endelion13@gmail.com

Ricardo S. Silvestre — Federal University of Campina Grande.
ricardoss@ufcg.edu.br

Candida de Sousa Melo — Federal University of Paraíba.
candida.jaci@gmail.com

Daniel Vanderveken — Université du Québec à Trois-Rivières and Federal University of Rio Grande do Norte.
daniel.vanderveken@gmail.com

Sheila R. M. Veloso — Computing and System Enginneering Department, Engineering Faculty, State University of Rio de Janeiro (UERJ).
sheila.murgel.veloso@gmail.com

Paulo A. S. Veloso — Computing and System Enginneering Program, COPPE, Federal University of Rio de Janeiro (UFRJ).
pasveloso@gmail.com

Vera Vidal — Philosophy Department, Federal University of Rio de Janeiro and Research Department, FIOCRUZ.
veravidal2000@hotmail.com

Preface

This volume is a homage to the scientist, philosopher and our friend Tarcísio Haroldo Cavalcante Pequeno. During the preparation of this volume, it became evident the extent to which Tarcísio has influenced the contributors that promptly joined this project. It is impressive the regard, respect and admiration our colleagues expressed to him.

Tarcísio was born in Fortaleza, the state capital of Ceará in Northeast Brazil. Under the epithet of "canelas pretas," due to the long-shaft farmer boots, his family, the Pequeno's, were spread through the cities of Icó and Crato in the country side of Ceará. His father, Haroldo Cipriano Pequeno, born in Icó, was agronomist and one of the first professors of the Faculty of Agronomy, one of the colleges that, together with the Faculty of Medicine and Faculty of Law, would latter join to compose the Federal University of Ceará under the guidance of the Antonio Martins Filho, the first rector. Haroldo taught physics, electricity, meteorology. He was responsible for the installation of several dozens of weather stations throughout Ceará, providing the first comprehensive amount of data about Ceará's climate.

Due to his interest in physics, Tarcísio joined the Federal University of Ceará in 1966 in the civil engineering course, the best place to study physics in Ceará at the time, completing the course in 1970. These were the first years of the military dictatorship that begun in 1964.

During this period, before finishing the engineering course, he started to work at IBM as an Autocoder programmer of a IBM 1401. This was the beginning of his work with computer science. In 1970, he was hired by the state phone company to run its data center, whose core was an IBM 360. In 1971 he joined the Federal University of Ceará as teaching assistant in the Physics Department. Due to his experience, he has been invited by the rector José Walter Cantídio to create and lead the university's data center and was responsible by the development of the first academic and administrative software systems.

In 1975, he joins the graduate course at the Pontifícia Universidade Católica of Rio de Janeiro (PUC-RJ). At this time, he met the professors Carlos José Pereira de Lucena, Paulo Veloso and Roberto Lins de Carvalho, who would later become his collaborators.

During the time in Rio, Tarcísio had the first contacts with logic, algebraic methods and their relationships with computer science. He worked with topics like automatic theorem proving, databases, semantics of programming languages and logical approaches to artificial intelligence. His PhD thesis was in the field of abstract data types.

At this time, the field of logic started to became mainstream due to the work or the prominent brazilian logician Newton da Costa on paraconsistent logic. In the next decades, Tarcísio would dedicate his research to the concept of rationality from a logical point of view, using paraconsistent and nonmonotonic logics as formalisms to capture rational behaviour. He was among the first (perhaps the first) to propose the use of paraconsistent logics to deal with the multiple extensions problems of default logic, which would allow merging several contradictory extensions in a single, inconsistent, but not trivial one.

In 1981, Tarcísio returns to Ceará where he founded the Department of Computer Science. In the following years, he would be responsible by the creation of the Master and PhD courses in the Computer Science Department.

Tarcísio dedicated his research to topics raging from artificial intelligence, philosophy, language and logic. His main contributions were in nonmonotonic and paraconsistent logics (the pair LEI and IDL), specially the study of negation, applications of intuitions from the game semantics to automatic theorem proving (in joint works with Hermann Haeusler), tableaux methods for paraconsistent logics (joint with Arthur Buchsbaum). It was also with Arthur that he developed a dialectic characterization of negation inspired in argumentation and refutation games. Also motivated by his interest in philosophy and the problem of demarcation of rationality, he studied the role of rule following and rule consciousness in cognition.

During the last forty years, Tarcísio oriented, inspired and collaborated with several researchers and established the field of logic as an active research area in the Federal University of Ceará. This book expresses the admiration of his colleagues and the recognition of his influence in their academic lives.

The contributions to this volume reflect the broad range of interests characteristics to Tarcísio's scientific interests, from computer science, artificial intelligence, language, logic and philosophy, broadly construed.

Hugo Macedo and Edward Hermann Haeusler's paper examines the different definitions of completeness in logic and particularly investigates Post completeness, relating it to Yoneda Lemma in category theory.

Paulo Veloso, Sheila Veloso, Mario Benevides and Isaque Lima's paper introduces a general graphical notation for the specification of model-checkers for specific logics.

Ana Teresa Martins and Francicleber Ferreira's paper briefly surveys the work on the LEI and IDL logics created by Tarcísio and further developed by the group of logic and Artificial Intelligence of the Federal University of Ceará. The paper also presents a Gentzen style proof calculus for IDL and investigates its proof theory.

Ricardo Silvestre's paper gives a philosophical justification for Tarcísio's axiomatization of the concept of plasibility, a central notion in Tarcísio's paraconsistent logic LEI.

Itala M. Loffredo D'Ottaviano and Rogério J. de R. da Silva Júnior's paper attacks Fitch's paradox in alethic-epistemic and alethic-doxastics logics and proposes a paraconsistent approach to deal with the problem.

Odinaldo Rodrigue's paper is dedicated to the comparison of SAT-based algorithms to the problem of preferred extension selection in abstract argumentation semantics.

Vládia Pinheiro, T. Pequeno and Vasco Furtado's paper presents an application of a pragmatist approach to natural language processing based on Robert Brandom's inferentialism.

Abel Lassalle Casanave and Luiz Carlos Pereira recall von Neumann's contribution to the proposals for the foundations of mathematics in the first half of 20th century, specially the formalist approach, that arouse as a response to the foundational crisis.

André Leclerc defends a contextualist approach to our linguistic practices and puts it in the context of the problem of manipulation of natural language by computers, one of the most important problems in Artificial Intelligence.

Marcos Silva's paper explores how Robert Brandom's logical expressivism can be used to support logical pluralism, putting forward the view that logical necessity is a kind of moral obligation.

Candida de Souza Melo and Daniel Vanderveken's work explores the limits of language and how those limits shape thought and mind.

Daniel Vanderveken presents a critique of standard deontic logic and lays the foundation of his *Principia Ethica*, which is intended to be a more suitable alternative and solve well known paradoxes.

Guido Imaguire's work defends a minimalist approach that amounts to a non-inflationist attitude towards ontology. He examines some cases to compare the minimalist and non-minimalist approaches and argues in favor of the superiority of the minimalist approach.

Also surfing the minimalist wave, Desidério Murcho argues in favor of eliminating modalities to explain logical truth.

Cícero Barroso's paper examines the contrasts between two views of logic: one strongly based on a metaphysical point of view, as advertised by Oswaldo Chateaubriand, and that defended by Tarcísio Pequeno, heavily founded on pragmatists grounds.

Manfredo Araújo de Oliveira explores Lima Vaz's metaphysical theory of truth and confronts it with two of the most prominent views of truth in contemporary philosophy.

Oswaldo Chateaubriand describes his account of meaning which derives from his view of Frege's theory of sense and Kripke's theory of reference combined with a descriptivist account of connotation.

Matthias Schirn's piece exposes and criticizes Frege's platonism as it appear in his *Grundlagen* and *Grundgesetze Der Arithmetik*.

João Branquinho argues against the claim that neo-fregean semantics for indexicals are notational variants of Millian's semantics (the Millian Notational Variance Claim).

David Miller's paper investigates the problem of the existence and multiplicity of finding minimal hidden assumptions that could be added to the premisses of an invalid argument to turn it into a valid one. He examines to what extent the so called *negative test* can accomplish this task.

Vera Vidal's paper examines the connections among theory, language and cognitive process in Quine's epistemology.

Jean-Yves Béziau closes this homage to Tarcísio with a philosophical account of *chance* and an analysis of its symbolic representation by dice throwing.

Last but not least, the editors would like to thank our colleagues that helped us serving as referees for the contributions: Carlos Brito, Guido Imaguire, André Leclerc, Luiz Carlos Pereira, Samy Sá, Marcos Silva, Paulo Veloso. We also would like to thank Robertty Freitas for the help with LaTeX and typesetting.

The Editors.

Yoneda's Embedding and Post-Completeness

HUGO D. MACEDO AND EDWARD H. HAEUSLER

1 Introduction

The motivation of any logic endeavor is the quest for veracity[1] of statements as: "$\sqrt{-1}$ is an imaginary number". Such veracity is established by classification of statements. To classify a statement one uses syntactical and semantical approaches (two sides of a coin). In the syntactical approach an artifact is used to attribute the same classification of already classified statements to an unclassified statement. In the semantical counterpart the statement is evaluated, i.e. it is mapped into an object of a model which becomes its truth value, thus classifying statements according to their truth values.

The syntactical tradition in logic is older than its semantical counterpart. Since the works of Aristotle, circa 400 BC, the logical deductive apparatus[2] has been extensively discussed and developed. In contrast, semantic investigations are rarely reported until the middle of XIX century, and the semantic analysis of logic is a fairly recent discussion. Although Aristotle himself started some seminal discussions on semantics in his *De Interpretatione*, such problematic did not last longer than a few centuries after. The first semantic aspects of modern logic seems to appear only in the XIX century. For instance truth tables, a cornerstone concept in logical semantics, became popular during the last years of the XIX century and its very hard to precisely name its first creator.

Nowadays, it seems clear that deductive systems derivation and truth-validation should be two sides of the same coin. For instance, to achieve the Poincaré principle in proof assistants, i.e. in between deduction steps, semantical evaluation can be used, to simplify the proof statements. Such coin with syntax and semantics on its sides spawns a meta-theoretical coin with soundness in one side and completeness on the other, which

---
Copyright © 2019 by Hugo D. Macedo and Edward H. Haeusler. All rights reserved.

[1] We choose this alternative to truth as the concept is connected to semantics.
[2] We remember that the word syllogism συλλογισμος means "inference".

is the result of reasoning at the meta level about syntax and semantics. For, given a syntactical artifact and a semantical evaluator for the same logic two questions appear: Is a syntactically derived statement also true in the model? (Soundness). And, conversely, can every semantical truth be derived syntactically? (Completeness). Such were among the meta-theoretical results that were studied by Hilbert, Bernays, Gödel, and made eternal in Post's thesis [5, 8, 7, 10, 1].

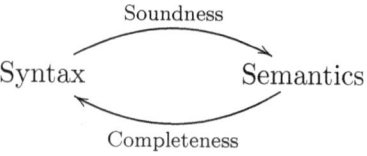

Concerning classical propositional logic, Post, in 1921, connected syntactical completeness (consistency) to functional truth-validation by showing that any Classical propositional logic sound deductive system is complete, iff, any proper consistent extension of it proves falsities. It is hard to say that this is the first completeness result ever. In fact it is stated in [1] that Bernays presents a proof of semantical completeness in 1918, before Post. It is even harder to say that it is a completeness result, since it was originally stated in a cumbersome manner.

The less obscure statement we follow in this work is due to Hilbert and states that if such logic was not complete then one could provide a substitution function that transforms a non-tautological statement into a tautology, deriving an inconsistent system. In this paper we show that the existence of such substitution may be constructed from a meta-mathematical result, the Yoneda lemma, but for that one needs to explore the naturality of inference rules.

Inference rules are designed in such a way that they are naturally used in many contexts. For instance, an inference rule such as modus ponens (when $\vdash A$ and $\vdash A \supset B$ hold infer $\vdash B$) can be used for any replacement of A and B by arbitrary formulas. Such context naturality is usually regarded as a main aspect of deductive systems and must be dealt with by the logic framework either by introducing machinery, e.g. substitutions functions, or by design of the formalism. In this paper we chose to use the framework of category theory (CT), a formalism where such naturality can be neatly achieved by design.

In category theory, naturality is precisely formalized by the concept of Natural Transformations. In fact the creators of CT state, in [3], that CT was created in order to mathematically define what is natural, at least in a formal system. In CT, mathematical concepts are framed and studied in the form of objects and morphisms (transformations) between objects. In particular, in the category where objects are the categories themselves and the morphisms between the categories are mappings preserving the categorical structure. Such structure preserving morphisms between categories are called functors in CT, and functors are the objects of the special structure preserving morphisms called natural transformations.

To show Hilbert's proof of Post-completeness is an instance of the Yoneda lemma we define a category **Form** where objects are the logical propositional formulas and a morphism embeds a formula into another. In **Form**, we are able to reframe logical formula semantical evaluation as a functor between the category of **Form** and the category of sets, written as **Set**. We also reframe formula substitution (and context naturality) as natural transformations. And after such reframing exercise, we are in conditions to establish the desired relation between the lemma and the proof.

The last step in establishing the result involves applying the lemma to the constructions of the **Form** framework. In such setting, the lemma establishes a bijection between the functor defining the logical semantical evaluation (the truth value producing functional) and the natural transformations providing the logical formula substitutions (inference rules). Thus obtaining a connection/duality between truth value and substitution, which in turn is the reasoning leading to the proof devised by Hilbert.

The remainder of the paper is organized as follows: We define rigorously the logic background concepts e.g. formula, derivation, substitution in Section 2. We discuss the different concepts of semantical completeness in Section 3. In Section 4 we define a categorical approach to propositional formulas. The Yoneda Lemma is introduced in Section 5. Section 6 contains the details on how to relate Hilbert's proof and the Yoneda lemma. And in Section 7 we finish our exposition with some concluding remarks.

2 Terminology and theoretical background

In this section we formalize, in the tradition of logic, the notions of formula, substitution, "syntactic" and "semantical" approach, and other concepts of the introduction that will be needed to formally define Post completeness.

Propositional formulas. We define the propositional language, $\mathcal{L}$, the set of formulas, of a propositional system as the set generated by recursion:

$$\varphi, \psi = \bot \mid p_i \mid \varphi \cup \psi \mid \varphi \cap \psi \mid \varphi \supset \psi \mid \neg \varphi,$$

Tautologies	Contingent	Contradictions
$A \supset A$	p_i	$A \cap \neg A$

Table 1. Partition of propositional classical formulas into three blocks

for p_i as an abbreviation of $p_1 \mid p_2 \mid \ldots$ where $i \in \mathbb{N}$ corresponding to propositional symbols.

Substitution. Given $\varphi \in \mathcal{L}$, and a mapping $\eta : Var \rightarrow \mathcal{L}$, we define the formula $\varphi[\eta]$ as the formula φ where every occurrence of the propositional variable x is replaced by $\eta(x)$.

Derivation rule. A derivation rule R produces a result (conclusion) formula φ from zero or more premises $\psi_1, \ldots, \psi_n$.

Propositional system. A propositional system is a tuple $(\mathcal{L}, \mathcal{R})$, where $\mathcal{L}$ is the set of formulas and $\mathcal{R}$ is a set of derivation rules.

Entailment system. In a system $(\mathcal{L}, \mathcal{R})$, $\Gamma \vdash \varphi$ denotes a relation where the formula φ is a consequence of the set of formulas Γ by a chain of rules of $\mathcal{R}$.

Semantics. A valuation is a mapping, $v : \mathcal{L} \rightarrow \{0, 1\}$, following the standard propositional logic interpretation rules (see p.18 of [2]).

Tautology. A formula φ is a tautology iff for every valuation v we conclude $v(\varphi) = 1$ and write $\models \varphi$.

Semantical consequence. We write $\Gamma \models \varphi$ iff for all valuation v, such that $v(\psi) = 1$ for every $\psi \in \Gamma$, then $v(\varphi) = 1$.

Theory. A logical theory is a set of formulas Γ s.t. if $\Gamma \vdash \varphi$, then $\varphi \in \Gamma$.

Consistency. A set of formulas Γ is consistent iff there is no φ such that $\Gamma \vdash \varphi$ and $\Gamma \vdash \neg\varphi$. The definition of a consistent theory is likewise.

Propositional logics. There are several systems of propositional logic depending on choice of formulas that characterize the particular theory. Such systems share the same language of formulas, but the partition of formulas into veracity classes is different. To make the partitioning notion clear, in Table 1 we depict the typical partition of formulas into the classes of formulas that are true for every valuation v of the model (Tautologies), formulas which can be true or false depending on the valuation chosen (Contingent), and formulas that are false for every valuation v (Contradictions).

For instance in classical logic $P \cup \neg P$ is part of the tautology class, which is not the case in intuitionistic logic. In Table 2 we present some propositional logic systems which are totally (linearly) ordered by the inclusions: Implication ⊂ Intuitionistic ⊂ Classical.

Logic	Rules
Implication	$N_\rightarrow$
Intuitionistic	NJ
Classical	NK

Table 2. Example propositional logic systems

In the following, we show how Post completeness allows to easily argue about the completeness of (at least some of) such logical systems.

3 Defining completeness

There are several meanings of completeness with respect to logical systems. Some of them are standard, others less known. For instance, Post himself proved several results of completeness besides the one we take as Post completeness. One such results is the folk functional completeness theorem [4], where a subset of logic connectors is proved to be enough to express all the other connectors.

In this section we provide the definition of Post completeness and we relate it to the popular definition (the one a student finds in a introductory logic course). The statement of the two definitions is displayed in Table 3, but to relate the two one needs to go beyond a pure statement. For that let us focus on the first definition that comes to our minds when one talks about completeness.

Semantical completeness. The most popular definition goes as follows: "All tautologies are syntactically derived" stating that every true statement in regarded to any model/valuation should appear as the conclusion of some syntactical derivation.

$$\models \varphi \text{ then } \vdash \varphi$$

We call such definition "Semantic completeness". Sometimes completeness induces a wrong intuition that a logic with a semantically complete system is decidable.

We now proceed to introduce Post completeness which goes beyond the coin syntax/semantics. Maybe because of that Post first introduces the notion using the word "closed":

> A system is "closed" if the addition of an unprovable formula makes all formulas provable

Variant	Description
Semantic completeness	Semantical truths (tautologies) are syntactically derived
Post Completeness	One cannot extend the logic without introducing inconsistency

Table 3. Semantical vs Post completeness definitions

We hope we can convey that Post completeness characterizes expressiveness of the logic system, beyond the syntax/semantics coin.

Post completeness. A logic L is Post complete iff L is consistent and L has no consistent proper extension. There are several logics that are Post complete and there is the concept of Post complete extensions. For instance, while the classical version of propositional logic is Post complete, the intuitionistic counterpart is not, because one can consistently properly extend intuitionistic logic by introducing the excluded middle axiom. For the case of Predicate logic, add as an axiom that our domain of discourse contains at most two elements. With this extra axiom the logic is extended without breaking its consistency. There are also more refined results, for instance in [6] it is shown why the modal logic S3 is incomplete by proving that it admits non-denumerable Post complete extensions.

Logic	Post Complete	Non-trivial extension contains
Minimal Implication	No	$P \supset (\neg P \supset Q)$
Intuitionistic Propositional	No	$P \cup \neg P$
Classical Propositional	Yes	-
Classical Predicate	No	$\forall y\, \exists x_1, x_2 : y = x_1 \lor y = x_2$

Table 4. Post completeness classification of some logics

The main difference between semantical completeness and Post completeness resides in the fact that the latter goes beyond providing a correspondence between semantics and syntax. It characterizes such correspondence in terms of the content of the axiomatic logical system (the theory) to argue about completeness itself. In Table 4 we present how the usage of Post completeness provides an argument on the completeness of traditional logical systems, including the ones referred in Table 2.

In our work we will start by focusing on Post completeness of Propositional Classical Logic, in fact the objective of Post was to prove the completeness of Principia Mathematica Propositional logic.

Hilbert's formulation. A logic L is Post complete iff L is consistent and it has no consistent proper extension. For classical logic ($\mathbb{CL}$) such property can be restated as follows (cf. Hilbert):

$$\forall \varphi \notin \mathbb{CL} \text{ then there } \exists \text{ a substitution } s \text{ such that } (\varphi[s] \supset p) \in \mathbb{CL}$$
$$\text{for } p \text{ atomic not occurring in } \varphi. \quad (1)$$

The point made in such restatement is that the addition of a formula φ that is not a tautology introduces the possibility of devising an artifact s that transforms φ into an artifact ($\varphi[s]$) from which any arbitrary literal p is implied.

With such formulation Hilbert provides a non-constructive proof of the completeness of Classical logic, thus deviating from being forced to provide a translation between truth in the semantic part and derivations in the syntactic part. This works in Classical logic because it is in some sense a maximal logic among propositional logic systems.

4 Putting propositional logics in categorical terms

The main contribution of our work is to relate Post completeness to the Yoneda Lemma. In this section we give an overview of the framework and how the result follows from it. Traditionally mathematical structures can be translated into category theory in several ways, for instance the category of monoids and the monoid category [3].

In our work this is still the same, we can use category theory to model the compositional aspect of formula construction, where $\varphi = \mathbb{F}(q)$ for some functor $\mathbb{F}$ and some propositional formula q. Or we can use category theory as a model of logics using the traditional Curry-Howard-Kolmogorov-Lambek correspondence. In the latter, we can see morphisms as a consequence relation and thus Yoneda lemma entails a metatheorem proved in intuitionistic logic.

Defining propositional formulas as subformulas. It is usual to syntactically define a formula φ as a function depending on all of its propositional variables. We extend such notion by making φ depend on each of its subformulas. For example, the formula $P \supset (Q \vee (P \supset R))$ depends on $P, R, Q, P \supset R$ and $Q \vee (P \supset R)$, besides depending on itself. We show below that this definition is convenient to capture syntactical substitution using well-known categorical concepts.

Any formula has a core that is invariant by substitution of variables by arbitrary formulas. The core of a formula does not change under substitutions of its propositional letters by other propositional formulas. This view of formulas is in accordance with our way of formalizing the relationship between Yoneda's lemma and Post completeness.

Given a formula φ, the set $[\![\varphi]\!] = \{\psi \mid \psi \text{ is subformula of } \varphi\}$, of its subformulas, may be identified with the formula ϕ itself.

A category of formulas. Let **Form** be the category where objects are $[\![\varphi]\!]$ and morphisms are provided by the set-theoretical inclusion, that is a morphism $A \to B$ means A is a subformula of B. Sometimes we use a formula φ itself in order to denote the set:

$$\{\psi : \exists \sigma, \text{ such that, } \sigma(\psi) \text{ is subformula of } \varphi\}^3.$$

The objects of **Form** are our first account for formulas.

A property of the category **Form** is that given a formula φ, the functor $Hom_{\textbf{Form}}(-, [\![\varphi]\!]) : \textbf{Form}^{op} \to \textbf{Set}$ can be identified with the formula φ.

As an example formula $P \supset (Q \vee (P \supset R))$ will be depicted categorically as:

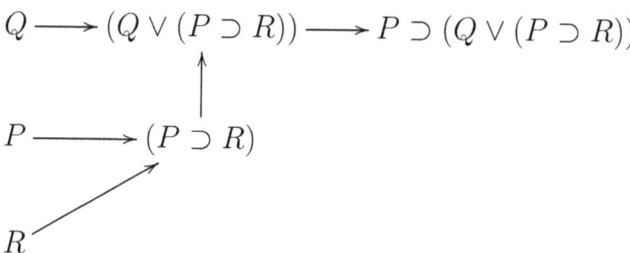

Formulas as functors. This identification of formulas with the Hom functor is quite convenient. When we fix the [co]-variant argument of it we obtain a family of functors $Hom_{\textbf{Form}}(X, -)$, one for each $X \in obj(\textbf{Form})$. We write $\varphi(X)$ as an abbreviate form for this functor. Precisely, the substitution of Y for X in φ can be seen as the φ component of the natural transformation between $Hom_{\textbf{Form}}(X, -)$ and $Hom_{\textbf{Form}}(Y, -)$, as shown below.

A natural transformation σ from $Hom(X, -)$ into $Hom(Y, -)$ is a collection of mappings σ_φ, $[\![\varphi]\!] \in obj(\textbf{Form})$, mapping each $\varphi(X)$ to $\varphi(Y)$. Notice that X and Y are arbitrary objects of **Form**, and, since Hom is a functor, this substitution operation happens as if X were atomic anyway.

Substitution as natural transformations $[Y/X]$. Given any formula φ, syntactical substitutions of Y for X in φ are represented by the component φ of the natural transformation between $Hom(X, \varphi)$ and $Hom(Y, \varphi)$.

---

[3] This could be made precise by an adjoint situation between a category of concrete formulas and **Form**, but this is useless for our present purpose.

$$\begin{array}{ccc}
\varphi & Hom(X,\varphi) \xrightarrow{\eta_\varphi} Hom(Y,\varphi) & \quad (2)\\
\downarrow h & \downarrow{\scriptstyle Hom(X,h)} \quad \downarrow{\scriptstyle Hom(Y,h)} & \\
\psi & Hom(X,\psi) \xrightarrow{\eta_\psi} Hom(Y,\psi) &
\end{array}$$

The diagram establishes the folk law for substitutions given that $\psi = h(\varphi)$:

$$\eta_\psi \cdot Hom(X,h) = Hom(Y,h) \cdot \eta_\varphi$$

In the above definition, $Hom_{\mathbf{Form}}(X,\varphi)$ can be identified with the formula $\varphi(X)$, that is $Hom_{\mathbf{Form}}(X,\varphi)$, whenever X occurs in $\varphi(X)$. In fact the substitution components maps $\varphi(X)$ into $\varphi(Y)$ as formulas. However, it is important to emphasize that the object $\varphi \in Obj(\mathbf{Form})$ is already representing both $\varphi(X)$ and $\varphi(Y)$.

5 Yoneda Lemma

In this section we provide the fragment of category theory we need to provide our categorical view on Post completeness. The result we use is a lemma that relates two fundamental concepts of the theory, namely functors and natural transformations.

The contravariant Yoneda Lemma. Let C be a locally small category, that is a category where hom-sets are proper sets. Consider $\mathbb{F}$ a contravariant functor from C into **Set**, and A an object in C. By the Yoneda lemma there is a bijection $\mathcal{Y}$,

$$\mathcal{Y}: Nat(Hom(-,A), \mathbb{F}) \approx \mathbb{F}(A), \quad (3)$$

sending each natural transformation $\eta : Hom(-,A) \to \mathbb{F}$ to $\eta_A\, id_A$ the image of the identity $A \to A$.

The lemma expresses that the set resulting of applying to the object A a 2-level morphism $\mathbb{F}$ from the category C into the category of sets (**Set**) is in bijection with the set of possible 3-level morphisms from arrows with A as codomain to the resulting set of applying $\mathbb{F}$ to the departure object. Therefore the set of natural transformations:

$$\begin{array}{ccc}
Y & Hom(X,A) \xrightarrow{\eta_X} \mathbb{F}X \\
\downarrow s & \downarrow{\scriptstyle Hom(s,A)} \quad \downarrow{\scriptstyle \mathbb{F}s} \\
X & Hom(Y,A) \xrightarrow{\eta_Y} \mathbb{F}Y
\end{array}$$

is in bijection with $\mathbb{F}(A)$.

The Yoneda lemma in general establishes that it is equivalent to study the object of a category A or to study morphisms on the category of sets and functions.

In our setting we have that when X is a subformula of Y, applying the function that transforms functions that build A from Y into the value of $\mathbb{F}(Y)$ after building formula Y is equal to applying the function that transforms functions that build A from X into the value of $\mathbb{F}(X)$ and then functorially build Y from X.

The possible results of $\mathbb{F}A$ are in bijection with the set of functions that transform subformulas of A into the result by F of the subformula in a structural way!

Yoneda embedding. A corollary of the lemma arises when we make a particular choice of functor: $\mathbb{F} = Hom(-, B)$. In such setting the lemma states:

$$\mathcal{Y} : Nat(Hom(-, A), Hom(-, B)) \approx Hom(A, B). \tag{4}$$

Note that applying the Yoneda lemma and embedding to the **Form** category we obtain for every substitution $[Y/X]$ a natural transformation in the shape of the diagram depicted in (2).

6 Relating Yoneda embedding to Post completeness

Using the category **Form** we build a contravariant functor from **Form** into **Set**. If we consider the well-known Boolean connectives forming formulas in **Form**, and, we add a semantical layer $\mathbb{F}$, such that $\mathbb{F} : \mathbf{Form}^{op} \longrightarrow \mathbf{Set}$ is the evaluation functor from formulas to truth values. In such setting, each truth assignment function $\sigma_\varphi : Hom(X, \varphi) \longrightarrow \mathbb{F}(X)$ associating φ to its truth value is a component of a natural transformation.

Rewriting the Yoneda lemma in the scope of the functor just defined we obtain:

$$\mathcal{Y} : Nat(Hom(-, \varphi), \mathbb{F}) \approx \mathbb{F}(\varphi), \tag{5}$$

an isomorphism between the possible evaluations of φ and the natural transformations between subformulas X of φ and the corresponding truth value. Such reading of the Yoneda just states that the truth values for a formula are in bijection with the various ways of transforming of its sub-formulas into the truth values of its sub-formulas.

Regarding Post completeness, Hilbert observed that if φ is not a tautology then there is a functor $\mathbb{F}$, such that $\mathbb{F}(\varphi) = \mathbb{F}(p)$ for a propositional variable p not in $[\![\varphi]\!]$. Using (5), and given that $\mathbb{F}(\varphi)$ is not empty, there is at least a natural transformation mapping to each subformula X of φ its truth $\mathbb{F}(X)$.

The substitution s is then derived component-wise for each subformula X, and by instantiating the natural transformation diagram we obtain for every atomic propositional variable q in the roots of φ, thus $q \to X$, and by naturality of η_X the following diagram commutes:

$$\begin{array}{ccc} q & Hom(X,\varphi) \xrightarrow{\eta_X} & \mathbb{F}X \\ {\scriptstyle s_X}\downarrow & {\scriptstyle Hom(s_X,\varphi)}\downarrow & \downarrow{\scriptstyle \mathbb{F}s_X} \\ X & Hom(q,\varphi) \xrightarrow{\eta_q} & \mathbb{F}q \end{array}$$

Thus, the inspection of the commutation entails whenever the truth value of φ can be put in terms of the truth value q, that is, the composition $\mathbb{F}s_X \cdot \eta_X$ is valid due to a non-emptiness of $\mathbb{F}(X)$, then there is a substitution $Hom(s_X, \varphi)$ arriving at the same $\mathbb{F}(q)$ and therefore $\varphi[s] \supset p$.

Notice in case $\mathbb{F}(\varphi)$ is not transformable into $\mathbb{F}(q)$, because either φ is a tautology or a contradiction, the result $\varphi[s] \supset p$ holds trivially. In the case of a contradiction apply the principle of explosion. In the case φ is a tautology its truth value is independent of its atomic subformulas truth values, therefore it is not possible to map $\mathbb{F}(\varphi)$ into $\mathbb{F}(q)$.

7 Conclusion

In this article we survey two well known definitions of logical system completeness. Beyond standard logical completeness, Post completeness relates inferential aspects and truth-functional aspects in a strong way, i.e. in addition to semantical completeness a Post completeness system is also maximal for some logical system ordering.

We started this work as an exercise on the extraction of a substitution operator from the Yoneda lemma, and ended up realizing that the Yoneda lemma also entails a meta-theorem for logics concerning the maximality of tautologies which may be a starting point for future work.

We show how the substitution appearing in Hilbert's proof of Post completeness for the classical propositional logic is a consequence of the bijection coming from the Yoneda lemma which puts the Functors into **Set**, defining semantical evaluation, into correspondence with the natural transformations providing substitutions in a category of propositional formulas.

In future revisions and additions to this work, the bijection between $Nat(Hom(-,\varphi), \mathbb{F}) \approx \mathbb{F}(\varphi)$ must provide the substitution s in a completely constructive way, making possible a comparison with the procedure found in textbooks that prove Post completeness. Also, we would like to continue the exploration our categorical framework in a more general metalogical framework, whether the syntactical completeness [10, 9] is naturally related to truth-validation in broader logical systems.

References

[1] W. A. Carnielli. "Paul Bernays and the eve of non-standard models in logic". In: *Universal logic: an anthology* (2012).

[2] D. van Dalen. *Logic and structure (3. ed.)* Universitext. Springer, 1994. ISBN: 978-3-540-57839-0.

[3] S. Mac Lane. *Categories for the working mathematician.* Springer, 1998.

[4] F. Pelletier and N. Martin. "Post's functional completeness theorem". In: *Notre Dame Journal of Formal Logic* 31.3 (1990), pp. 462–475.

[5] E. L. Post. "Introduction to a General Theory of Elementary Propositions". In: *American Journal of Mathematics* 43.3 (1921), pp. 163–185. ISSN: 00029327. URL: http://www.jstor.org/stable/2370324.

[6] K. Segerberg. "Post Completeness in Modal Logic". English. In: *The Journal of Symbolic Logic* 37.4 (1972), pp. 711–715. ISSN: 00224812. URL: http://www.jstor.org/stable/2272418.

[7] W. Sieg. "Hilbert's programs: 1917–1922". In: *Bulletin of Symbolic Logic* 5.01 (1999), pp. 1–44.

[8] T. Skura. *Post Completeness in Multiple-Conclusion Logic.* To appear in Logica Universalis. 2016.

[9] H. C. Wasserman et al. "Admissible rules, derivable rules, and extendible logistic systems". In: *Notre Dame Journal of Formal Logic* 15.2 (1974), pp. 265–278.

[10] R. Zach. "Completeness before Post: Bernays, Hilbert, and the development of propositional logic". In: *Bulletin of Symbolic Logic* (1999), pp. 331–366.

On General Graphical Model-Checkers

Sheila R. M. Veloso, Paulo A. S. Veloso, Mario R. F. Benevides, and Isaque M. S. Lima

1 Introduction

We introduce a general graphical approach to specification and construction of model-checkers. This graphical approach gives general model-checkers that can be instantiated to model-checkers for specific logics.

A model-checker for a logic receives (representations of) a finite model $\mathfrak{M}$, a formula φ and state a of $\mathfrak{M}$ and decides whether model $\mathfrak{M}$ satisfies formula φ at state a (see, e. g. [2]).

The graphical approach gives general model-checkers that can be instantiated to model-checkers for a specific logic. This graphical approach provides two components: a logic-dependent converter and a general model checker; the former receives a formula φ and converts it to a graphical expression E (see Fig. 1, p. 13), whereas the latter, upon receiving expression E, a model $\mathfrak{M}$ and a state a of $\mathfrak{M}$, decides whether $\mathfrak{M}$ satisfies E at a (see Fig. 2, p. 14).

Figure 1. Graphical converter Cnv

Formula φ ⟶ [Cnv] ⟶ Graphical expression E
↑
Logic Lg

The user will be concerned only with the graphical description of the semantics of the specific logic (see Fig. 3, p. 14). The converter will eliminate the symbols of the formula according to this description. This approach has crucial issues concerning formulation of general rules and handling expressions; graphical concepts are very convenient for them.

Copyright © 2019 by Sheila R. M. Veloso, Paulo A. S. Veloso, Mario R. F. Benevides, and Isaque M. S. Lima. All rights reserved.

Figure 2. General model-checker GMC

Graphical expression E → [GMC] → Boolean sort **Bln**

with State a ↓ going into GMC and Model $\mathfrak{M}$ ↑ coming up into GMC.

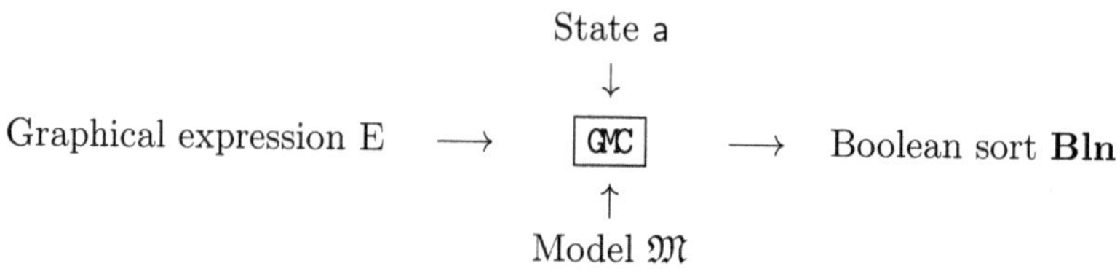

Figure 3. General model-checker GMC with converter Cnv

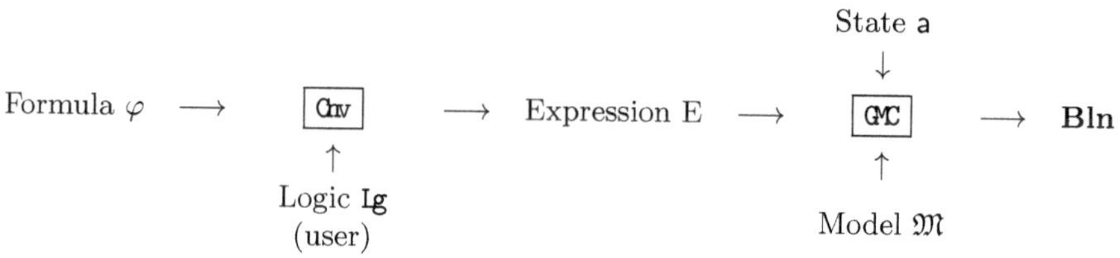

Formula φ → [Cnv] → Expression E → [GMC] → **Bln**

with Logic Lg (user) ↑ feeding Cnv, State a ↓ feeding GMC, and Model $\mathfrak{M}$ ↑ feeding GMC.

The structure of this paper is as follows. In the remainder of this section, we will recall some concepts of modal logics, to fix notation. Sct. 2 introduces basic ideas of our graphical approach: graphical concepts and constructions. In Sct. 3, we examine graphical formulations of semantics. In Sct. 4, we introduce our general graphical converter and model-checker and examine their instances. Sct. 5 presents some concluding remarks about our approach.

In this sequel, we will consider modal languages, with a set PL of *propositional letters*, each one characterized by its 0-ary, 1-ary and 2-ary connectives (†, ∇ and †) and modalities ($\mu \in \Xi$).[1] Such a language **M** has set Φ of *formulas* generated by the following grammar:

$$\varphi ::= p \mid \dagger \mid \nabla \varphi \mid \varphi' \bullet \varphi'' \mid \mu \varphi \qquad (p \in PL) \qquad (1)$$

A *model* $\mathfrak{M}$, over universe $M \neq \emptyset$, assigns subsets and relations as follows: a subset $\varphi^{\mathfrak{M}}$ of M, to each formula $\varphi \in \Phi$, and a 2-ary relation $\mu^{\mathfrak{M}}$ on M, to each modality μ. A model $\mathfrak{M}$ for a simple logic is often characterized by the subsets $p^{\mathfrak{M}}$, for $p \in PL$, and the relations $r^{\mathfrak{M}}$, for each relation name $r \in RN$ (see Examples 2 and 3 on classical and intuitionistic modal models, p. 17).

---

[1] We will mention richer languages in Sct. 5 (in connection with extensions).

2 Graphical Concepts and Constructions

We now introduce some basic ideas involved in our graphical approach: graphical concepts (p. 15) and constructions (p. 16).[2]

Our graphical objects involve nodes and arcs. Arcs may be unary or binary (represented differently for better visualization). A 1-ary arc is meant to capture the fact that a formula holds at a state; we represent that formula φ holds at node w by a dashed line from w to φ: w - - -⊰φ . A 2-ary arc stands for accessibility between states; we represent that node v is accessible from node u by the relation of t by a solid arrow labelled t from u to v: u $\xrightarrow{t}$ v . (See also Fig. 4: Arcs unary and binary, p. 15.)

We now introduce some concepts: draft, page, book and expression.

(Δ) A *draft* consists of finite sets N, of nodes, and A, of arcs. An example of a 3-node draft is Δ = p⊱ - - - u $\xrightarrow{r}$ v $\xrightarrow{s}$ w - - -⊰q .

(P) A *page* consists of an underlying draft together with a link node (marked ̂). E. g. P = p⊱ - - - $\hat{u}$ $\xrightarrow{r}$ v $\xrightarrow{s}$ w - - -⊰q is a page with link u and underlying draft Δ above.

(B) A *book* is a finite set of (alternative) pages. Examples of books are { P, $\hat{z}$ } and the empty book { }, with no page.

(E) The *expressions* are the formulas, the pages, the books and their complements (represented by an overbar); they will represent sets of states. We also allow 1-ary arcs with expressions: w - - -⊰E .

Figure 4. Arcs unary and binary

1-ary   w - - -⊰E   2-ary   u $\xrightarrow{\mu}$ v

Graphical semantics is as follows; for a given model $\mathfrak{M}$, over universe M.

(E) *Set of expression*: for a formula $\varphi \in \Phi$, $(\varphi)_\mathfrak{M} := \varphi^\mathfrak{M}$; if expression E is a page or a book, then $(E)_\mathfrak{M} := [E]_\mathfrak{M}$ (see below); for complementation: $(\overline{E})_\mathfrak{M} := M \setminus (E)_\mathfrak{M}$.

(⊩) *Satisfaction* under assignment $g : N \to M$ ($w \in N \mapsto w^g \in M$).

---

[2] For more details about the graphical approach see, e. g. [3, 6, 7].

(a) Arcs: $g \Vdash_{\mathfrak{M}} w \dashv E$ iff $w^g \in (E)_{\mathfrak{M}}$ and
$g \Vdash_{\mathfrak{M}} u \xrightarrow{\mu} v$ iff $(u^g, v^g) \in \mu^{\mathfrak{M}}$.

(Δ) Draft: $g \Vdash_{\mathfrak{M}} \Delta$ iff $g \Vdash_{\mathfrak{M}} a$, for every arc a of Δ.

(Pg) The *behaviour of page* P, with link node w and underling draft $\underline{P}$, is the set of all values $w^g \in M$ for the assignments g satisfying $\underline{P}$, namely: $[P]_{\mathfrak{M}} := \{w^g \in M \,/\, g \Vdash_{\mathfrak{M}} \underline{P}\}$.

(Bk) *Behaviour of book*: $[B]_{\mathfrak{M}} := \bigcup_{P \in B} [P]_{\mathfrak{M}}$.

($\equiv$) Expressions E and F are *equivalent* in a class K of models iff, for every model $\mathfrak{M} \in K$: $(E)_{\mathfrak{M}} = (F)_{\mathfrak{M}}$.

We introduce graphical constructions for expression set $F = \{F_1, \ldots, F_h\}$. These constructions aim at capturing simultaneous and alternative satisfaction, as well as change under transition (see Lemma 1, p. 16).

(PG) *Page of expression set* $PG(F) := \widehat{x} \dashv F_1 \quad Pg(E) := PG(\{E\})$.
$\vdots$
F_h

(BK) *Book of expression set*

$$BK[F] := \left\{ \begin{array}{c} Pg(F_1) \\ \vdots \\ Pg(F_h) \end{array} \right\} = \left\{ \begin{array}{c} \widehat{x} \dashv F_1 \\ \vdots \\ \widehat{x} \dashv F_h \end{array} \right\}.$$

(FP) *Follow-page* ($\mu \in \Xi$) $FP[\mu, F] := \widehat{x} \xrightarrow{\mu} y \dashv F_1$.
$\vdots$
F_h

We use the abbreviation:

$fp[r, E] := FP[\langle r \rangle, \{E\}] = \widehat{x} \xrightarrow{r} y \dashv E$.

These concepts will be used for formulating semantical clauses in the sequel (see Sct. 3: Graphical Formulations of Semantics.)

LEMMA 1 (Graphical constructions). *For an expression* E, *we have:* $[Pg(E)]_{\mathfrak{M}} = (E)_{\mathfrak{M}}$. *For a finite expression set* F, *we have:*
($\cap$) $[PG(F)]_{\mathfrak{M}} = \bigcap_{F \in F} (F)_{\mathfrak{M}}$; ($\cup$) $[BK[F]]_{\mathfrak{M}} = \bigcup_{F \in F} (F)_{\mathfrak{M}}$, *and*
($\exists$) $[FP[\mu, F]]_{\mathfrak{M}} = \{a \in M \,/\, \exists b \in M \,[\, (a, b) \in \mu^{\mathfrak{M}} \wedge b \in [PG(F)]_{\mathfrak{M}}\,]\}$. ⊣

Proof. By graphical semantics, p. 15. ∎

We also use *neat assertions* and *neat-assertion sets*, jointly generated as follows:

$$F ::= p \mid \overline{F} \mid PG(\boldsymbol{F}) \mid BK[\boldsymbol{F}] \mid FP[\mu, \boldsymbol{F}] \qquad p \in PL$$
$$\boldsymbol{F} ::= \emptyset \mid \boldsymbol{F} \cup \{F\} \qquad (\boldsymbol{F} = \{F_1, \ldots, F_h\}) \tag{2}$$

We can also use draft-like diagrams (with wiggly arrows) to represent models.

EXAMPLE 2 (Classical modal model). Consider a model $\mathfrak{C}$ for classical modal logic KM with 4-state universe $M = \{a, b, c, d\}$; subsets $p^{\mathfrak{C}} = \{d\}$ and $q^{\mathfrak{C}} = \{b\}$; and relations $r^{\mathfrak{C}} = \{(a, b), (a, c)\}$ and $s^{\mathfrak{C}} = \{(c, d)\}$. We can visualize this model $\mathfrak{C}$ as shown in Fig. 5: Classical modal model $\mathfrak{C}$ (p. 17).

Figure 5. Classical modal model $\mathfrak{C}$

q⟩⤳ b ⟵ʳ a ⤳ʳ c ⤳ˢ d ⤳⟨p

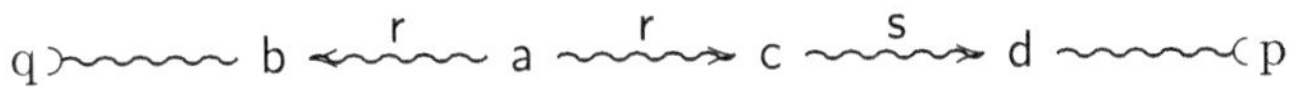

EXAMPLE 3 (Intuitionistic modal model). Consider a model $\mathfrak{J}$ for intuitionistic modal logic JM as follows. Model $\mathfrak{J}$ has universe $M = \{a, b, c, d, e, f\}$; subsets $p^{\mathfrak{J}} = \{f\}$ and $q^{\mathfrak{J}} = \{b, e\}$; and relations $r^{\mathfrak{J}} = \{(a, b), (a, c), (a, d), (a, e)\}$, $s^{\mathfrak{J}} = \{(c, f), (e, f)\}$ and $\prec^{\mathfrak{J}} = \{(b, d), (c, e)\}$ (we use $\prec$ for 'strictly precedes': "precedes and different".) We can visualize this model $\mathfrak{J}$ as shown in Fig. 6: Intuitionistic modal model $\mathfrak{J}$ (p. 17).

Figure 6. Intuitionistic modal model $\mathfrak{J}$

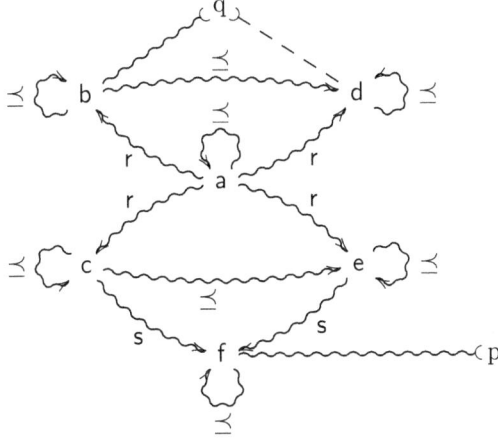

3 Graphical Formulations of Semantics

We now introduce graphical formulations of semantics.

We will consider some *simple* modal languages (cf. grammar (1), p. 14):

$$\dagger := \bot \qquad \nabla := \neg \qquad \bullet \in \{\wedge, \vee, \rightarrow\}$$
$$\Xi := \{\langle r \rangle \,/\, r \in \mathsf{RN}\} \cup \{[r] \,/\, r \in \mathsf{RN}\}.$$

We will illustrate how to formulate graphically the Kripke semantics of modal logics: classical KM (cf., e. g. [1]) and intuitionistic JM (à la Simpson [5]).

We begin with classical modal logic KM. We first recall the notion of satisfaction of a formula at a state of a model and then formulate it graphically.

EXAMPLE 4 (Classical modal semantics). Classical satisfaction clauses. Satisfaction of formula φ at state $\mathbf{a}$ of model $\mathfrak{C}$ is recursively defined [1].

1. Propositional fragment $\qquad\qquad\qquad\qquad\qquad p \in \mathsf{PL}, \bot, \neg, \wedge, \vee, \rightarrow$

    (p) For a propositional letter $p \in \mathsf{PL}$: $\mathbf{a} \Vdash_{\mathfrak{C}} p \quad$ iff $\quad \mathbf{a} \in p^{\mathfrak{C}}$.

    ($\bot$) For $\bot$: $\quad \mathbf{a} \nVdash_{\mathfrak{C}} \bot$.

    ($\neg$) For negation: $\mathbf{a} \Vdash_{\mathfrak{C}} \neg \varphi \quad$ iff $\quad \mathbf{a} \nVdash_{\mathfrak{C}} \varphi$.

    ($\wedge$) For conjunction: $\mathbf{a} \Vdash_{\mathfrak{C}} \psi \wedge \theta \quad$ iff $\quad \mathbf{a} \Vdash_{\mathfrak{C}} \psi$ and $\mathbf{a} \Vdash_{\mathfrak{C}} \theta$.

    ($\vee$) For disjunction: $\mathbf{a} \Vdash_{\mathfrak{C}} \psi \vee \theta \quad$ iff $\quad \mathbf{a} \Vdash_{\mathfrak{C}} \psi$ or $\mathbf{a} \Vdash_{\mathfrak{C}} \theta$.

    ($\rightarrow$) For conditional: $\mathbf{a} \Vdash_{\mathfrak{C}} \psi \rightarrow \theta \quad$ iff $\quad \mathbf{a} \nVdash_{\mathfrak{C}} \psi$ or $\mathbf{a} \Vdash_{\mathfrak{C}} \theta$.

2. Modal fragment $\qquad\qquad\qquad\qquad\qquad\qquad$ modalities $\langle r \rangle, [r]$

    ($\langle \rangle$) For $\langle r \rangle$: $\mathbf{a} \Vdash_{\mathfrak{C}} \langle r \rangle \varphi \quad$ iff $\quad$ for some $\mathbf{b}$ with $(\mathbf{a}, \mathbf{b}) \in r^{\mathfrak{C}}$: $\mathbf{b} \Vdash_{\mathfrak{C}} \varphi$.

    ([]) For $[r]$: $\mathbf{a} \Vdash_{\mathfrak{C}} [r] \varphi \quad$ iff $\quad$ for every $\mathbf{b}$ with $(\mathbf{a}, \mathbf{b}) \in r^{\mathfrak{C}}$: $\mathbf{b} \Vdash_{\mathfrak{C}} \varphi$;
    i. e. $\qquad\qquad\quad$ there is no $\mathbf{b}$ such that $(\mathbf{a}, \mathbf{b}) \in r^{\mathfrak{C}}$ and $\mathbf{b} \nVdash_{\mathfrak{C}} \varphi$.

Derived clause for $\top$: $\quad \mathbf{c} \Vdash_{\mathfrak{B}} \top$. $\qquad\qquad\qquad\qquad\qquad$ b

We can now formulate classical modal semantics in a graphical manner (cf. Sct. 2: Graphical Concepts and Constructions, p. 15).

EXAMPLE 5 (Graphical KM semantics). Satisfaction and expressions.

($\bot$) $\quad \mathbf{a} \nVdash_{\mathfrak{C}} \bot \quad$ iff $\quad \mathbf{a} \in (\{\ \})_{\mathfrak{C}} \qquad\qquad\qquad\qquad$ empty book

($\neg$) $\quad \mathbf{a} \nVdash_{\mathfrak{C}} \varphi \quad$ iff $\quad \mathbf{a} \in (\overline{\varphi})_{\mathfrak{C}} \qquad\qquad\qquad$ complemented expression

($\wedge$) $\quad \mathbf{a} \Vdash_{\mathfrak{C}} \psi$ and $\mathbf{a} \Vdash_{\mathfrak{C}} \theta \quad$ iff $\quad \mathbf{a} \in (\psi \succ --\widehat{x}-- \prec \theta)_{\mathfrak{C}} \quad$ 1-node page

(∨) a ⊩$_{\mathfrak{C}}$ ψ or a ⊩$_{\mathfrak{C}}$ θ   iff   a ∈ $\left(\left\{\begin{array}{c} \psi \!\succ\!-\!-\!-\widehat{x} \\ \widehat{x}\!-\!-\!-\!\prec\theta \end{array}\right\}\right)_{\mathfrak{C}}$ 2-page book

(→) a ⊮$_{\mathfrak{C}}$ ψ or a ⊩$_{\mathfrak{C}}$ θ   iff   a ∈ $\left(\left\{\begin{array}{c} \overline{\psi}\!\succ\!-\!-\!-\widehat{x} \\ \widehat{x}\!-\!-\!-\!\prec\theta \end{array}\right\}\right)_{\mathfrak{C}}$ 2-page book

(⟨⟩) ∃ b ((a, b) ∈ r$^{\mathfrak{C}}$ & b ⊩$_{\mathfrak{C}}$ φ)   iff   a ∈ ($\widehat{x} \xrightarrow{r} y\!-\!-\!-\!\prec\varphi$)$_{\mathfrak{C}}$
2-node page

([ ]) ∄ b ((a, b) ∈ r$^{\mathfrak{C}}$ & b ⊮$_{\mathfrak{C}}$ φ)   iff   a ∈ $\overline{(\widehat{x} \xrightarrow{r} y\!-\!-\!-\!\prec\overline{\varphi})}_{\mathfrak{C}}$
complemented 2-node page

For instance: (⟨r⟩φ)$^{\mathfrak{C}}$ = ($\widehat{x} \xrightarrow{r} y\!-\!-\!-\!\prec\varphi$)$_{\mathfrak{C}}$, for every KM-model $\mathfrak{C}$. ♭

By combining Examples 4 and 5, we obtain Tabs. 1 and 2.

Table 1. Connectives, graphical expressions and constructs for KM

Formula	Expression	Construct
⊥	{ }	BK[∅]
¬φ	$\overline{\varphi}$	$\overline{\varphi}$
ψ ∧ θ	ψ≻ − − − $\widehat{x}$ − − − ≺θ	PG({ψ, θ})
ψ ∨ θ	$\left\{\begin{array}{c} \widehat{x}\!-\!-\!-\!\prec\psi \\ \widehat{x}\!-\!-\!-\!\prec\theta \end{array}\right\}$	BK[{ψ, θ}]
ψ → θ	$\left\{\begin{array}{c} \widehat{x}\!-\!-\!-\!\prec\overline{\psi} \\ \widehat{x}\!-\!-\!-\!\prec\theta \end{array}\right\}$	BK[{$\overline{\psi}$, θ}]

Table 2. Modalities, graphical expressions and constructs for KM

Formula	Expression	Construct
⟨r⟩φ	$\widehat{x} \xrightarrow{r} y\!-\!-\!-\!\prec\varphi$	fp[r, φ]
[r]φ	$\widehat{x} \xrightarrow{r} y\!-\!-\!-\!\prec\overline{\varphi}$	$\overline{\text{fp}[r, \overline{\varphi}]}$

Thus, we obtain the *elimination rules* for classical modal logic KM shown in Tabs. 3 (p. 20) and 4 (p. 20).

Table 3. Elimination rules for the classical propositional connectives

$\vec{\bot}_{KM}$	:=	BK[∅]	empty book	{ }
$\vec{\neg}_{KM}[\varphi]$	:=	$\overline{\varphi}$	expression	$\overline{\varphi}$
$\vec{\wedge}_{KM}[\psi, \theta]$	:=	PG({ψ, θ})	1-node page	$\hat{x} \dashv \psi, \theta$
$\vec{\vee}_{KM}[\psi, \theta]$	:=	BK[{ψ, θ}]	2-page book	{ $\hat{x} \dashv \psi$, $\hat{x} \dashv \theta$ }
$\vec{\rightarrow}_{KM}[\psi, \theta]$	:=	BK[{$\overline{\psi}, \theta$}]	2-page book	{ $\hat{x} \dashv \overline{\psi}$, $\hat{x} \dashv \theta$ }

Table 4. Elimination rules for classical modalities ⟨r⟩ and [r]

$\vec{\langle r \rangle}_{KM}[\varphi]$	:=	fp[r, φ]	follow-page	$\hat{x} \xrightarrow{r} y \dashv \varphi$
$\vec{[r]}_{KM}[\varphi]$	:=	$\overline{\text{fp}[r, \overline{\varphi}]}$	complemented follow-page	$\hat{x} \xrightarrow{r} y \dashv \overline{\varphi}$

Tab. 5 (p. 21) specifies classical modal logic KM. We thus have a converter from ML-formulas to KM-equivalent neat expressions (see Example 8, p. 24).

We now consider intuitionistic modal logic JM. [5]

EXAMPLE 6 (Intuitionistic modal semantics). Consider an intuitionistic modal model $\mathfrak{J}$ with world precedence $\preceq^{\mathfrak{J}}$.

Satisfaction of formula φ at state **a** of $\mathfrak{J}$ is recursively defined as follows.

(⋆) For p, ⊥, ∧, ∨ and ⟨⟩: as in classical modal semantics (cf. Example 4, p. 18).

(¬) For negation: a $\Vdash_{\mathfrak{J}} \neg \varphi$ iff for every b with (a, b) $\in \preceq^{\mathfrak{J}}$: b $\nVdash_{\mathfrak{J}} \varphi$; i. e. there is no b such that (a, b) $\in \preceq^{\mathfrak{J}}$ and b $\Vdash_{\mathfrak{J}} \varphi$.

(→) For →: a $\Vdash_{\mathfrak{J}} \psi \rightarrow \theta$ iff for all (a, b) $\in \preceq^{\mathfrak{J}}$: if b $\Vdash_{\mathfrak{J}} \psi$ then b $\Vdash_{\mathfrak{J}} \theta$; i. e. there is no b with (a, b) $\in \preceq^{\mathfrak{J}}$ such that b $\Vdash_{\mathfrak{J}} \psi$ and b $\nVdash_{\mathfrak{J}} \theta$.

([]) For modality [r]: a $\Vdash_{\mathfrak{J}} [r] \varphi$ iff whenever (a, b) $\in \preceq^{\mathfrak{J}}$ and (b, c) $\in r^{\mathfrak{J}}$: c $\Vdash_{\mathfrak{J}} \varphi$; i. e. there are no b, c with (a, b) $\in \preceq^{\mathfrak{J}}$, (b, c) $\in r^{\mathfrak{J}}$ and

Table 5. Graphical specification for classical modal logic KM

1. Propositional: KM-elimination rules for $\bot, \neg, \wedge, \vee, \rightarrow$ (cf. Tab. 3, p. 20):

$$\bot := \{\} \quad \neg\varphi := \overline{\varphi} \quad \psi \wedge \theta := PG(\{\psi, \theta\})$$

$$\psi \vee \theta := BK[\{\psi, \theta\}] \quad \psi \rightarrow \theta := BK[\{\overline{\psi}, \theta\}]$$

2. Modal: KM-elimination rules for $\langle r \rangle, [r]$ (cf. Tab. 4, p. 20):

$$\langle r \rangle \varphi := fp[r, \varphi] \qquad [r]\varphi := \overline{fp[r, \overline{\varphi}]}$$

$c \not\Vdash_J \varphi$.

We now provide the graphical formulation.

($\neg$) $\neg \varphi$ JM-equivalent to   complemented 2-node page $\hat{x} \xrightarrow{\preceq} y \dashrightarrow \prec \varphi$.

($\rightarrow$) $\psi \rightarrow \theta$ JM-equivalent to   complemented 2-node page $\hat{x} \xrightarrow{\preceq} y \dashrightarrow \prec \psi$, $\searrow \prec \overline{\theta}$

([]) $[r]\varphi$ JM-equivalent to   complemented 3-node page $\hat{x} \xrightarrow{\preceq} y \xrightarrow{r} z \dashrightarrow \prec \overline{\varphi}$.

For $\bot, \wedge, \vee$ and $\langle r \rangle$, as in classical modal semantics (cf. Example 5).

Tab. 6 (p. 21) shows the *elimination rules* for the intuitionistic connectives $\neg$ and $\rightarrow$ and for the intuitionistic modality $[r]$.

Table 6. Elimination rules for intuitionistic $\neg, \rightarrow$ and $[r]$

$\vec{\neg}_{\mathbb{P}}[\varphi]$	:=	$\overline{fp[\preceq, \varphi)}$	complemented follow-page
$\vec{\rightarrow}_{\mathbb{P}}[\psi, \theta]$	:=	$\overline{fp[\preceq, PG(\{\psi, \overline{\theta}\}))}$	complemented follow-page
$\vec{[r]}_{\mathbb{M}}[\varphi]$	:=	$\overline{fp[\preceq, (fp[r, \overline{\varphi}]))}$	complemented follow-page

Table 7. Graphical specification for intuitonistic modal logic 𝕁M

1. Propositional fragment:

    (a) 𝕁M-elimination rules for $\bot, \wedge, \vee$ (cf. Tab. 3, p. 20):

    $$\bot := \{\} \quad \psi \wedge \theta := \mathsf{PG}(\{\psi, \theta\}) \quad \psi \vee \theta := \mathsf{BK}[\{\psi, \theta\}]$$

    (b) 𝕁M-elimination rules for $\neg, \rightarrow$ (cf. Tab. 6, p. 21):

    $$\neg \varphi := \overline{\mathsf{fp}[\preceq, \varphi)} \qquad \psi \rightarrow \theta := \overline{\mathsf{fp}[\preceq, \mathsf{PG}(\{\psi, \overline{\theta}\}))}$$

2. Modal fragment   𝕁M-elimination rule for $\langle r \rangle$ and $[r]$:

    (a) For modality $\langle r \rangle$ (cf. Tab. 4, p. 20): $\langle r \rangle \varphi := \mathsf{fp}[r, \varphi)$
    (b) For modality $[r]$ (cf. Tab. 6, p. 21):
    $[r] \varphi := \overline{\mathsf{fp}[\preceq, (\mathsf{fp}[r, \overline{\varphi})))}$

Tab. 7 (p. 22) specifies intuitonistic modal logic 𝕁M. We thus have a converter from 𝕄L-formulas to 𝕁M-equivalent neat expressions (see Example 9, p. 24).

PROPOSITION 7 (Graphical specifications). *Consider the specifications in Tabs. 5 (p. 21), for 𝕂M, and 7 (p. 22), for 𝕁M. For every elimination rule $\varphi := E$: formula φ is equivalent to expression E in the corresponding class of models.*

Proof. By graphical constructs (cf. Sct. 2, p. 16) and each semantics. ■

In general, graphical formulation of semantics (as illustrated in the preceding examples) involves two steps as follows.

1. Translate the satisfaction clauses to graphical expressions; cf. Examples 5 (p. 18) and 6 (p. 20).

2. Formulate the resulting expressions by means of expression constructs (cf. Sct. 2 (p. 16): simultaneous PG, alternatives BK and change FP); cf. the graphical specifications in Tabs. 5 (p. 21) for 𝕂M, and 7 (p. 22) for 𝕁M.

4 General Converter and Model-Checker

We now examine general converter (p. 23) and model-checker (p. 25).

We first consider general and specific graphical converters (see Fig. 7: General and specific converters, p. 23.)

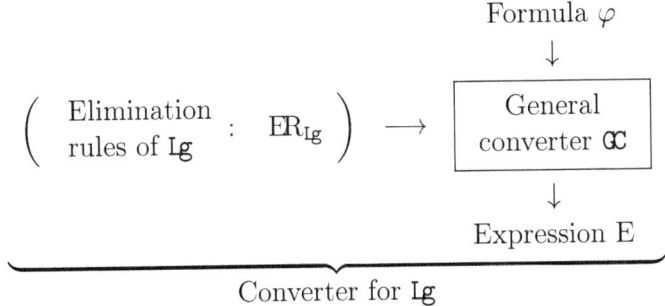

Figure 7. General and specific converters

$$\left(\begin{array}{c}\text{Elimination}\\\text{rules of Lg}\end{array} : \text{ER}_{\text{Lg}}\right) \longrightarrow \boxed{\begin{array}{c}\text{General}\\\text{converter } \mathcal{C}\end{array}}$$

Formula φ ↓ ... ↓ Expression E

Converter for Lg

A general converter generates, for each logic Lg, a converter $\mathcal{C}_{\text{Lg}}$, which eliminates logical symbols producing neat expressions (cf. grammar (1), in Sct. 1: Introduction, p. 14). We define our *general converter* $\mathcal{C}$ as follows (see also Tab. 8: General graphical converter $\mathcal{C}$, p. 24):

$$\mathcal{C} : \text{Lg} : \text{Logic} \to (\text{Formulas} \to \text{Expressions})$$

(p) $\mathcal{C}_{\text{Lg}}(p) := p$ (convert propositional letter p ∈ PL)

(†) $\mathcal{C}_{\text{Lg}}(\dagger) := \vec{\dagger}_{\text{Lg}}$ (eliminate 0-ary †)

(∇) $\mathcal{C}_{\text{Lg}}(\nabla \varphi) := \vec{\nabla}_{\text{Lg}}[\mathcal{C}_{\text{Lg}}(\varphi)]$ (eliminate 1-ary ∇)

(•) $\mathcal{C}_{\text{Lg}}(\psi \bullet \theta) := \vec{\bullet}_{\text{Lg}}[\mathcal{C}_{\text{Lg}}(\psi), \mathcal{C}_{\text{Lg}}(\theta)]$ (eliminate 2-ary •)

(μ) $\mathcal{C}_{\text{Lg}}(\mu \varphi) := \vec{\mu}_{\text{Lg}}[\mathcal{C}_{\text{Lg}}(\varphi)]$ (eliminate modality μ)

We now consider instantiated graphical converters (see Fig. 8: Instantiated graphical converter, p. 23) for KM and JM: from formulas to neat expressions.

Figure 8. Instantiated graphical converter

Formula φ ⟶ $\boxed{\text{ER}_{\text{Lg}}} \to \boxed{\mathcal{C}}$ ⟶ Expression E

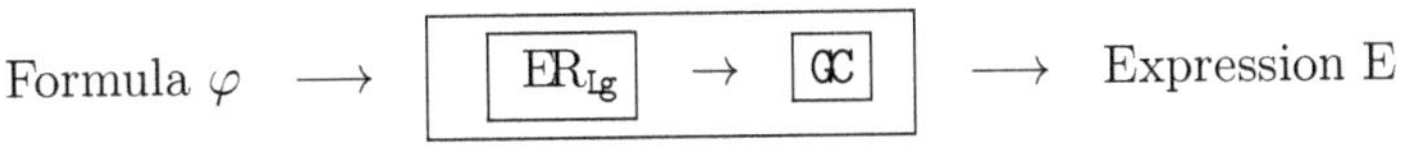

Table 8. General graphical converter ₵

function ₵ (Lg : Logic , φ : Formula) → Expression
    case[φ]
        $\varphi \in$ PL   **return** φ
        $\varphi = \dagger$ **return** $\vec{\dagger}_{\text{Lg}}$
        $\varphi = \nabla \theta$ **return** $\vec{\nabla}_{\text{Lg}}[₵(\theta)]$
        $\varphi = \psi \bullet \theta$ **return** $\vec{\bullet}_{\text{Lg}}[₵(\psi), ₵(\theta)]$
        $\varphi = \mu \psi$ **return** $\vec{\mu}_{\text{Lg}}[₵(\psi)]$
    esac(φ)

end-function ₵.

EXAMPLE 8 (₭₥ converter). ₵$_{\text{₭₥}}$: ₥L-formula → Neat expression (cf. Tab. 5: Graphical specification for classical modal logic ₭₥, p. 21; see also Fig. 9: Instantiated graphical converter ₵$_{\text{₭₥}}$, p. 25).

(p)   ₵$_{\text{₭₥}}$(p) := p   propositional letter;

($\bot$)   ₵$_{\text{₭₥}}$($\bot$) = $\vec{\bot}_{\text{₭₥}}$ = { }   empty book ₿₭[∅];

($\neg$)   ₵$_{\text{₭₥}}$($\neg \varphi$) = $\vec{\neg}_{\text{₭₥}}[₵_{\text{₭₥}}(\varphi)]$ = $\overline{₵_{\text{₭₥}}(\varphi)}$ complemented expression;

($\wedge$)   ₵$_{\text{₭₥}}$($\psi \wedge \theta$) = $\vec{\wedge}_{\text{₭₥}}[₵_{\text{₭₥}}(\psi), ₵_{\text{₭₥}}(\theta)]$ = PG($\{₵_{\text{₭₥}}(\psi), ₵_{\text{₭₥}}(\theta)\}$)   page;

($\vee$)   ₵$_{\text{₭₥}}$($\psi \vee \theta$) = $\vec{\vee}_{\text{₭₥}}[₵_{\text{₭₥}}(\psi), ₵_{\text{₭₥}}(\theta)]$ = ₿₭$[\{₵_{\text{₭₥}}(\psi), ₵_{\text{₭₥}}(\theta)\}]$ book;

($\rightarrow$)   ₵$_{\text{₭₥}}$($\psi \rightarrow \theta$) = $\vec{\rightarrow}_{\text{₭₥}}[₵_{\text{₭₥}}(\psi), ₵_{\text{₭₥}}(\theta)]$ = ₿₭$[\{\overline{₵_{\text{₭₥}}(\psi)}, ₵_{\text{₭₥}}(\theta)\}]$ book;

($\langle r \rangle$)   ₵$_{\text{₭₥}}$($\langle r \rangle \varphi$) = $\vec{\langle r \rangle}_{\text{₭₥}}[₵_{\text{Lg}}(\varphi)]$ = fp[r, ₵$_{\text{₭₥}}(\varphi)$)   follow-page;

([r])   ₵$_{\text{₭₥}}$([r]φ) = $\vec{[r]}_{\text{₭₥}}[₵_{\text{Lg}}(\varphi)]$ = $\overline{fp[r, \overline{₵_{\text{₭₥}}(\varphi)})}$ complemented
     follow-page.

For instance, ₵$_{\text{₭₥}}$($\langle r \rangle$ [s] $\neg$p) = $\overline{fp[r, \overline{(fp[s, \overline{\overline{p}}]))}}$.

EXAMPLE 9 (₥ converter). ₵$_{\text{₥}}$: ₥L-formula → Neat expression (cf. Tab. 7: Graphical specification for intuitonistic modal logic ₥, p. 22; see also Fig. 10: Instantiated graphical converter ₵$_{\text{₥}}$, p. 25).

($\star$)   For p, $\bot$, $\wedge$, $\vee$ and $\langle r \rangle$:   as for classical modal logic (cf. Example 8);

($\neg$)   ₵$_{\text{₧}}$($\neg \varphi$) = $\vec{\neg}_{\text{₥}}[₵_{\text{₥}}(\varphi)]$ = $\overline{fp[\preceq, ₵_{\text{₥}}(\varphi))}$;

($\rightarrow$)   ₵$_{\text{₧}}$($\psi \rightarrow \theta$) = $\vec{\rightarrow}_{\text{₥}}[₵_{\text{₥}}(\psi), ₵_{\text{₥}}(\theta)]$ = $\overline{fp[\preceq, PG(\{₵_{\text{₥}}(\psi), \overline{₵_{\text{₥}}(\theta)}\}))}$.

Table 9. Instantiated graphical converter $\mathbb{CC}_{KM}$

function $\mathbb{CC}_{KM}$ (φ : **ML**-formula) $\rightarrow$ Neat expression

    case[φ]

	$\varphi \in \mathbf{PL}$	**return** φ
	$\varphi = \bot$	**return** $\{\,\}$
	$\varphi = \neg \theta$	**return** $\overline{\mathbb{CC}_{KM}(\theta)}$
	$\varphi = \psi \wedge \theta$	**return** $\mathsf{PG}(\{\mathbb{CC}_{KM}(\psi), \mathbb{CC}_{KM}(\theta)\})$
	$\varphi = \psi \vee \theta$	**return** $\mathsf{BK}[\{\mathbb{CC}_{KM}(\psi), \mathbb{CC}_{KM}(\theta)\}]$
	$\varphi = \psi \rightarrow \theta$	**return** $\mathsf{BK}[\{\overline{\mathbb{CC}_{KM}(\psi)}, \mathbb{CC}_{KM}(\theta)\}]$
	$\varphi = \langle r \rangle \psi$	**return** $\mathsf{fp}[r, \mathbb{CC}_{KM}(\psi))$
	$\varphi = [r] \psi$	**return** $\mathsf{fp}[r, \overline{\mathbb{CC}_{KM}(\psi)})$

    esac(φ)

end-function $\mathbb{CC}_{KM}$.

([r]) $\mathbb{CC}_{\mathbb{M}}([r]\varphi) = [\vec{r}]_{\mathbb{M}}[\mathbb{CC}_{\mathrm{lg}}(\varphi)] = \overline{\mathsf{fp}[\preceq, (\mathsf{fp}[r, \overline{\mathbb{CC}_{\mathbb{M}}(\varphi)})))}$.

For instance, $\mathbb{CC}_{\mathbb{M}}(\langle r \rangle [s] \neg \mathrm{p}) = \mathsf{fp}[r, \overline{(\mathsf{fp}[\preceq, \overline{(\mathsf{fp}[s, \overline{\mathsf{fp}[\preceq, \mathrm{p}]}))})}))$.

Table 10. Instantiated graphical converter $\mathbb{CC}_{\mathbb{M}}$

function $\mathbb{CC}_{\mathbb{M}}$ (φ : **ML**-formula) $\rightarrow$ Neat expression

    case[φ]

	$\varphi \in \mathbf{PL}$	**return** φ
	$\varphi = \bot$	**return** $\{\,\}$
	$\varphi = \neg \theta$	**return** $\mathsf{PG}(\{\mathbb{CC}_{\mathbb{M}}(\psi), \mathbb{CC}_{\mathbb{M}}(\theta)\})$
	$\varphi = \psi \wedge \theta$	**return** $\mathsf{PG}(\{\mathbb{CC}_{\mathbb{M}}(\psi), \mathbb{CC}_{\mathbb{M}}(\theta)\})$
	$\varphi = \psi \vee \theta$	**return** $\mathsf{BK}[\{\mathbb{CC}_{\mathbb{M}}(\psi), \mathbb{CC}_{\mathbb{M}}(\theta)\}]$
	$\varphi = \psi \rightarrow \theta$	**return** $\overline{\mathsf{fp}[\preceq, \mathsf{PG}(\{\mathbb{CC}_{\mathbb{M}}(\psi), \mathbb{CC}_{\mathbb{M}}(\theta)\}))}$
	$\varphi = \langle r \rangle \psi$	**return** $\mathsf{fp}[r, \mathbb{CC}_{\mathbb{M}}(\psi))$
	$\varphi = [r] \psi$	**return** $\overline{\mathsf{fp}[\preceq, (\mathsf{fp}[r, \overline{\mathbb{CC}_{\mathbb{M}}(\psi)})))}$

    esac(φ)

end-function $\mathbb{CC}_{\mathbb{M}}$.

We now consider our general graphical model-checker.

We will use a Boolean sort **Bln**, having 2 values **false** and **true**, with the usual 2-ary operations **and** and **or**, which we extend naturally to sets as follows:

$$\underline{\mathrm{and}}(\{v_1, \ldots, v_k\}) := v_1 \underline{\mathrm{and}} \ldots \underline{\mathrm{and}} v_k$$
$$\underline{\mathrm{or}}[\{v_1, \ldots, v_k\}] := v_1 \underline{\mathrm{or}} \ldots \underline{\mathrm{or}} v_k.$$

We set $\underline{and}(\emptyset) := \mathbf{true}$ and $\underline{or}[\emptyset] := \mathbf{false}$. We also have the usual 1-ary operation $\underline{not}$, which we extend naturally to sets:

$$\underline{not}[V] := \{\underline{not}\, v\, /\, v \in V\}$$

We have the usual properties (like De Morgan).

We will also use *reachable sets* consisting of the states reached under a transition: $[a^\mu)_\mathfrak{M} := \{b \in M\, /\, (a, b) \in \mu^\mathfrak{M}\}$. See Tab 11 (p. 26).

Table 11. Reachable set $[a^\mu)_\mathfrak{M}$ for some modalities

Modality ⟨t⟩		Special modalities [ ]	
Relation name r ∈ RN	$\{b \in M\, /\, (a, b) \in r^\mathfrak{M}\}$		
Constant null 0	∅	Universal D	M
Constant square ⋈	M		
Constant identity ι	{ a }	Difference E	M \ { a }
Constant diversity ∂	M \ { a }		

A general model checker receives a model $\mathfrak{M}$, a state a of $\mathfrak{M}$ and a neat expression E, and checks whether $\mathfrak{M}$ satisfies expression E at state a. We define our *general graphical model-checker* as follows (cf. grammar (2), in Sct. 2: Graphical Concepts and Constructions, p. 17).

Function GMc : (Model $\mathfrak{M}$, State a : Neat expression E] → Bln

We use simply GMc(a : E], when model $\mathfrak{M}$ is clear (see also Tabs. 12, 13).

(p) Expression E is a propositional letter $p \in PL$ (direct check):

$$\text{GMc}(a : E] := \mathbf{true} \text{ iff } a \in p^\mathfrak{M}.$$

(¬) Expression E is a complemented expression $\overline{F}$ (reduce by $\underline{not}$):

$$\text{GMc}(a : E] := \underline{not}\, \text{GMc}(a : F].$$

(PG) Expression is an expression-page PG($\boldsymbol{F}$):   (reduce by $\underline{and}$):

$$\text{GMc}(a : E] := \underline{and}(\{\text{GMc}(a : F]\, /\, F \in \boldsymbol{F}\}).$$

(BK) Expression is an expression-book BK[$\boldsymbol{F}$]:   (reduce by $\underline{or}$):

Table 12. General graphical model-checker GMc

function GMc (a : State , E : Neat expression) → **Bln**

   case[E]

$E \in PL$	if $a \in E^{\mathfrak{M}}$ then return true else return false fi
$E = \overline{F}$	return <u>not</u> GMc(a : F]
$E = PG(F)$	return <u>and</u>({GMc(a : F] / F $\in$ F})
$E = BK[F]$	return <u>or</u>[{GMc(a : F] / F $\in$ F}]
$E = FP[\mu, F]$	return <u>or</u>[{GMc(b : PG(F)] / b $\in$ [a$^\mu$)$_\mathfrak{M}$}]

   esac(E)

end-function GMc.

Table 13. Special case of graphical model-checker GMc

function GMc (a : State , E : Neat expression) → **Bln**

   case[E]

$E \in PL$	if $a \in E^{\mathfrak{M}}$ then return true else return false fi
$E = \overline{F}$	return <u>not</u> GMc(a : F]
$E = PG(\{F_\leftarrow, F_\rightarrow\})$	return GMc(a : F$_\leftarrow$] <u>and</u> GMc(a : F$_\rightarrow$]
$E = BK[\{F_\leftarrow, F_\rightarrow\}]$	return GMc(a : F$_\leftarrow$] <u>or</u> GMc(a : F$_\rightarrow$]
$E = fp[t, F]$	return <u>or</u>[{GMc(b : Pg(F)] / (a, b) $\in$ t$^{\mathfrak{M}}$}]

   esac(E)

end-function GMc.

$$\mathfrak{Mc}(\mathsf{a}:E] := \underline{\text{or}}[\{\mathfrak{Mc}(\mathsf{a}:F] \,/\, F \in \boldsymbol{F}\}].$$

($\overset{\mu}{\mapsto}$) Expression is a follow-page (reduce to $[\mathsf{a}^\mu)_\mathfrak{M}$):

$$\mathfrak{Mc}(\mathsf{a}:E] := \underline{\text{or}}[\{\mathfrak{Mc}(\mathsf{b}:\mathrm{PG}(F)] \,/\, \mathsf{b} \in [\mathsf{a}^\mu)_\mathfrak{M}\}].$$

We can instantiate our general graphical model-checker $\mathfrak{Mc}$ by a logic Lg (cf. Fig. 3: General model-checker $\mathfrak{MC}$ with converter $\mathfrak{Cnv}$, p. 14). We obtain a *graphical model-checker for logic* Lg, defined as follows.

$\mathfrak{Mc}_{\mathsf{Lg}}(\mathfrak{M}, \mathsf{a}:\varphi] := \mathfrak{Mc}(\mathfrak{M}, \mathsf{a}:\mathfrak{CC}_{\mathsf{Lg}}(\varphi)]$ (cf. $\mathfrak{CC}_{\mathsf{Lg}}$, p. 23, and $\mathfrak{Mc}$, p. 25).

We use simply $\mathfrak{Mc}(\mathsf{a}:\varphi]$, when logic Lg and model $\mathfrak{M}$ are clear.

EXAMPLE 10 (Graphical model-checker for KM). Consider model $\mathfrak{C}$ of Example 2: Classical modal model, p. 17. We wish to check whether formula $\varphi = \langle r \rangle [s] \neg p$ holds at state a of model $\mathfrak{C}$. Model-checker $\mathfrak{Mc}(\mathsf{a}:\varphi]$ (short for $\mathfrak{Mc}_{\mathsf{KM}}(\mathfrak{C}, \mathsf{a}:\varphi]$) operates as follows (see also Fig. 9: Operation of graphical model-checker for KM, p. 29).

1. $\mathfrak{Mc}(\mathsf{a}:\varphi] = \mathfrak{Mc}(\mathsf{a}:\mathrm{fp}[r, \overline{(\mathrm{fp}[s, \overline{\overline{p}})})])]$ as

$$\mathfrak{CC}_{\mathsf{KM}}(\langle r \rangle [s] \neg p) = \mathrm{fp}[r, \overline{(\mathrm{fp}[s, \overline{\overline{p}}]))]};$$

(cf. Example 8: KM converter, p. 24).

2. $\mathfrak{Mc}(\mathsf{a}:\mathrm{fp}[r, \overline{(\mathrm{fp}[s, \overline{\overline{p}}]))]}] \overset{(\overset{r}{\mapsto})}{=} \underline{\text{or}} \left[\left\{ \begin{array}{l} \mathfrak{Mc}(\mathsf{b}:\overline{\mathrm{fp}[s, \overline{\overline{p}}]}) \\ \mathfrak{Mc}(\mathsf{c}:\overline{\mathrm{fp}[s, \overline{\overline{p}}]}) \end{array} \right\} \right]$ as $[\mathsf{a}^r)_{\mathfrak{C}} = \{\mathsf{b}, \mathsf{c}\}$.

   (a) $\mathfrak{Mc}(\mathsf{b}:\overline{\mathrm{fp}[s, \overline{\overline{p}}]}) \overset{(-)}{=} \underline{\text{not}}\, \mathfrak{Mc}(\mathsf{b}:\mathrm{fp}[s, \overline{\overline{p}}])$.

   (b) $\mathfrak{Mc}(\mathsf{b}:\mathrm{fp}[s, \overline{\overline{p}}]) \overset{(\overset{s}{\mapsto})}{=} \mathrm{false}$ as $[\mathsf{b}^s)_{\mathfrak{C}} = \emptyset$.

   So, $\mathfrak{Mc}(\mathsf{b}:\overline{\mathrm{fp}[s, \overline{\overline{p}}]}) = \underline{\text{not}}\, \mathrm{false} = \mathrm{true}$.

   (a) $\mathfrak{Mc}(\mathsf{c}:\overline{\mathrm{fp}[s, \overline{\overline{p}}]}) \overset{(-)}{=} \underline{\text{not}}\, \mathfrak{Mc}(\mathsf{c}:\mathrm{fp}[s, \overline{\overline{p}}])$.

   (b) $\mathfrak{Mc}(\mathsf{c}:\mathrm{fp}[s, \overline{\overline{p}}]) \overset{(\overset{s}{\mapsto})}{=} \mathfrak{Mc}(\mathsf{d}:\overline{\overline{p}})$ as $[\mathsf{c}^s)_{\mathfrak{C}} = \{\mathsf{d}\}$.

   (c) $\mathfrak{Mc}(\mathsf{d}:\overline{\overline{p}}) \overset{(-)^2}{=} \underline{\text{not not}}\, \mathfrak{Mc}(\mathsf{d}:p] = \mathfrak{Mc}(\mathsf{d}:p]$.

Figure 9. Operation of graphical model-checker for 𝕂𝕄

$$\mathfrak{Mc}(a:\varphi) \;=\; \mathfrak{Mc}(a:\mathrm{fp}[r,(\overline{\mathrm{fp}[s,\overline{\mathrm{p}}]})])$$

$$\|\;(\stackrel{r}{\mapsto})$$

| $\mathfrak{Mc}(b:\overline{\mathrm{fp}[s,\overline{\mathrm{p}}]})$ | **or** | $\mathfrak{Mc}(c:\overline{\mathrm{fp}[s,\overline{\mathrm{p}}]})$ |

$\|\;(\overline{})$ $\|\;(\overline{})$

not $\mathfrak{Mc}(b:\mathrm{fp}[s,\overline{\mathrm{p}}])$ **not** $\mathfrak{Mc}(c:\mathrm{fp}[s,\overline{\mathrm{p}}])$

$\|\;(\stackrel{s}{\mapsto})$ $\|\;(\stackrel{s}{\mapsto})$

                                     not $\mathfrak{Mc}(d:\overline{\overline{\mathrm{p}}})$

                                     $\|\;(\overline{})^2$

not false   ⋮   **not** $\mathfrak{Mc}(d:\mathrm{p})$

$\|$ $\|\;(\mathrm{p})$

                                      not true

true   **or** $\|$

                    $\|$ **false**

                true

(d) $\mathfrak{Mc}(d:\mathrm{p}) \stackrel{(\mathrm{p})}{=}$ **true**   as $d \in \mathrm{p}^{\mathfrak{M}}$.

So, $\mathfrak{Mc}(c:\overline{\mathrm{fp}[s,\overline{\mathrm{p}}]}) =$ **not true** = **false**.

Thus, $\mathfrak{Mc}(a:\varphi) =$ **true or false** = **true**. ♭

EXAMPLE 11 (Graphical model-checker for 𝕁𝕄). Consider model 𝔍 of Example 3: Intuitionistic modal model, p. 17. We wish to check whether formula $\varphi = \langle r \rangle [s] \neg \mathrm{p}$ holds at state a of model 𝔍. Model-checker $\mathfrak{Mc}(a:\varphi)$ (short for $\mathfrak{Mc}_{\mathbb{JM}}(\mathfrak{J}, a:\varphi)$) operates as follows.

- First, $\mathfrak{Mc}(a:\varphi) = \mathfrak{Mc}(a:\mathrm{fp}[r,(\mathrm{fp}[\preceq,(\mathrm{fp}[s,\overline{\mathrm{fp}[\preceq,\mathrm{p}]}])])])$; as $\mathfrak{CC}_{\mathbb{JM}}(\langle r \rangle [s] \neg \mathrm{p}) = \mathrm{fp}[r,(\mathrm{fp}[\preceq,(\mathrm{fp}[s,\overline{\mathrm{fp}[\preceq,\mathrm{p}]}])])]$ (cf. Example 9, p. 24).

- Then, $\mathfrak{Mc}(a:\mathrm{fp}[r,(\mathrm{fp}[\preceq,(\mathrm{fp}[s,\overline{\mathrm{fp}[\preceq,\mathrm{p}]}])])]) =$ **true** (by applications of $(\stackrel{r}{\mapsto})$, $(\overline{})$, $(\stackrel{\preceq}{\mapsto})$, $(\stackrel{s}{\mapsto})$ and (p).)

Thus, $\mathfrak{Mc}(a:\varphi) =$ **true**. ♭

5 Concluding Remarks

We now comment on our approach and on some extensions and variations.

We have introduced a general graphical approach to specification and construction of model-checkers, which gives general model-checkers that can be instantiated to model-checkers for specific logics. Our approach involves two components: a logic-dependent converter and a general model checker; the user is concerned only with the former (cf. Fig. 3, p. 14), which can be obtained by instantiating a general converter (cf. Fig. 8, p. 23). We have illustrated the application of this approach to two modal logics: classical KM and intuitionistic JM (which clearly applies to their propositional fragments). Our approach, however, already handles some relational constants and special modalities, like D and E [1] (cf. Tab. 11: Reachable set $[a^\mu]_{\mathfrak{M}}$ for some modalities, p. 26).

We now consider the extension of our graphical approach to structured modalities (as in PDL [4]): it suffices to describe the semantics of the structured relations. For instance, consider composition. A composed relation r;s can be described by two consecutive arrows, one labelled r followed by one labelled s; so we pre-process the model: whenever we have two consecutive arrows a $\xrightarrow{r}$ b and b $\xrightarrow{s}$ c, we add a new arrow a $\xrightarrow{r;s}$ c. We can similarly expand a model for each relation operation ⊔, ⊓, ⌣ and iteration ∗, as well as for the constants (see, e. g. [6]). Tests cause no problem: we can eliminate $\langle\psi?\rangle\,\theta$ and $[\psi?]\,\theta$ to the 1-node page $\psi \succ\text{-}\text{-}\text{-}\widehat{x}\text{-}\text{-}\text{-}\prec\theta\ =\ \text{PG}(\{\psi,\theta\})$.

We now consider some variations of our approach. In this paper, the converter eliminates all logical symbols once and for all and the model-checker proceeds in a forward direction by simplifying the resulting expression. One could use stepwise elimination, instead. Some model-checkers (as for CTL [2]) use a backward approach. One could annotate the model, starting with the given information and adding more complex expressions (much as above for structured relations).

We also have the flexibility of combining these variations:

Converter/Model-checker	Forward checking	Backward checking
Initial elimination	Initial Forward	Initial Backward
Stepwise elimination	Stepwise Forward	Stepwise Backward

We finally comment on the user's task: graphical specification of semantics. As illustrated in Sct. 3, with classical and intuitionistic modal logics, this is not too difficult. We can visualize their models by draft-like diagrams (cf. Figs. 5, p. 17, and 6, p. 17). One often reasons intuitively about their semantics also in a graphical manner. Our graphical approach formalizes such intuitive arguments.

The main feature of our approach is its generality, which provides flexibility without loss of efficiency. It could also be used for theorem provers.

References

[1] P. Blackburn, M. de Rijke, and Y. Venema. *Modal Logic*. Cambridge, England: Cambridge University Press, 2001. ISBN: 0 521 52714 7 (pbk).

[2] E. M. Clarke, O. Grumberg, and D. A. Peled. *Model Checking*. Cambridge, Massachusetts: The MIT Press, 1999. ISBN: 0262032708.

[3] S. Curtis and G. Lowe. "Proofs with graphs". In: *Science of Computer Programming* 26.1–3 (May 1996), pp. 197–216.

[4] D. Harel, D. Kozen, and J. Tiuryn. *Dynamic Logic*. Cambridge, Massachusetts: The MIT Press, 2000. ISBN: 0262082896 (hbk).

[5] A. K. Simpson. "The Proof Theory and Semantics of Intuitionistic Modal Logic". PhD thesis CST-114-94. Laboratory for Foundations of Computer Science, Dept. of Computer Science, Univ. of Edinburgh, Aug. 1994.

[6] P. A. S. Veloso, S. R. M. Veloso, and M. R. F. Benevides. "PDL for structured data: a graph-calculus approach". In: *Logic Journal of the IGPL* 22.5 (2014), pp. 737–757.

[7] P. A. S. Veloso, S. R. M. Veloso, and M. R. F. Benevides. "On Graph Calculi for Multi-modal Logics". In: *Electr. Notes Theor. Comput. Sci* 312 (2015), pp. 231–252.

A Proof Theory for the Inconsistent Default Logic

ANA TERESA MARTINS AND FRANCICLEBER FERREIRA

1 Introduction

Reasoning in practical situations involves drawing conclusions from incomplete knowledge. In this context, we usually assume facts that may be later discharged in face of new evidences. If a computer program is intended to be artificially intelligent, it must be flexible enough to cope with the lack of knowledge and uncertainties.

Part of the work of Artificial Intelligence has been devoted to reasoning formalisation and automatisation in order to solve complex problems. In particular, logical systems have been used as an extremely attractive tool, not only for providing a formal language which allows a non-ambiguous representation of knowledge, but also for suggesting precise and powerful ways to manipulate knowledge, e.g. the logical inferences.

The classical logic, for instance, has been traditionally used either in argumentation analysis or in the mathematical foundation of deductive sciences. Within Artificial Intelligence, logic has been used for capturing patterns of reasoning in order to automatise it [29]. Nevertheless, it is useful to model ideal situations, such as those found in mathematics or in philosophical argumentations, where the knowledge at hand is precise and complete enough to perform inferences whose conclusions have a status of certainty (modulo the assumed premises). In actual situations, where the knowledge is incomplete and imprecise and it is necessary to assume plausible suppositions, possibly refutable *a posteriori* by the acquisition of new knowledge, the classical logic is not sufficient. Non-monotonic logics have been proposed in order to give a more suitable modelling of practical reasoning.

Copyright © 2019 by Ana Teresa Martins and Francicleber Ferreira. All rights reserved.

⋆ As a tribute to Tarcísio Pequeno, this paper reproduces some previously published works: Sections 1 and part of Section 2 are from [36], and Sections 2.1, 3–6 were previously published in [45] with some minor changes.

The monotonicity of a logical system means that, if a conclusion is entailed from some set of premises, no additional premise will invalidate this conclusion. However, in actual situations, it is usual to draw reasonable conclusions based on our experience and on what we think as plausible, which may be refutable in face of new knowledge. The canonical example, which involves inheritance of prototypical properties, is about birds which flies. Whenever one discovers that Tweety is a bird, it is legitimate to assume that it flies. Nevertheless, the additional information that Tweety is a penguin, makes one to revise his/her previous belief about Tweety's capacity of flying, although keeping the belief that it is usual flying birds.

Practical reasoning has a strong non-monotonic component. It has been largely defended that almost all practical inferences are from this nature. *Ergo*, it is perfectly sound to argue for intelligent automatic systems which are capable of performing some kind of non-monotonic inferences as the ones above explained.

Historically, non-monotonic logics were officially launched in 1980 in a special issue of *Artificial Intelligence* journal [2]. Among the approaches to non-monotonic reasoning formalisation, we may cite Default Logic. It was introduced in [52], a seminal paper that points out many interesting features and problems related to non-monotonic reasoning. One of them is the seemingly always unavoidable *multiple extension problem*, discussed in [51] through the *Nixon's Diamond problem*. A somewhat general investigation about inheritance of properties in hierarchies with exceptions was presented in [13] involving the use of seminormal default rules. They intended to avoid the multiple extension problem whenever the inheritance of values in the hierarchy allows one to prefer one extension instead of others. Some default logic variants were also proposed to cope with related problems [35],[6, 5], [56, 55].

One of the most serious problems associated to formal approaches of non-monotonic inferences is that, in general, they are undecidable, i.e., the set of all non-monotonic theorems is not recursively enumerated. The application of a default rule, for instance, require the consistency check of the conclusion, In other words, it is necessary to show that any attempt to prove the negation of the conclusion will fail. The lack of an algorithm for deciding non-theoremicity for first-order formulas makes non-monotonic logic undecidable. This feature carries out two consequences. The first one is that the implementation of non-monotonic systems tend to be extremely slow by the necessity of repeatedly check for consistency. The second consequence is that these systems frequently give contradictory conclusions that may be a problem if it is supposed to interact with classical first-order theorem provers. A solution to this problem may be the use of paraconsistent first-order theorem provers, instead of classical ones.

Some approaches to non-monotonic reasoning, as circumscription, avoid contradictions discharging conflicting conclusions. Other approaches, as Reiter's default logic, split contradictory conclusions into multiple extensions, each internally consistent. We will present below a somewhat credulous alternative, the Inconsistent Default Logic, IDL for short, which keep all contradictions into a single extension to be paraconsistently tolerated.

The generation of multiple extensions also carries out another problem: the *anomalous extension problem*. Anomalous extensions are those that are not intuitively intended. They reflect the weakness of several non-monotonic logics to properly deal with preferences among non-monotonic rules. This problem was first revealed in [58, 30] through a temporal projection problem, the famous *Yale Shooting Problem*. This paper inspired the development of several proposals for the formalisation of non-monotonic inferences aiming to avoid such problem as the pointwise circumscription [34], the logic of chronological ignorance [57], the use of nonnormal default rules to avoid anomalous extensions [48], the logic IDL to be here presented, the Cumulative Default Logic with filters [5], [27], among others.

In this paper, we address the proof theory for IDL. In Section 2, we briefly describe the Brazilian Academic Scenario in which the IDL system was developed. In Section 3, we introduce IDL more formally. The sequent calculus for IDL is introduced in Section 4 and, in Section 5, we discuss the cut elimination for this calculus. We summarize our results in the last section.

2 The Brazilian Academic Scenario in the 1990s

Since the publication of the special issue of *Artificial Intelligence* journal specially dedicated to non-monotonic reasoning at the beginning of the 80s [2], the non-monotonic area has grown up impressively around the world. A large number of researchers joined this area, even encouraging the creation of several groups in Brazil working in this topic (and others related to it). A great number of international conferences has been dedicated to this question as the leading conference for *Principles of Knowledge Representation and Reasoning*, with proceedings published by the AAAI Press.

With respect to the Brazilian group, it began at PUC/RJ and then it attracted researchers from several Brazilian universities such as: UFRJ, Unicamp, UFC and UECE. The last two universities founded the Laboratory of Artificial Intelligence (LIA) in 1990.

In Brazil, the starting point to the investigation on the formalisation and automatisation of the non-monotonic reasoning was the critics with respect to the generation of anomalous extensions and the way contradictory conclusions in practical reasoning was treated. It was within this context that the logics IDL (for Inconsistent Default Logic) [31], LEI (for Logic of Epistemic Inconsistency) [33] and DLEF (for Defeasible Logic with Exceptions First) [49], [50] were proposed as early results of our group.

The pair of logics IDL & LEI aims to support two complementary aspects of practical reasoning: non-monotonicity and paraconsistency. The argument in favour of this complement is so resumed [33]: non-monotonic reasoning comes from situations where we are asked to take decisions based on partial knowledge. This knowledge may eventually carry out conflicts which reflects not an inconsistency of the world itself but a lack of knowledge about it. This epistemic inconsistency occurs even under the hypothesis of coherence of reality. Such an inconsistency may eventually be avoided by a deep analysis of the knowledge at hand or by additional information which may allow the solution of the conflict. But there are cases where this is not possible since the available knowledge is not really sufficient to decide what to do. In this cases, the inconsistency is, at least, temporarily unavoidable and must be tolerated until new evidences comes to solve it.

Logics which support inconsistency without trivialising the deductive systems (in the sense of explosion of theorems) are called paraconsistent logics. Paraconsistent logics are logics where the *ex falso sequitur quodlibet principle* does not holds. LEI is a paraconsistent logic proposed to make precise the notion of epistemic inconsistency. It is the monotonic basis of the non-monotonic logic IDL, dealing classically with monotonic conclusions and paraconsistently with plausible conclusions that come from IDL default rules.

After IDL initial proposal in [31], an axiomatic system was designed to LEI with a recursive semantics. In fact, a whole non classical semantic framework was proposed in [1]. We may also cite other approaches to LEI and IDL semantics based on a notion of plausible worlds [**wcp97**, 37, 44, 41].

Based on Buchsbaum's framework [1], a family of paraconsistent, paracomplete (used to model situations where the principle of excluded-middle does not hold) and non-alethic systems (systems that are simultaneously paraconsistent and paracomplete) were developed, beyond LEI and LSR (the Logic for Skeptical Reasoning) which deal with plausible knowledge. Abstract methods of (meta) proofs were also proposed in [1] (see, also, [14]). With such generic methods, it was possible to simultaneously prove important (meta) theorems as completeness and the deduction theorem for the whole family of logics above mentioned.

Proof methods for these logics have also been investigated. In [7], a generalised *tableau* which allows the development of tableau systems for non classical logics was presented. Specific proof methods were developed for IDL & LEI, also dealing with non-monotonic aspects of reasoning [8], [9]. A generic resolution method, which may be used as a proof method for LEI and other paraconsistent logics, was presented in [59]. A natural deduction system, a sequent calculus and a proof theory for LEI was also developed in [53, 42, 43, 38, 39, 46, 47].

Inspired on these works, subsequent results of our research group on non-monotonic logics other than IDL were also published in [18, 16, 17, 23, 15, 22, 21, 20, 19, 24].

2.1 In the route to a sequent calculus

Beyond tableau, resolution and natural deduction systems, a proof theory for IDL and LEI through sequent calculus was also investigated. In spite of sequent calculus being a very appealing instrument for the formal description of logical systems, its use along with non-monotonic logics has not been an usual tool at that time. Some examples of attempts in that direction can be found in the literature [10, 3, 4]. There are, of course, good reasons for that. In order to implement a non-monotonic behaviour, a context sensitive mechanism is mandatory. This mechanism must appeal to global properties, being able to detect changes in the theory that may prevent the application of previously applicable rules. This is the essence of non-monotonicity and it seems to be in frontal opposition to the principle of locality and context independence which is such an essential feature of sequent calculus rules. So, *prima facie*, they just look incompatible with each other.

Nevertheless, the difficulty of making the sequent calculus method applicable to non-monotonic logic has deprived the area of the benefits it is so praised for. Among them, the modularity in expressing the connectives behaviour, enabling a great deal of clarity in their characterisation, the possibility of standardisation in the construction of proofs and a bunch of results and techniques for the analysis of logical systems. Above all, a sequent calculus characterisation of a non-monotonic logic provide it with a much more tractable definition to be used in the proof of meta properties [32], comparing with fixed-point characterisations or semantic ones usually available.

The presentation of a sequent calculus formulation for IDL, with LEI as its monotonic basis, is the subject of the following sections. The sequent calculus for LEI was previously presented in [36, 42, 53, 43]. Here we focus on a sequent calculus for IDL which was published in [45]. The strategy adopted to apply sequent calculus for IDL without deforming it too much has been to confine the disturbing effect of globality, intrinsic to non-monotonicity, to some restricted versions of weakening rules to be used in a precise point within a non-monotonic normal proof. As a result, we have got a formulation holding most of the praised properties such as cut elimination and proof normalisation.

3 The Inconsistent Default Logic

The main idea behind the conception of IDL is the realisation that the disposition to perform inferences under partial evidences lead to contradictions by the occurrence of partial evidences supporting opposite conclusions, in cases where the available knowledge does not provide the means for a definite decision between them. Of course, this is not a novel insight in the field of non-monotonic reasoning, but a well-known feature. What is new is the approach adopted in IDL of assimilating all competing conclusions into a single extension to be paraconsistently treated. By doing so, instead of trying to conceal this inconsistency problem, we have opted out for giving to it a full logical treatment, by recognising the introduction of inconsistencies as an essential feature of non-monotonic reasoning and an unavoidable consequence of the disposition to enlarge deduction.

Another essential feature of IDL is that it marks the conclusions introduced using default rules by a question mark '?' in order to distinguish them from the ones deductively inferred. It is intended to distinguish, within the formalism, the different epistemic status they really have. The marked formulas are defeasible and may be removed by the incoming of fresh knowledge while the others are irrevocable ones. This distinction is crucial for the paraconsistent treatment. Contradictions are allowed in an IDL theory, without trivialising it, only among question marked formulas (?-formulas). We will call such contradictions '$\beta? \wedge \neg\beta?$' a *weak contradiction*.

The basic element of IDL, its default rule, has about the same format of Reiter's one with some changes in its semantics: $\frac{\alpha:\beta;\gamma}{\beta?}$. The peculiarity of IDL relies on the way this rule is interpreted. There is a different treatment given to the normal part 'β' and to the seminormal part 'γ' of the justification of the default rule. To inhibit the application of an IDL rule you must either derive a strong conclusion '$\neg\beta$', or merely a plausible conclusion '$\neg\gamma?$'. The first restriction allows to tolerate and assimilate conflicting defeasible conclusions into a same extension since a weak contradiction '$\beta? \wedge \neg\beta?$' does not prevent the application of the rule. The second restriction reflects a policy of prioritising exceptions called the *exception-first principle* [49], that is, when you want to apply a rule, first check if its exception is not plausible. It is a crucial feature to allow IDL discriminative skills, making it able to successfully avoid the anomalous extension problem as the *Yale Shooting Problem*.

Here we are concerned in equipping IDL with a sequent calculus formulation, one which goes uniformly from its deductive part to its non-monotonic one. In our formulation, we had to use a kind of structural rule not really orthodox in the sequent calculus practise. The employment of this kind of non constructive rule is really unavoidable for the very essence of the non-monotonic reasoning discussed above. We try to minimise the hazardous effects that they may provoke by encapsulating their use in a precise point during the construction of the proofs. We end this section with the definitions of an IDL theory, IDL extension and IDL consequence operator.

Before presenting IDL consequence operator, we would like to compare IDL with Reiter's default logic. IDL differs from Reiter's default logic in the way default rules are tested for their applicability and in how conflicts among default conclusions are solved. The special treatment given to the exception condition part of an IDL rule naturally induces an order among default rules, from which we take advantage here. However, in order to do so, some restrictions in the construction of IDL theories must be introduced. They come from the fact that, although IDL has been designed to assimilate all conclusions, being them contradictory or not, in a single extension, its formalism is powerful enough even to fully express Reiter's rule[1], whose application would lead to multiple extension. On the other hand, badly written specification may lead to zero extension IDL theories. Both are, in fact, ill formulated IDL theories. In order to avoid them, some usage guidelines must be respected [54] and we will present then in Section 4.1. Moreover, in order to make default rules more akin to sequent calculus treatment, the traditional default format $\frac{\alpha:\beta;\gamma}{\beta?}$ is translated into a labelled implicational default, of the form $[\alpha \rightarrow \beta;\gamma] / (\alpha? \rightarrow \beta?)$, where $[\alpha \rightarrow \beta;\gamma]$ is referred as the (supporting) label. Labelled formulas belong to the object language and they obey the same operational rules of the sequent calculus. Having the default rules within the object language, it will be possible to reason about them, to perform reasoning by cases and contraposition [11], obviously respecting label conditions. For details, see [54].

An IDL default theory is a pair $\Delta = <\mathcal{W},\mathcal{D}>$, where $\mathcal{W}$ is the set of facts and $\mathcal{D}$ is the set of default rules. Two languages will be used to present an IDL theory: $\mathcal{L}_?$ and $\mathcal{LL}_?$, for formulas in $\mathcal{W}$ and $\mathcal{D}$, re-

---

[1]Reiter's rule $\frac{\alpha:\gamma}{\beta}$ could be translated to the IDL rule $\frac{\alpha:\beta;\beta\wedge\gamma}{\beta?}$

spectively. $\mathcal{L}_?$ is the LEI language defined as the union of the classical language $\mathcal{L}$ with the set of all formulas in $\mathcal{L}$ suffixed by a question mark ?. $\mathcal{LL}_?$ will be the language used for default rules written as labelled formulas. Cn stands for LEI consequence operator (defined by the same operational, identity and structural rules given in Section 4 but the weakening rules which are the same as the classical ones).

DEFINITION 1 (IDL extension and IDL consequence operator). Let $\Delta = <\mathcal{W}, \mathcal{D}>$ be an IDL theory over $\mathcal{L}_?$ and C an operator on $\mathcal{L}_?$. $C(<\mathcal{S}, \mathcal{D}>)$, for some $\mathcal{S} \in \mathcal{L}_?$, is the minimal set of sentences which satisfies the following properties:

1. $\mathcal{W} \subseteq C(<\mathcal{S}, \mathcal{D}>)$

2. $Cn(C(<\mathcal{S}, \mathcal{D}>)) = C(<\mathcal{S}, \mathcal{D}>)$

3. If $[\alpha \to \beta; \gamma] / (\alpha? \to \beta?) \in \mathcal{D}$, $\neg(\alpha \to \beta) \notin Cn(\mathcal{W})$ and $(\neg\gamma)? \notin \mathcal{S}$, then $\alpha? \to \beta? \in C(<\mathcal{S}, \mathcal{D}>)$.

A set of sentences $\mathcal{E}$ is an extension for Δ iff $C(<\mathcal{E}, \mathcal{D}>) = \mathcal{E}$. C will be understood as the IDL consequence operator.

4 The Sequent Calculus

In this section, the rules will be introduced in Sections 4.2 and 4.3, and the notion of well-presented IDL theory is given in Section 4.1.

4.1 Well-presented IDL Theories

In order to precisely characterise the notion of well-presented IDL theories, some relations must be introduced before.

DEFINITION 2 ($\lessapprox$, $\prec$, $\prec_e$ and $\ll$). Let $\Delta = <\mathcal{W}, \mathcal{D}>$ be an IDL default theory. Without loss of generality, assume that all formulas in $\mathcal{W}$ are in clausal form [36]. Partial relations $\lessapprox$, $\prec$, $\prec_e$ and $\ll$ on literals are defined as follows:

1. If $\alpha \in \mathcal{W}$, then $\alpha = \alpha_1 \vee \ldots \vee \alpha_n$. For all $\alpha_i, \alpha_j \in \{\alpha_1, \ldots, \alpha_n\}$, if $\alpha_i \neq \alpha_j$, let $\sim(\alpha_i) \lessapprox \alpha_j$.[2]

2. If $\delta \in \mathcal{D}$, then $\delta = [\alpha \to \beta; \gamma]/\alpha? \to \beta?$. Let $\alpha_1, \ldots, \alpha_r$, $\beta_1, \ldots, \beta_s$, and $\gamma_1, \ldots, \gamma_t$ be literals obtained from the clausal forms of α, β, and γ, respectively. Then:

    (a) If $\alpha_i \in \{\alpha_1, \ldots, \alpha_r\}$ and $\beta_j \in \{\beta_1, \ldots, \beta_s\}$, let $\alpha_i \lessapprox \beta_j$.

    (b) If $\alpha_k \in \{\alpha_1, \ldots, \alpha_r\}$, $\beta_j \in \{\beta_1, \ldots, \beta_s\}$, $\gamma_i \in \{\gamma_1, \ldots, \gamma_t\}$, let $(\neg\gamma_i)? \prec (\sim \alpha_k)$ and $(\neg\gamma_i)? \prec \beta_j$.

---

[2] $\sim \alpha =_{def} \alpha \to (B \wedge \neg B)$, which behaves as classical negation in LEI [33]

(c) Also, $\beta = \beta_1 \wedge \ldots \wedge \beta_m$, for some $m \geq 1$. For each $i \leq m$, $\beta_i = (\beta_{i,1} \vee \ldots \vee \beta_{i,m_i})$, where $m_i \geq 1$. Thus, if $\beta_{i,j}, \beta_{i,k} \in \{\beta_{1,1}, \ldots, \beta_{m,m_m}\}$ and $\beta_{i,j} \neq \beta_{i,k}$, let $\sim \beta_{i,j} \underline{\ll} \beta_{i,k}$.

3. The following relationships holds for $\underline{\ll}$, $\prec$, $\prec_e$ and $\ll$:

    (a) If $\alpha \underline{\ll} \beta$ and $\beta \underline{\ll} \gamma$, then $\alpha \underline{\ll} \gamma$.
    (b) If $(\alpha \prec \beta$ and $\beta \underline{\ll} \gamma)$ or $(\alpha \underline{\ll} \beta$ and $\beta \prec \gamma)$, then $\alpha \prec \gamma$.
    (c) If $\alpha \prec \beta$, then $\alpha \prec_e \beta$.
    (d) If $\alpha \prec \beta$, $\beta \prec \gamma$ and $\gamma \prec_e \delta$, then $\alpha \prec_e \delta$.
    (e) If $\alpha \prec_e \beta$, $\beta \prec \gamma$ and $\gamma \prec \delta$, then $\alpha \prec_e \delta$.
    (f) If $\alpha \prec \beta$, then $\alpha \ll \beta$.
    (g) If $\alpha \ll \beta$ and $\beta \ll \gamma$, then $\alpha \ll \gamma$.

The intuitive meaning of '$\alpha \underline{\ll} \beta$' ($\alpha$ may help β), '$\alpha \prec \beta$' (α is a direct exception of β), '$\alpha \prec_e \beta$' (α is an exception of β) and '$\alpha \ll \beta$' (α may be an exception of β) is that 'α' appears somehow in the inference of 'β'. The relation '$\underline{\ll}$' is used to link two literals whenever there is a possibility of one appear in a proof of the other. It is a transitive relation since if α may help the proof of β and β may help the proof of γ then α may help the proof of γ. The relation '$\prec$' is used to link the seminormal part (the exception condition) of a default rule to the consequent of the same rule. '$\prec$' and '$\underline{\ll}$' are related in the following way: if α is a direct exception of β and β may help the proof of γ, then α may be a direct exception of γ, or γ may be a direct exception of α. Since the exception of an exception is not an exception, $\prec$ is not transitive. However, it is possible a sort of chain of exceptions and it is defined through the definition of $\prec_e$. The relation '$\prec_e$' is used to link two literals whenever one may be used to block the proof of the other, directly or indirectly, using other formulas in $\mathcal{W}$ or $\mathcal{D}$. Relation '$\ll$' will be used further, in section 4.3. Relations '$\underline{\ll}$' and '$\ll$' are based on the ones defined by Etherington for Reiter's default logic [12] and are used there to build up Reiter's extensions. Here, we use them, in addition to '$\prec$' and '$\prec_e$', to precisely identify the IDL theories we are interested in and for which we will provide a sequent calculus.

Incoherent (zero-extensional) theories are characterised by the occurrence of a literal α such that '$\alpha \prec_e \alpha$'. In cases of theories with multiple extensions, the characterisation comes from the existence of literals α and β such that '$\alpha \prec_e \beta$' and '$\beta \prec_e \alpha$', simultaneously. Acyclic IDL theories are those where such problems do not occur.

DEFINITION 3 (Well-presented IDL theories). The well-presented IDL theories are the acyclic ones.

It is also possible to prove a nice result that summarize our goals:

THEOREM 4 (Uniqueness of Extension). *A well-presented IDL theory has always a single extension [54].*

The *uniqueness of extension* is a much desired result since IDL, with LEI paraconsistent basis, has no need to pull apart conflicts. We will not loose theorems, if compared to Reiter's default theories, but we will keep all extensions into a single one. From a computational point of view, we neither need to keep all extensions nor use extra logical mechanisms to change contexts from one extension to another. The uniqueness of extension does not belong to the policies that guide us in our rational criteria used to represent practical knowledge but it is a very nice and expected result. In fact, it yields to a finer characterisation of IDL consequence operator in the same spirit of deductive consequence operators where only one theory, or extension in the default jargon, is taken into account.

4.2 Operational and Identity Rules

The IDL sequent calculus was designed for the consequence operator given in Definition 1. For the sake of uniformity, we will write any formula α from $\mathcal{W}$ also as a labelled formula, but an empty-labelled one, $[]/\alpha$ ($[]/\Gamma$ is a short form for $\{[]/\gamma \mid \gamma \in \Gamma\}$). The signature of the IDL sequent calculus relation will be $\hspace{0.5em}\sim$: $2^{\mathcal{LL}?} \times 2^{\mathcal{LL}?}$ since it deals with labelled formulas in each side of the sequent.

Some restrictions may be adopted in the way we build up IDL proofs: an IDL sequent calculus proof goes from bottom to top, starting with $[]/\mathcal{W}, \mathcal{D} \hspace{0.5em}\sim\hspace{0.5em} []/\alpha$ until identities are reached where $\Delta = <\mathcal{W}, \mathcal{D}>$ is the default theory in focus and $[]/\alpha$ is the formula to be checked out as an IDL theorem. In the first part of the proof, operational rules, the cut rule and structural rules, but $LW_4\&RW_4$, $LW_5\&RW_5$ and $LW_6\&RW_6$ (which we will call the *global rules*) may be all applied. When their application is no longer necessary, we will reach to what we will call as the *mid-sequent*. This is the appropriate time to apply the global rules which are responsible to find out all the default rules applicable in the extension and to discharge the others. After all, we may not consider the labels and we can treat all remaining formula as irrefutable ones. Therefore, the calculus will be reduced to the monotonic case where any LEI sequent rule is applicable. Recall that the monotonic case is, in fact, LEI sequent calculus [53] since IDL sequent calculus is a proper extension of it. This has been referred as the *supradeductibility* property[3].

The global rules $LW_4\&RW_4$, $LW_5\&RW_5$ and $LW_6\&RW_6$, are the critical ones. They are used to rule out non-applicable default rules, the ones that are not used to build up the extension, and to clear the labels of those taken as the applicable ones. It is not easy to find out such distinction since restrictions for the application of a default must be checked against the final extension, unknown by the time we are searching for the applicable default rules. A default may be considered as an applicable one only if it has no exceptions as theorems. If we discharge a default without checking if its exceptions are provable, we will not obey the exception-first principle. In order to do so, default rules will be ordered by using those partial relations of definition 2 and some auxiliary functions of section 4.3.

By *context* we mean the set of formulas that appear in each side of the sequent. Two-premises rules, such as $L\vee$, keep almost the same left and right contexts Γ and Λ. Operational rules preserve the same label of the main formula and its subformulas. Keeping left and right contexts invariant, as much as possible, assures that the very same IDL theory $<\mathcal{W},\mathcal{D}>$ will be taken into account when exceptions are checked out as theorems, even though we will now consider simpler subformulas.

We will begin by showing IDL operational and identity rules. For the sake of simplicity, we will not present rules for quantifiers.

$$L?\,\frac{\Gamma,\ell/\alpha\mathrel{\mid\!\sim}\Lambda}{\Gamma,\ell/\alpha?\mathrel{\mid\!\sim}\Lambda}(*) \qquad \frac{\Gamma\mathrel{\mid\!\sim}\ell/\alpha,\Lambda}{\Gamma\mathrel{\mid\!\sim}\ell/\alpha?,\Lambda}R?$$

(*) where all formulas in Γ and Λ are ?-closed or α is ?-closed[4].

Restrictions to the application of rule $L?$ may be better understood by making an analogy of ? with S5 possibility $\Diamond$, although they have been shown to be distinct operators [25]. Question mark rules $L?$ and $R?$ reflect LEI axioms for ?, namely: $\alpha \to \alpha?$, $\alpha?? \to \alpha?$, $(\alpha? \to \beta?)? \to (\alpha? \to \beta?)$ beyond others [33].

$$L\wedge\,\frac{\Gamma,\ell/\alpha,\ell/\beta\mathrel{\mid\!\sim}\Lambda}{\Gamma,\ell/(\alpha\wedge\beta)\mathrel{\mid\!\sim}\Lambda} \qquad \frac{\Gamma\mathrel{\mid\!\sim}\ell/\alpha,\Lambda\quad\Gamma\mathrel{\mid\!\sim}\ell/\beta,\Lambda}{\Gamma\mathrel{\mid\!\sim}\ell/(\alpha\wedge\beta),\Lambda}\,R\wedge$$

$$L\vee\,\frac{\Gamma,\ell/\alpha\mathrel{\mid\!\sim}\Lambda\quad\Gamma,\ell/\beta\mathrel{\mid\!\sim}\Lambda}{\Gamma,\ell/(\alpha\vee\beta)\mathrel{\mid\!\sim}\Lambda} \qquad \frac{\Gamma\mathrel{\mid\!\sim}\ell/\alpha,\ell/\beta,\Lambda}{\Gamma\mathrel{\mid\!\sim}\ell/(\alpha\vee\beta),\Lambda}\,R\vee$$

---

[3] Actually, in the literature, supradeductibility is stated as supraclassicality, assuming that classical logic is the underlying monotonic logic. We believe that our generalisation to supradeductibility is on the right spirit.

[4] A formula is ?-closed iff all occurrences of atomic formulas are under the scope of ?.

Note that the rules for $\wedge$ and $\vee$ differ from the rules for these connectives in classical logic, and also in LEI. The classical rules for these connectives embeds a form of monotonicity that may not be generally allowed in IDL.

$$L\rightarrow \quad \frac{\Gamma \mathrel{\mid\!\sim} \ell/\alpha, \Lambda \quad \Gamma, \ell/\beta \mathrel{\mid\!\sim} \Lambda}{\Gamma, \ell/(\alpha\rightarrow\beta) \mathrel{\mid\!\sim} \Lambda} \qquad \frac{\Gamma, \ell/\alpha \mathrel{\mid\!\sim} \ell/\beta, \Lambda}{\Gamma \mathrel{\mid\!\sim} \ell/(\alpha\rightarrow\beta), \Lambda} \quad R\rightarrow$$

Two-premise rules rules, such as $R\wedge$, $L\vee$ and $L\rightarrow$ differ from the correspondent classical ones since we here keep the same set of formulas Γ and Λ on each premise, to maintain contexts invariant as much as possible.

$$L\neg \quad \frac{\Gamma \mathrel{\mid\!\sim} \ell/A, \Lambda}{\Gamma, \ell/(\neg A) \mathrel{\mid\!\sim} \Lambda} \qquad \frac{\Gamma, \ell/\alpha \mathrel{\mid\!\sim} \Lambda}{\Gamma \mathrel{\mid\!\sim} \ell/(\neg\alpha), \Lambda} \quad R\neg$$

Notice that $L\neg$ is restricted to ?-free formulas, expressed by 'A'. This is the key point to achieve the paraconsistent effect of inhibiting the use of *reductio ad absurdum* axiom for ?-formulas, that is, only $(\alpha \rightarrow B) \rightarrow ((\alpha \rightarrow \neg B) \rightarrow \neg\alpha)$ is provable, where 'B' is a ?-free formula.

The second group of rules to be presented is the Identity Rules.

$$[]/\alpha \mathrel{\mid\!\sim} []/\alpha \quad Identity$$

$$\frac{\Gamma \mathrel{\mid\!\sim} \ell/\alpha, \Lambda \quad \Gamma, \ell/\alpha \mathrel{\mid\!\sim} \Lambda}{\Gamma \mathrel{\mid\!\sim} \Lambda} \quad Cut$$

By the restrictions in the way we build up IDL proofs, the Identity rule will only be applied to empty-labelled formulas. In fact, it will be used just at the end of the proof when all non-applicable default rules were ruled out and we find ourselves in a monotonic context where all LEI rules hold (recall again that IDL sequent calculus satisfies supradeductibility).

4.3 Structural Rules

Let us first present some of the well-known structural rules: exchange, contraction and weakening rules, respectively.

$$LX \quad \frac{\Gamma, \ell_1/\alpha, \ell_2/\beta, \Gamma' \mathrel{\mid\!\sim} \Lambda}{\Gamma, \ell_2/\beta, \ell_1/\alpha, \Gamma' \mathrel{\mid\!\sim} \Lambda} \qquad \frac{\Gamma \mathrel{\mid\!\sim} \Lambda, \ell_1/\alpha, \ell_2/\beta, \Lambda'}{\Gamma \mathrel{\mid\!\sim} \Lambda, \ell_2/\beta, \ell_1/\alpha, \Lambda'} \quad RX$$

$$LC \quad \frac{\Gamma, \ell/\alpha, \ell/\alpha \mathrel{\mid\!\sim} \Lambda}{\Gamma, \ell/\alpha \mathrel{\mid\!\sim} \Lambda} \qquad \frac{\Gamma \mathrel{\mid\!\sim} \ell/\alpha, \ell/\alpha, \Lambda}{\Gamma \mathrel{\mid\!\sim} \ell/\alpha, \Lambda} \quad RC$$

$$LW_1 \quad \frac{[]/\Gamma \mathrel{\mid\!\sim} []/\Lambda}{[]/\Gamma, []/\alpha \mathrel{\mid\!\sim} []/\Lambda} \qquad \frac{[]/\Gamma \mathrel{\mid\!\sim} []/\Lambda}{[]/\Gamma \mathrel{\mid\!\sim} []/\alpha, []/\Lambda} \quad RW_1$$

Since there is no underlying meaning attached to the order in which formulas appear in a sequent, any permutation is allowed among those which appear in the same side. This is what is expressed by exchange rules. Contraction rules, on the other hand, merge a double occurrence of a formula into a single one. Therefore, changing places of formulas or duplicating them do not modify left/right contexts of an IDL sequent.

The first two weakening rules, LW_1 and RW_1, deal with empty-labelled formulas. In such context, all monotonic rules are allowed by the fact that empty-labelled formulas are those irrefutable. These rules are, in fact, the very same weakening rules of LEI sequent calculus [53].

$$LW_2 \quad \frac{\Gamma \mid\!\sim \Lambda \quad \Gamma \mid\!\sim []/\alpha}{\Gamma, []/\alpha \mid\!\sim \Lambda}$$

LW_2 shows that an IDL theorem preserves extensions when added to the set Γ of premises as they do not change the set of applicable default rules. This rule reflects the property of Cautious Monotonicity, proved in [36].

$$LW_3 \quad \frac{\Gamma \mid\!\sim \Lambda}{\Gamma, \ell/\alpha \mid\!\sim \Lambda}$$

where $(l_{min}(\ell/\alpha) > l_{max}(\Lambda),\ l_{min}(\Gamma) \leq l_{min}(\ell/\alpha))$.

LW_3 states that, if a labelled formula' ℓ/α does not interfere in the proof of Λ, then it can be added to the left context. This is expressed by the notion of order between labelled formulas and a labelled formulas is greater than another if the former does not enter in the exception chain of the latter. This is expressed by the functions l_{min} and l_{max} presented below.

$$LW_4 \quad \frac{\Gamma \mid\!\sim \Lambda \quad W \vdash \neg(\alpha \to \beta)}{\Gamma, [\alpha \to \beta, \gamma]/\delta \mid\!\sim \Lambda} \qquad \frac{\Gamma \mid\!\sim \Lambda \quad W \vdash \neg(\alpha \to \beta)}{\Gamma \mid\!\sim [\alpha \to \beta, \gamma]/\delta\, \Lambda} \quad RW_4$$

where $l_{min}([\alpha \to \beta, \gamma]/\delta) \geq l_{min}(\Gamma)$ and $l_{min}([\alpha \to \beta, \gamma]/\delta) \geq l_{min}(\Lambda)$

$$LW_5 \quad \frac{\Gamma \mid\!\sim \Lambda \quad \Gamma \mid\!\sim []/(\neg\gamma)?}{\Gamma, [\alpha \to \beta, \gamma]/\delta \mid\!\sim \Lambda} \qquad \frac{\Gamma \mid\!\sim \Lambda \quad \Gamma \mid\!\sim []/(\neg\gamma)?}{\Gamma \mid\!\sim [\alpha \to \beta, \gamma]/\delta\, \Lambda} \quad RW_5$$

where $l_{min}([\alpha \to \beta, \gamma]/\delta) \geq l_{min}(\Gamma)$ and $l_{min}([\alpha \to \beta, \gamma]/\delta) \geq l_{min}(\Lambda)$

$$LW_6 \quad \frac{\Gamma, []/\delta \mid\!\sim \Lambda \quad W \not\vdash \neg(\alpha \to \beta) \quad \Gamma \not\mid\!\sim []/(\neg\gamma)?}{\Gamma, [\alpha \to \beta, \gamma]/\delta \mid\!\sim \Lambda} \qquad \frac{\Gamma \mid\!\sim []/\delta, \Lambda \quad W \not\vdash \neg(\alpha \to \beta) \quad \Gamma \not\mid\!\sim []/(\neg\gamma)?}{\Gamma \mid\!\sim [\alpha \to \beta, \gamma]/\delta\, \Lambda}$$
$$RW_6$$

where $l_{min}([\alpha \to \beta, \gamma]/\delta) \geq l_{min}(\Gamma)$ and $l_{min}([\alpha \to \beta, \gamma]/\delta) \geq l_{min}(\Lambda)$.

Since IDL sequent calculus deals with default rules on both sides of the sequent, we must be careful about what are those that belongs to the extension (applicable default rules) and those that do not (non-applicable default rules). The global rules LW_4 & RW_4, LW_5 & RW_5 and LW_6 & RW_6 are the critical rules. Before reaching identities, all labelled formulas must have empty labels. LW_4 & RW_4 and LW_5 & RW_5 are the ones used to rule out non-applicable default rules and rules LW_6 & RW_6 are used to clear labels of the applicable ones. Obviously, these rules are really hard to compute. The purpose of IDL sequent calculus is not the impossible task of building a (semi)decidable calculus to non-monotonic reasoning but to identify what are (some of) the atomic steps performed in the course of reasoning upon incomplete and inaccurate knowledge.

LW_3 and the last three pairs of weakening rules use functions l_{min} and l_{max} defined upon the partial order $\ll$ of Definition 2. Let $\mathbb{N}$ be the set of natural numbers, Δ the IDL default theory in focus and $literals(\Delta)$ is the set of literals of such theory. With these functions, we can discover and discharge all non-applicable default rules. Function $l : literals(\Delta) \mapsto \mathbb{N}$ encodes the length of longest chain of default rules which could figure in a proof of a literal, that is: (i) let $\alpha, \beta \in literals(\Delta)$, if $\alpha \ll \beta$, then $l(\alpha) \leq l(\beta)$ and if $\alpha \ll \beta$, then $l(\alpha) < l(\beta)$; (ii) if $\beta \in literals(\Delta)$ and for no $\alpha \in literals(\Delta)$ is $(\alpha \ll \beta)$ then $l(\beta) = 0$ and (iii) if $n \in \mathbb{N}, \beta \in literals(\Delta)$ and $l(\beta) > n$, then there is an $\alpha \in literals(\Delta)$ s.t. $(\alpha \ll \beta)$ and $l(\alpha) = n$.

Let ℓ/α be any labelled formula and $\alpha_1, ..., \alpha_n$ its literals. Functions l_{min} and l_{max} are applied to α taking, respectively, the minimum and the maximum result of function l applied to $\alpha_1, ..., \alpha_n$. These functions are inspired on the ones proposed by Etherington for Reiter's default logic [12]. Now, back to the restriction on LW_3, we say that, if it is impossible for ℓ/α be an exception to Λ, since the longest chain of exceptions for ℓ/α is already greater than the longest chain of exceptions for Λ, and since we are complying with the exception-first principle, we can add ℓ/α to the set of premises Γ without interfering in the proof of Λ. In the case of global rules, the same reasoning is adopted: we can introduce a default, no matter whether it is applicable or not, in any context, if it is not an exception to any other. The whole proof of correctness and completeness for this calculus in relation to Definition 1 is presented in [36].

5 Cut Elimination

The notion of a normal proof for a sequent calculus [26] does not imply in a standardisation of a proof procedure such as: 'use left (operational) rules before applying right (operational) rules'. We have only to avoid the use of the cut rule which introduces redundant formulas. The notion of normal IDL sequent proof is:

DEFINITION 5 (Normal IDL sequent proof). A normal IDL sequent proof is a proof where there is no application of the cut rule and the bottom-up process is divided in three parts: (i) the lower part: all operational rules, the cut rule and some structural rules (but the global rules) are successively applied; (ii) the central part: all non-applicable default rules are ruled out and all remaining formulas have their labels cleared; (iii) the upper part: it is of monotonic nature and all LEI rules may be applicable.

THEOREM 6 (Normalisation theorem). *It is always possible to transform an IDL sequent proof to a normal IDL sequent proof [36].*

The proof of such theorem is based on the following ones: (i) LEI cut elimination [53]; (ii) IDL is a LEI conservative extension (see below); (iii) all well-presented IDL theories have an extension, a single one [54]; (iv) IDL satisfies a restricted form of semimonotonicity (see below).

The first two theorems allow us to inherit LEI proof of cut elimination since the IDL cut rule, even if applied to the lower part, may be pushed up to the upper part. If cut occur in the upper part where the context is monotonic then, by the fact that IDL is a LEI conservative extension, that is, $\Gamma \vdash \alpha$ iff $[]/\Gamma \hspace{2pt}\mid\hspace{-4pt}\sim\hspace{2pt} []/\alpha$, it is facultative the choice of LEI or IDL rules at this point. Therefore, the proof of cut elimination for LEI implies the proof of cut elimination for IDL.

The last two theorems assure that the process of discharging all applicable default rules finishes with success. This restricted form of semimonotonicity states that it will always be possible to build up an IDL extension for well-presented theories step-by-step, by checking the application of default rules obeying the exception-first principle. Hence, it will always be possible to identify applicable and non-applicable default rules in a well-presented IDL theory and build up an IDL sequent proof in a normal form. Some main consequences of this theorem are: (i) a proof of the *consistency of the calculus*; (ii) the *subformula property*, an important result for automated deduction since we need just to look for predictable end-sequent subformulas, and (iii) *the inversion principle*: it states that each left rule for a logical constant must have a right rule that introduces the same symbol at the right side. The cut elimination is one way to check if the operational rules are conservative definitions of logical constants. The rewriting process behind cut elimination makes explicit the interaction of left and right rules reflecting the deep symmetries among them and assuring the inversion principle [28].

6 Conclusion

The Brazilian community that works with logic and computer science is very grateful to the pioneer work of Tarcísio Pequeno in non-monotonic and paraconsistent logis, and on philosophical aspects of logic. In this paper, we have tried to give a partial and brief retrospective of the work of his research group in the use of logics to formalize practical reasoning. Here, we focused on the presentation of a proof theory for the Inconsistent Default Logic IDL.

To offer a tool to perform practical computation in IDL or even to provide a decision procedure for the IDL consequence operator which could be useful in supporting efficient computations was clearly not our purpose here. As a direct vehicle to support real life non-monotonic inference, the sequent calculus we have developed is certainly not very suited, keeping untouched all the intrinsic hardness of the problem of performing formal reasoning under non-monotonic environments. Instead of that, by giving a full description of the IDL/LEI system in a uniform fashion, taking the advantages of the elegance and technical skills of sequent calculus, we aimed to expose in a clear way or, at least, from a different perspective, the subtleties of non-monotonic reasoning in general and of IDL in particular and to provide an effective instrument for the formal analysis of its properties.

One of the welcomed features of the sequent calculus style is its ability in providing individualised description of the behaviour of each connective. With operational rules, we can give local definitions of logical symbols by showing, in atomic steps, how they are used. One of the policies adopted in designing operational rules was the one of minimise changes in context, in order of not loosing track of the extension being constructed. Naturally, the application of operational rules may change contexts by inserting and/or deleting formulas in each side of the sequent. So, we must take care of the preservation of the overall context we are dealing with.

Another worthy feature of a sequent calculus is the possibility of dealing with structural aspects of a logic in the object language and not in the meta language, as usually done. Structural rules for non-monotonic logics have a crucial task of discovering which of the default rules have effectively been used in building up the extension. This is not a simple problem, due to the global features inherent to the non-monotonic reasoning. We face such difficult by introducing a partial order among default rules: a lower default may be considered as an exception to a higher one. Since the theory is well-presented, such order will be well-defined. Following this order, a constructive definition of extension is then possible. A default will only be considered as applicable if no lower default, a possible exception, is provable.

Besides its usefulness as a tool to identify and clarify some essential features of practical reasoning, this calculus was used to prove some meta properties of IDL in a very elegant and, sometimes, surprisingly simple way. In fact, some rules are an almost straightforward expression of some (meta) properties such as OR ($L\vee$), S ($R \rightarrow$) and Cumulativity (Cut and LW_2) [32]. In a subsequent paper [40] (and [36]), we prove, by using IDL sequent rules, that IDL consequence operator satisfies the minimal properties required to a cumulative non-monotonic relation, namely: *reflexivity, cut, cautious monotonicity, right weakening* and *left logical equivalence*. Moreover, we show that IDL also satisfies *supradeductibility, OR* and *S*. Such characterisation makes of IDL a preferential logic.

References

[1] A.R. Buchsbaum. "Lógicas da Inconsistência e Incompletude: Semântica, Axiomatização e Automatização". Orientator: T.Pequeno. PhD thesis. Rio de Janeiro: Departamento de Informática, Pontifícia Universidade Católica, 1995.

[2] D.G. Bobrow. "Special issue on nonmonotonic logic". In: *Artificial Intelligence* 13 (1980).

[3] P. A. Bonatti. "Sequent Calculi for Default and Autoepistemic Logics". In: *Proceedings of TABLEAUX'96, LNCS 1071*. Berlin: Springer-Verlag, 1996, pp. 127–42.

[4] P. A. Bonatti and N. Olivetti. "A Sequent Calculus for Skeptical Default Logic". In: *Proceedings of TABLEAUX'97, LNCS 1227*. Berlin: Springer-Verlag, 1997, pp. 107–21.

[5] G. Brewka. "Cumulative Default Logic: in defense of nonmonotonic inference rules". In: *Artificial Intelligence* 50.2 (1991), pp. 183–205.

[6] G. Brewka. "Reasoning About Priorities in Default Logic". In: *Proceeding of AAAI*. 1994.

[7] A. Buchsbaum and T. Pequeno. "O Método dos Tableaux Generalizados e sua Aplicação ao Raciocínio Não Monotônico". In: *O que nos faz pensar: Cadernos do Departamento de Filosofia da PUC-Rio* (1990).

[8] M.S. Corrêa, A.R. Buchsbaum, and T. Pequeno. "Sensible Inconsistent Reasoning: A Tableau System for LEI". In: *Technical Notes of AAAI Fall Symposium on Automated Deduction in Non-Standard Logics*. 1993.

[9] M.S. Corrêa and T. Pequeno. "Raciocínio Automático com Conhecimento Inconsistente II: Um Método Computacional para IDL". In: *Nono Simpósio Brasileiro de Inteligência Artificial.* suportado pela Sociedade Brasileira de Computação. Out, 02 1992, 297–310.

[10] G. Crocco and F. Cerro. "Structure Consequence Relation". In: *What is a Logical System ? Studies in Logic and Computation.* Princeton: Oxford Science Publication, 1994.

[11] J.P. Delgrand and W.K. Jackson. "Default Logic Revisited". In: *Proceedings of the 2nd International Conference on the Principles of Knowledge Representation and Reasoning.* Ed. by J.A.Allen, R.Fikes, and E.Sandewall. San Mateo: Morgan Kaufmann Publishers Inc., Apr. 1991, pp. 118–27.

[12] D.W. Etherington. "Formalizing Nonmonotonic Reasoning Systems". In: *Artificial Intelligence* 31 (1987), pp. 41–85.

[13] D.W. Etherington and R. Reiter. "On Inheritance Hierarchies with Exceptions". In: *Proceedings of the Third National Conference on Artificial Intelligence.* 1983, pp. 104–108.

[14] A.R. Buchsbaum and T. Pequeno. *Uma Família de Lógicas Paraconsistentes e/ou Paracompletas com Semânticas Recursivas.* Monografias em Ciência da Computação 5/91. Rio de Janeiro: Departamento de Informática da PUC-Rio, 1991.

[15] F. M. Ferreira and A. T. Martins. "Expressiveness and Definability in Circunscription". In: *The Bulletin of Symbolic Logic (Also in Annals of the CLE 30 Years XV Brazilian Logic Conference and XIV Latin-American Symposium on Mathematical Logic, p.144, 2008)* 15 (2009), pp. 354–354.

[16] F. M. Ferreira and A. T. Martins. "Expressiveness and definability in circumscription". In: *Manuscrito* 34 (2011), pp. 233–266.

[17] F. M. Ferreira and A. T. Martins. "Recursive definitions and fixedpoints on well-founded structures". In: *Theoretical Computer Science* 412 (2011), pp. 4893–4904.

[18] F. M. Ferreira and A. T. Martins. "Expressible preferential logics". In: *Journal of Logic and Computation* 22 (2012), pp. 1125–1143.

[19] F. M. Ferreira and A. T. C. Martins. "Minimalidade e Hierarquia de Expressividade". In: *XIV Encontro Brasileiro de Lógica, 2006, Itatiaia, Caderno de Resumos.* CLE - UNICAMP, 2006.

[20] F. M. Ferreira and A. T. C. Martins. "The Predicate-Minimizing logic MIN". In: *10th IBERAMIA and 18th SBIA - 20th International Joint Conference, 2006, Ribeirão Preto, Lecture Notes in Artificial Intelligence: Advances in Artificial Intelligence - IBERAMIA - SBIA 2006*. Vol. 4140. Berlin: Springer-Verlag, 2006, pp. 582–591.

[21] F. M. Ferreira and A. T. C. Martins. "On Minimal Models". In: *Logic Journal of the IGPL (Print)* 15 (2007), pp. 503–526.

[22] F. M. Ferreira and A. T. C. Martins. "Minimal Models and Expressiveness Hierarchy". In: *Anais do XXVIII Congresso da Sociedade Brasileira de Computação, 2008, Belém*. Porto Alegre: SBC, 2008, pp. 89–96.

[23] F. M. Ferreira and A. T. C. Martins. "Recursive Definitions and Fixed-Points". In: *Electronic Notes in Theoretical Computer Science* 247 (2009), pp. 19–37.

[24] D. F. Frota et al. "An ALC Description Default Logic with Exceptions-First". In: *Proceedings ofthe Brazilian Conference on Intelligent Systems, 2014, São Carlos. (Also in:Proceedings of Brazilian Conference on Intelligent Systems BRACIS 2014*. Vol. 1. Conference Publishing Services of IEEE Computer Society, 2014, pp. 172–179.

[25] A.T.C. Martins and T. Pequeno. "Paraconsistency and Plausibility in the Logic of Epistemic Inconsistency". Submitted to the organization of the 1st World Congress on Paraconsistency, University of Gent, Belgium, 12 pg, to be published in a journal. 1997.

[26] G. Gentzen. "Investigations into Logical Deductions". In: *The Collected Papers of Gerhard Gentzen*. Ed. by M. E. Szabo. Studies in Logic and The Foundations of Mathematics. Amsterdam: North-Holland, 1969, pp. 68–131.

[27] G. Zaverucha. "On Cumulative Default Logic with Filters". In: *Sixth International Workshop on Nonmonotonic Reasoning*. Timberline, June 1996.

[28] J.-Y. Girard, Y. Lafont, and P. Taylor. "Proofs and Types". In: *Number 7 in Cambridge Tracts in Theoretical Computer Science*. Cambridge University Press, 1988.

[29] D. Gabbay, C.J. Hogger, and J.A. Robinson, eds. *Handbook of Logic in Artificial Intelligence and Logic Programming, Nonmonotonic Reasoning and Uncertain Reasoning*. Vol. 2. Oxford: Clarendon Press, 1994.

[30] S. Hanks and D. McDermott. "Default Reasoning, Nonmonotonic Logis, and the Frame Problem". In: *Proceedings of the Fifth National Conference on Artificial Intelligence*. 1986, 328–33.

[31] T.H.C. Pequeno. *A Logic for Inconsistent Nonmonotonic Reasoning*. Tech. rep. 90/6. London: Department of Computing, Imperial College, 1990.

[32] S. Kraus, D. Lehmann, and M. Magidor. "Nonmonotonic reasoning, preferential models and cumulative logics". In: *Artificial Intelligence* 44 (1990), pp. 167–207.

[33] T.H.C. Pequeno and A.R. Buchsbaum. "The Logic of Epistemic Inconsistency". In: *2nd International Conference on Principles of Knowledge Representation and Reasoning*. Boston, 1991.

[34] V. Lifschitz. *Pointwise Circumscription*. Technical Report. Stanford: Stanford University, 1987.

[35] W. Lukaszewicz. "Considerations on default logic — an alternative approach". In: *Computational Intelligence* 4 (1988), pp. 1–16.

[36] A. T. Martins. "A Syntactical and Semantical Uniform Treatment for the IDL & LEI Nonmonotonic System". PhD thesis. Department of Informatics. Federal University of Pernambuco, 1997.

[37] A. T. C.. Martins. "An Algebraic Semantics for a Nonmonotonic and Paraconsistent Logic". In: *Proceedings of The International Meeting on Language, Logic and Artificial Intelligence*. Fortaleza: Universidade Federal do Ceará, 1998, pp. 15–16.

[38] A. T. C. A Martins. "Proof Theory for the Paraconsistent Logic of Epistemic Inconsistency". In: *Science, Truth and Consistency* 1 (2009), pp. 75–77.

[39] A. T. C. Martins, L. R. Martins, and F. F. Morais. "Natural Deduction and Weak Normalization for the Paraconsistent Logic of Epistemic Inconsistency". In: *Handbook of Paraconsistency: Studies in Logic and Cognitive Systems*. Amsterdam: Elsevier/North-Holland v.9, p, 2007, pp. 355–382.

[40] A. T. C. Martins, M. Pequeno, and T. Pequeno. "Some Characteristics of the Inconsistent Default Logic Reasoning Style". In: *Journal of the Interest Group in Pure and Applied Logics* 4.3 (June 1996), pp. 517–19.

[41] A. T. C. Martins, M. Pequeno, and T. H. C. A Pequeno. "Multiple Worlds Semantics to a Paraconsistent Nonmonotonic Logic". In: *PARACONSISTENCY: THE LOGICAL WAY TO THE INCONSISTENCY*. New York: Marcel Dekker Inc, 2002, pp. 187–211.

[42] A. T. C. Martins and T. H. C. Pequeno. "Proof-Theoretical Considerations about the Logic of Epistemic Inconsistency". In: *Proceedings of the Conference on Philosophical Logic, 1994, Gent*. Gent: University of Gent, 1994, pp. 1–2.

[43] A. T. C. Martins and T. H. C. Pequeno. "Proof-Theoretical Considerations about the Logic of Epistemic Inconsistency". In: *Logique et Analyse* 143.4 (1996), pp. 245–260.

[44] A. T. C. Martins and T. H. C. A Pequeno. "Paraconsistent Nonmonotonic Logic through its Semantics In: , 2000," in: *Proceedings of the Second World Congress on Paraconsistency*. Campinas: CLE- UNICAMP, 2000, pp. 66–67.

[45] A. T. Martins, T. Pequeno, and M. Pequeno. "A Sequent Calculus for a Paraconsistent Default Logic". In: *Proc. of the 6th Workshop on Logic, Language, Information and Computation*. Itatiaia, 1999, pp. 139–149.

[46] F. F. Morais and A. T. C. Martins. "Weak and Strong Normalization for the Paraconsistent Logic of Epistemic Inconsistency". In: *XIII Encontro Brasileiro de Lógica, 2003, Campinas*. Campinas: Centro de Lógica, Epistemologia e História da Ciência, 2003, pp. 74–76.

[47] F. F. Morais and A. T. C. Martins. "Weak Normalization for the Logic of Epistemic Inconsistency In: Third World Congress on Paraconsistency". In: *Proceedings of the Third World Congress on Paraconsistency, WCP III, 2003*. Toulouse: IRIT (Institut de Recherche en Informatique de Toulouse), 2003, pp. 53–53.

[48] C.G. Morris. "The Anomalous Extension Problem in Default Reasoning". In: *Artificial Intelligence* 35 (1988), pp. 383–99.

[49] M. Pequeno. "Defeasible Logic with Exception First". Supervisor: D.Gabbay. PhD thesis. London: Imperial College, 1994.

[50] M. Pequeno. "Pruning Multi-Extensions via Exceptions, II". In: *Dutch/German Workshop on Nonmonotonic Reasoning*. Utrecht, Mar. 1995.

[51] R. Reiter and G. Criscuolo. "On interacting defaults". In: *Proceedings of the Seventh International Joint Conference on Artificial Intelligence*. 1981, pp. 270–276.

[52] R. Reiter. "A Logic of default reasoning". In: *Artificial Intelligence* 13 (1980), pp. 81–132.

[53] A.T.C. Martins and T. Pequeno. "A Sequent Calculus for the Logic of Epistemic Inconsistency". In: *Proceedings of the 11th Brazilian Symposium on Artificial Intelligence*. Federal University of Ceará. Fortaleza, Oct. 1994.

[54] A.T.C. Martins, M. Pequeno, and T. Pequeno. "Well-Behaved IDL Theories". In: *Lecture Notes in Artificial Intelligence, 1159:11-20*. Proceedings of the 13th Brazilian Symposium on Artificial Intelligence. Curitiba: Springer-Verlag, Oct. 1996.

[55] T. Schaub. "Assertional Default Theories: a semantical view". In: *Proceedings of the Second International Conference on the Principles of Knowledge Representation and Reasoning*. Ed. by R.Fikes J.A.Allen and E.Sandewall. San Mateo: Morgan Kaufmann Publishers Inc., Apr. 1991, pp. 496–506.

[56] T. Schaub. "Considerations on Default Logics". PhD thesis. Darmstadt: Technische Hochschule Darmstadt, 1992.

[57] Y. Shoham. "Chronological Ignorance". In: *Proceedings of the Fifth National Conference on Artificial Intelligence*. 1986, pp. 389–93.

[58] S. Hanks and D. McDermott. "Nonmonotonic Logic and Temporal Projection". In: *Artificial Intelligence* 33 (1987), pp. 27–39.

[59] J.M. Coelho Filho. "Um Ambiente Genérico de Dedução Automática Baseado em Resolução". Orientator: T.Pequeno. MA thesis. Departamento de Informática, Pontifícia Universidade Católica, 1993.

Modality, Paraconsistency and the Concept of Inductive Plausibility

RICARDO S. SILVESTRE

1 Combining Modality and Paraconsistency: The Logic of Epistemic Inconsistency

The practical side of the problem of combining logics [6] is one of its most appealing aspects. Take knowledge representation for example. An agent able to interact with its external environment has to represent not only its beliefs about the external world and its internal states but also how these beliefs change during time. Supposing that the agent representation mechanism is a logic-based one, we have then to face the problem of combining doxastic logic with temporal logic. If besides this the agent is equipped with information about obligations and permissions, we will have the further task of adding a deontic component to the combined system.

Paraconsistent logics are logics able to formalize inconsistent but non-trivial theories [9]. They have been advertised as important tools in applications such as knowledge representation, multi-agent systems and database management. In the situation above described, for example, the agent might have evidences both to believe and not to believe something, or its normative component might both require and prohibit something, in cases of which a paraconsistent mechanism might be required. Therefore it seems quite natural to combine paraconsistent logic with logics able to represent modalities; this has been done in the context of relevant logics (which is one type of paraconsistent logic) [12] [25] [34], da Costa's systems [10] [22] and "truth-value gluts" paraconsistent logics [14].

Something remarkable about modal logic and paraconsistent logic, and consequently about the enterprise of combining them, is the alleged relation that exists between the two classes of systems [2] [4] [5] [24]. Take the following definition of paraconsistency: a paraconsistent negation is a unary operator that does not satisfy the principle of explosion (for any formulas α and β, $\{\alpha, \neg\alpha\} \vdash \beta$) and has enough properties to be called

Copyright © 2019 by Ricardo S. Silvestre. All rights reserved.

a negation; a paraconsistent logic then is a logic having a paraconsistent negation [4]. If we define $\sim$ as $\Diamond\neg$, we will have that S5 (meant as a consequence relation), for instance, contains a unary operator that does not satisfy the principle of explosion and has enough properties to be called a negation, which would entitle us to classify it as a paraconsistent logic [4].

This is of course due to the logical properties of $\Diamond$ when used along with classical negation; or, more specifically, to the fact that $\Diamond\neg\alpha$ can get along with α and $\Diamond\alpha$ without trivializing the theory. This fact, which we know is the basis of Jaśkowski's calculus for contradictory deductive systems [19], has led some to speak of a subtler sort of paraconsistency, named hertian [3] and conceptual [35] paraconsistency. Of course we could speak of a true or formal paraconsistency (regarding the primitive symbol $\neg$ in connection with $\Diamond$) if we had, for some α and β, something like $\{\neg\Diamond\alpha, \Diamond\alpha\} \nvdash \beta$; this would certainly make stronger the claim that modal logic is paraconsistent.

In [31] a paraconsistent modal logic called LEI (*Logic of Epistemic Inconsistency*) was proposed to serve as the monotonic base for a version of Reiter's default logic [33]. This gave rise to a whole family of paraconsistent modal systems and a quite significant contribution to the field of non-classical logic [1] [27] [28] [35] [7] [26] [38] [39]. The original idea was to mark formulas obtained through the use of defaults with a modal operator – the symbol ?, which was used in a post-fixed notation – so that they might be treated paraconsistently. α? was read as "α is plausible"; LEI was therefore a sort of logic of plausibility or, if you will, an attempt to formalize the concept of plausibility. It was presented both proof-theoretically, through Hilbert's axiomatic method, and semantically, through a Kripkean-like model theoretic framework.

From the proof-theoretical point of view, the paraconsistency was obtained by a slight modification on the *reductio ad absurdum* axiom:

(1) $\quad (\alpha \rightarrow B) \rightarrow ((\alpha \rightarrow \neg B) \rightarrow \neg\alpha)$

, wherein B is a ?-free formula. In this way, formulas such as β? and $\neg(\beta?)$ are not able to trivialize the theory. However, because the negative ?-marked formulas introduced through default rules are of the form $(\neg\alpha)$?, in order for the paraconsistency to be really required, there had to be some way to transform $(\neg\alpha)$? into $\neg(\alpha?)$. Therefore the following axiom:

(2) $\quad \neg(\alpha?) \leftrightarrow (\neg\alpha)?$

There were some interesting things about this axiom. First, it allows us to go from conceptual paraconsistency, say, to formal paraconsistency.

From a semantic point of view, ? corresponds to the ◊ operator of traditional modal logic: α? is true iff α is true in at least one member of a set of worlds (which in this case might be called plausible worlds). Therefore it shares with ◊ the property of tolerating contradictions of the form $\{\alpha?, (\neg\alpha)?\}$: there is a model which satisfies both α? and $(\neg\alpha)$?. But since from $(\neg\alpha)$? we get $\neg(\alpha?)$, we have that this model also satisfies α? and $\neg(\alpha?)$, which is the same as saying that its paraconsistency also applies to true contradictions of the form $\{\alpha?, \neg(\alpha?)\}$.

Second, the existence of a ◊-like operator suggests a □-like operator with an axiom corresponding to (2). Letting this operator be represented by ! (also used in a post-fixed notation), we would have the following formula as the aforementioned axiom:

(3) $\neg(\alpha!) \leftrightarrow (\neg\alpha)!.$

Since ! is to be interpreted like □ - α! is true iff α is true in all plausible worlds – we have that ! can be said to represent a sort of strong plausibility, with ? representing a weak plausibility. Second, there shall be a semantic model in which neither $(\neg\alpha)$! nor α! are true. But since by (3) $(\neg\alpha)$! implies $\neg(\alpha!)$, this model shall not satisfy $\neg(\alpha!)$ either. Therefore, we have a model in which neither α! nor $\neg(\alpha!)$ are satisfied; this entails, for instance, that the excluded middle principle is not universally valid.

The property a logic might have of not satisfying the principle of excluded middle, which has been named paracompleteness [21], is usually taken as the dual of paraconsistency [3]. Logics which are both paraconsistent and paracomplete have been called paranormal or non-alethic logics[1]. To have then such a !-operator would 'generate' a logic in which the modal operators ! and ? when taken along with ¬ would exhibit, respectively, the dual properties of paracompleteness and paraconsistency. Besides this, if this logic were presented in the closest possible way to standard modal logic, in special along with a standard Kripkean possible world semantics, we would have an instance of combining logics which would possibly shed some light on the relation between paraconsistency, paracompleteness and modality. Such a logic could be fairly called *paranormal modal logic*[2].

---

[1]This "para" notation is due to Miró Quesada [3].

[2]The term "paranormal modal logic" might certainly be misleading, for "normal modal logic" has already an established meaning in modal logic literature – a normal modal logic is a modal logic in which K ($\Box(\alpha \to \beta) \to (\Box\alpha \to \Box\beta)$) is valid and has *modus ponens*, generalization (from α conclude $\Box\alpha$) and the rule of uniform substitution as inference rules [17] – and the "para" in "paranormal" does not apply

Now, despite the fact that all resources, I may say, for this paranormal modal logic were present in [31] and subsequent works, some things prevented it from arising. First, despite clearly using resources of modal logic, most systems of the LEI family were presented in a non-standard way, so that the fact that they were a sort of modal logic, or more generally, a combination of modal logic with something else was not sufficiently clear. Second, except for the quite trivial fact that in modal logic the right-to-left side of (2), $(\neg\alpha)? \to \neg(\alpha?)$, does not hold, very few was said about the formal relations between LEI systems and traditional modal logic. Thirdly, no satisfactory philosophical justification for the right-to-left side of (2) was given. This last point is important for if we take ? as representing the notion of plausibility (or weak plausibility), we have to give very convincing arguments that (2) is an important feature of this notion, indeed one of the features which would distinguish it from the notion of possibility[3].

In connection to this it is worthy noticing that in their latest published work about the subject [7], Pequeno and Buchsbaum use as the monotonic base of their nonmonotonic system a version of LEI in which neither the right-to-left side of (2) nor the left-to-right side of (3) are valid. In other words, they present a so-called logic of plausibility which, from the structural point of view, is indistinguishable from the logic of possibility. This, we might say, was a consequence of their failure to find a suitable justification for the odd sides of (2) and (3). In its turn, this was a consequence of a regrettable lack of emphasis on the conceptual and philosophical side of LEI, which in its turn prevented it from being classified, I may say, as a genuine attempt to explain the concept of plausibility.

2 Paranormal Modal Logic

In [35] a first attempt was made to fill in these gaps and present a *paranormal modal logic* [4]. It was presented both proof-theoretically and semantically, being these two forms of presentation equivalent to each other (the system is sound and complete). Only the propositional case was presented. From the semantic point of view, it was used a model-

---

to it. However, the other term used to designate logics that are both paraconsistent and paracomplete, "non-alethic", has also a well-established meaning in modal logic literature. In the lack then of a better term, I have decided to stick to this 'para' tradition of logical notation.

[3]The same can be said about the left-to-right side of (3) and the notions of strong plausibility and necessity

[4]As far as I know, the only work which resembles what I am calling paranormal modal logic is [30]; it gets to paranormal analogues to K by taking intuitionistic modal versions of K as starting points.

theoretical structure as closest as possible to the Kripkean structures used in traditional modal logic. In special, the semantic model uses the same three elements of propositional modal logic: a set of worlds, in this case called plausible words, an accessibility relation and a truth valuation function. Later on in [37], [38] and [39] this paranormal modal logic was presented inside a general framework in which a wide range of logics, including classical logic and traditional normal modal logic, could be defined. This had a couple of advantages.

First of all, paranormal modal logic could be now seen not as an individual logic, but as a family of logics akin to the family of normal modal logics. In the same way that the system K can be extended into D, T, B, S4, S5, etc., the most basic paranormal modal logic $K_?$ can be extended into corresponding paranormal modal systems: add $\alpha! \to \alpha?!$ (axiom $D_?$) to $K_?$ and you have system $D_?$; add $\alpha! \to \alpha$ (axiom $T_?$) to $K_?$ and you have system $T_?$; add $\alpha \to \alpha?!$ (axiom $B_?$) to $T_?$ and you have system $B_?$; add $\alpha! \to \alpha!!$ (axiom $4_?$) to $T_?$ and you have system $S4_?$; add axiom $B_?$ to $S4_?$ and you have $S5_?$. Below there is the axiomatics of this most basic of paranormal modal logics, $K_?$:

Positive Classical Axioms

P1: $\alpha \to (\beta \to \alpha)$

P2: $(\alpha \to (\beta \to \varphi)) \to ((\alpha \to \beta) \to (\alpha \to \varphi))$

P3: $\alpha \wedge \beta \to \alpha$

P4: $\alpha \wedge \beta \to \beta$

P5: $\alpha \to (\beta \to \alpha \wedge \beta)$

P6: $\alpha \to \alpha \vee \beta$

P7: $\beta \to \alpha \vee \beta$

P8: $(\alpha \to \beta) \to ((\varphi \to \beta) \to (\alpha \vee \varphi \to \beta))$

P9: $((\alpha \to \beta) \to \alpha) \to \alpha$

Paranormal Classical Axioms

A1: $(\alpha \to \beta) \to ((\alpha \to \neg\beta) \to \neg\alpha)$, wherein β is ?-free and α is !-free

A2: $\neg\alpha \to (\alpha \to \beta)$, wherein α is ?-free

A3: $\alpha \vee \neg\alpha$, wherein α is !-free

Non-Positive Additional Classical Axioms

N1: $\neg(\alpha \to \beta) \leftrightarrow (\alpha \wedge \neg\beta)$

N2: $\neg(\alpha \wedge \beta) \leftrightarrow (\neg\alpha \vee \neg\beta)$

N3: $\neg(\alpha \vee \beta) \to (\neg\alpha \wedge \neg\beta)$

N4: $\neg\neg\alpha \leftrightarrow \alpha$

Paranormal Modal Axioms

K1: $\alpha? \leftrightarrow \sim((\sim\alpha)!)$

K2: $(\neg\alpha)! \leftrightarrow \neg(\alpha!)$

K3: $(\neg\alpha)? \leftrightarrow \neg(\alpha?)$

Modal Axioms

$K_?$: $(\alpha \to \beta)! \to (\alpha! \to \beta!)$

Rules of Inference

MP: $\alpha, \alpha \to \beta / \beta$

$N_?$: $\alpha/\alpha!$

As one would expect, from the semantic point of view the differences between these paranormal systems are the same as between their normal counterparts: while the accessibility relation of $K_?$ has no restriction at all, $D_?$'s models are serial, $T_?$'s are reflexive, $B_?$'s are reflexive and symetric, $S4_?$'s are reflexive and transitive and $S5_?$'s are reflexive, transitive and symmetric.

Second, as it shall be already clear from what I have said above, by using this general framework it was possible to have a more precise comparative analysis between paranormal modal logic and and other logics. In special, it was easier to determine the specific features of paranormal modal logic which make it to depart from traditional modal logics. It was also easier to see the similarities between these two families of logics. For instance, despite contrary appearances, it was shown that in a very important sense paranormal modal logic and normal modal logic are equivalent to each other. Letting $\Im_\Diamond$ be normal modal logic language (with $\Box$ and $\Diamond$ as modal operators) and $\Im_?$ paranormal modal logic language (with ! and ? used in post-fixed notation as modal operators) we define the following pairs of functions (Π and $\amalg$ and Δ and ∇):

(i) $\Pi(p) = \amalg(p) = p$;

(ii) $\Pi(\alpha?) = \Diamond\Pi(\alpha)$;

(iii) $\amalg(\alpha?) = \Box\amalg(\alpha)$;

(iv) $\amalg(\alpha!) = \Box\Pi(\alpha)$;

(v) $Ш(α!) = ◊Ш(α)$;
(vi) $Π(¬α) = ¬Π(α)$;
(vii) $Ш(¬α) = ¬Ш(α)$;
(viii) $Π(α ⊕ β) = Π(α) ⊕ Π(β)$, where $⊕ ∈ \{∧, ∨, →\}$;
(ix) $Ш(α ⊕ β) = Ш(α) ⊕ Ш(β)$, where $⊕ ∈ \{∧, ∨\}$;
(x) $Ш(α → β) = Π(α) → Ш(β)$.

(i) $Δ(p) = ∇(p) = p$;
(ii) $Δ(◊α) = Δ(α)?$;
(iii) $∇(◊α) = ∇(α)!$;
(iv) $Δ(□α) = Δ(α)!$;
(v) $∇(□α) = ∇(α)?$;
(vi) $Δ(¬α) = ¬∇(α)$;
(vii) $∇(¬α) = ¬Δ(α)$;
(viii) $Δ(α ⊕ β) = Δ(α) ⊕ Δ(β)$, where $⊕ ∈ \{∧, ∨, →\}$;
(ix) $∇(α ⊕ β) = ∇(α) ⊕ ∇(β)$, where $⊕ ∈ \{∧, ∨\}$;
(x) $∇(α → β) = Δ(α) → ∇(β)$.

Let $A ⊆ \mathfrak{F}_?$ and $B ⊆ \mathfrak{F}_◊$.

(i) $Π(A) = \{Π(α) \mid α ∈ A\}$;
(ii) $Ш(A) = \{Ш(α) \mid α ∈ A\}$;
(iii) $Δ(B) = \{Δ(α) \mid α ∈ B\}$;
(iv) $∇(B) = \{∇(α) \mid α ∈ B\}$.

Letting then $⊢_K$ and $\models_K$ be, respectively, the deductive and logical consequence relations of normal modal logic K, and $⊢_{K?}$ and $\models_{K?}$ the deductive and logical consequence relations of paranormal modal logic $K_?$, the following theorems could be proved[5]:

$A ⊢_K α$ iff $Δ(A) ⊢_{K?} Δ(α)$
$A ⊢_{K?} α$ iff $Π(A) ⊢_K Π(α)$
$A \models_K α$ iff $Δ(A) \models_{K?} Δ(α)$
$A \models_{K?} α$ iff $Π(A) \models_K Π(α)$

---

[5] In the construction of the inference relations, [37] and [39] used two sets of premises – the so-called local and global premises – instead of just one. See [11].

Thirdly, this way of presenting things made it easier to deal with paranormal modal logic's aspect as a combination of modal logic with paraconsistent and paracomplete logics. It was straightforward, for instance, to combine in this framework paraconsistent and paracomplete logic with first-order modal logic and obtain a new family of logics: paranormal first-order modal logic [39]. In [39] it also introduced a logic resultant from the combination of paranormal modal logic with normal modal logic, resulting in logic with two pairs of modalities, one ($\lozenge$ and $\square$) behaving classically and other (? and !) having a paraconsistent and paracomplete behavior.

3 Plausibility and the Two Approaches to Induction

Paranormal modal logic was presented as a real attempt to formalize or conceptually explain the notion of plausibility. The philosophical motivation of such a project, which is described in detail in [37], relies on the problem of inductive inconsistencies [15] [16] [18] [32] [13] and the connection it might be shown to have with two different but complementary approaches: the *skeptical* and the *credulous* approaches to induction. Justifying the right-to-left side of (2) and the left-to-right side of (3) is an important part of such explanatory endeavor.

Let Δ be a consistent set of statements. Supposing the existence of some inductive (in the sense of a rational non-deductive [8]) mechanism of inference I (which might simply be a set of inductive inferential rules) to be applied to the members of Δ, we name the deductive closure of each consistent set of conclusions obtained from Δ an *inductive extension*. A trivial consequence of this definition is that the cases where contradictions are obtained from the application of I to Δ lead to more than one inductive extension. In these cases, we have at least two options at our disposal: to ignore contradictions and recognize as sound only those inductive conclusions belonging to the intersection of all extensions, or to take contradictions seriously and accept as authentic inductive conclusions all statements belonging to the union of all extensions. While the first option is a strict or skeptical approach which requires a great deal to accept an inductive conclusion as sound, the second is a tolerant or credulous approach which requires just the minimum to accept a statement as an authentic inductive conclusion.

This distinction between a skeptical approach and a credulous one is not new [29] [31] [23]. It has been used in the nonmonotonic literature, for instance, to classify some of the available formalisms to common sense reasoning [23]. What is new however is its being used to name two general approaches to inductive reasoning, which are in fact the end-product of a conceptual analysis to the notion of induction which takes seriously the phenomenon of inductive inconsistencies [36]. As far as the philosophical literature is concerned, even though the existence of these two approaches to induction has not been explicitly acknowledged, it is possible to identify isolated uses of them in several discussions related to the problem of inductive inconsistencies. It can be shown for instance how some of the main solutions given to the lottery paradox [20] can be seen as instances either of a skeptical approach or of a credulous one, and that when we recognize these approaches as complementary instead of competing, the whole controversy regarding the proper solution to the lottery paradox is dissolved [36].

Now, supposing that we can effectively infer new conclusions from our set of inductive inferential rules, it seems natural to qualify such conclusions so to distinguish them from deductively obtained conclusions, for example. The common attitude in philosophy has been to use some probability notion to do the job. In order to distinguish such concept of probability from other probability notions, in special from his notion of logical probability, Carnap [8] used the term "pragmatical probability". I shall use here the less controversial term "plausibility" or "inductive plausibility". What I have called so far inductive conclusions are thus the same as plausible conclusions, plausible statements or plausible hypotheses.

Notice that according the general idea being presented here we cannot speak of inductive conclusions *per se*. Instead, we must speak of inductive or plausible conclusions *according* to this or that approach: when α is true in all inductive extensions we say that α is plausible according to a skeptical approach, and when α is true in at least one extension we say that α is plausible according to a credulous approach. Trivially then, the skeptical and credulous approaches work as evaluation functions which assess in different ways the truthfulness of plausible statements, giving rise in fact to two plausibility notions: what we might call *skeptical plausibility* and *credulous plausibility*.

From a general point of view, we can say that the credulous and skeptical approaches represent, respectively, minimizing and maximizing strategies of truth assessing. If one adopts a credulous position, for example, he will not require too much to accept statement α as plausible. If we use 1 to represent truth and 0 to represent falsehood, this can be restated by saying that he will somehow try to *maximize* or bring close to 1 the truth-value of plausible statements. On the other hand, if one adopts a skeptical position he will be more demanding in the matter of accepting α as plausible, which means that he will try to *minimize* or bring close to 0 the truth-value of plausible statements. Therefore, while adopting a skeptical position means to be strict in the matter of accepting something as true, in our case the plausibility of sentences, adopting a credulous position means to be tolerant, not so demanding in the matter of taking something as truth.

What has been said so far can be quite fairly represented with the aid of a Kripkean semantic framework. First of all, each inductive extension might be naturally associated with a possible world, in this case of a special kind of world named *plausible world*. Second, following the notation introduced in the previous sections and representing our notions of skeptical and credulous plausibility with the help of the modal operators ! and ? – α! means "α is plausible according to a skeptical approach" and α? "α is plausible according to a credulous approach" – we have that while ! is interpreted alike to traditional modal operator $\Box$, ? is interpreted alike to $\Diamond$. In other words, α! is true iff α is true in all plausible worlds, and α? is true iff α is true in at least one plausible world[6].

Since the key semantic notion is the notion of plausible worlds, there will be important conceptual differences between a logic of plausibility so conceived and the logic of possibility and necessity as formalized, say, by S5. But there will be important similarities too. For instance, since every plausible world is a possible world, we might set the following relations between the notions of necessity, possibility, skeptical plausibility and credulous plausibility: $\Box \alpha \rightarrow \alpha!$, $\alpha! \rightarrow \alpha?$ and $\alpha? \rightarrow \Diamond \alpha$ [7]. More important however is that this logic of plausibility seems to have the same formal structure as traditional modal logic, so that the task of building a logic of plausibility would be reduced to the task of deciding which one of the normal modal systems, say, is more adequate to our needs. This in fact would be so if it were not for the following fact: *having a skeptical and a credulous approach to evaluate the truth value of plausible formulas causes the negation operator to behave in a way that traditional modal logic simply cannot handle.*

To start with, let us examine how the notion of implausibility would be represented inside our sketched framework. First of all, for all intends and purposes, the notion of implausibility might be seen simply as the negation of plausibility, so that "α is implausible" can be taken as an abbreviation to "it is not the case that α is plausible". But since here the concept of plausibility is being taken obligatorily according either to a credulous view or to a skeptical view, the same should be done to all notions derived from it, in special to the notion of implausibility. Therefore, we shall have something like (I) and (II) below:

---

[6]I am here neglecting the accessibility relation component which, as it was said, is an important part of paranormal modal logic's formal semantics.

(I)   it is not the case that α is plausible (α is implausible) according to a skeptical position.

(II)   it is not the case that α is plausible (α is implausible) according to a credulous position.

According to our notation, (I) and (II) are trivially represented as $\neg(\alpha!)$ and $\neg(\alpha?)$, respectively.

Note however that there is an ambiguity in the reading of these two sentences. Are we negating the plausibility of α according to such and such approach, or are we negating, according to that approach, the plausibility of α? This can be better seen with the help of brackets, where (i) or (ii) below correspond to each one of the two possible ways we can read (I) and (II):

(i)   it is not the case that [α is plausible according to a skeptical (credulous) position].

(ii)   [it is not the case that α is plausible] according to a skeptical (credulous) position.

In the skeptical case, for example, while (i) means that we were not able to take "α is plausible" as truth according to a rigid, strict posture, (ii) means that we *did* succeed in the task of attributing "true" to the sentence "α is not plausible" according to a posture that requires quite a lot to attach "true" to any sentence, including of course the sentence "α is not plausible." Similarly for the credulous case: while (i) means that adopting a tolerant posture concerning truth-assignment we were not able to classify "α is plausible" as true, all that (ii) says is that "α is not plausible" is true according to a posture whose goal is to maximize the truth of sentences or, we may say, to easily classify statements a true.

Now, (i) clearly involves a negation pretty much alike to the negation of traditional modal logic: (i) is true iff α is false in at least one world, in the case of the skeptical approach, and iff α is false in all worlds, in the case of the credulous one. Regarding (ii), however, the situation seems to be quite different: instead of denying that α is plausible according to a specific position, (ii) is in fact classifying the whole sentence "it is not the case that α is plausible" as true according to a specific position. Our question now is how to semantically analyze this sort of negation. According to what I have explained above, to evaluate "α is not plausible" according to a skeptical position means that we shall be very strict, requiring the maximum we can to classify "α is not plausible" as true. On the other hand, to evaluate "α is not plausible" according to a credulous position means exactly the opposite: that we shall be very tolerant, requiring the minimum we can to classify "α is not plausible" as true. Given the semantic framework sketched here, clearly to require the *maximum* we can to classify "α is not plausible" as true means to require α to be false in all plausible worlds, and to require the *minimum* we can to classify "α is not plausible" as true is tantamount to requiring α to be false in at least one world. This means that the skeptical version of (ii), or in symbols $\neg(\alpha!)$, is true iff α is false in all plausible worlds, and the credulous version of (ii), or in symbols $\neg(\alpha?)$, is true iff α is false in at least one plausible world.

As already mentioned, a trivial presupposition present in analyses such as the one I am doing here is that the notion of implausibility is to be analyzed, represented or described in terms of the concepts of negation and plausibility. As consequence of that, it can be claimed that a fundamental step in the task of formally disambiguating statements (I) and (II) involves having two different negations, one for each reading of (I) and (II). Let us use the symbol $\sim$ refer to the negation involved in (i) and $\neg$ to the negation involved in (ii), so that the first reading of (I) and (II) might be formally represented as $\sim(\alpha!)$ and $\sim(\alpha?)$, respectively, and the second reading of (I) and (II) as $\neg(\alpha!)$ and $\neg(\alpha?)$, respectively. While $\sim$ is a negation which interprets $\sim(\alpha!)$ and $\sim(\alpha?)$, respectively, in exactly the same way as $\sim \Box\alpha$ and $\sim \Diamond\alpha$ in traditional modal logic, $\neg$ has a different, non-classical behavior, according to which $\neg(\alpha!)$ is true iff α is false in all plausible worlds and $\neg(\alpha?)$ is true iff α is false in at least one plausible world. About the relations between these two negations, it is easy to see that neither $\neg\alpha \rightarrow \sim\alpha$ nor $\sim\alpha \rightarrow \neg\alpha$ are generally valid: even though $\neg(\alpha!) \rightarrow \sim(\alpha!)$ holds, $\neg(\alpha?) \rightarrow \sim(\alpha?)$ is not valid; and even though $\sim(\alpha?) \rightarrow \neg(\alpha?)$ holds, $\sim(\alpha!) \rightarrow \neg(\alpha!)$ is not valid.

One might say that there is still a third reading to (I) and (II):

(iii)   [it is not the case that α] is plausible according to a skeptical (credulous) position.

This, however, does not correspond to the implausibility of α, but to the plausibility of the negation of α, which is trivially represented as $(\approx\alpha)!$ and $(\approx\alpha)?$, where $\approx$ is any one of our two negations. Differently from $\approx(\alpha!)$ and $\approx(\alpha?)$, however, there is no ambiguity in the reading of $(\approx\alpha)!$ and $(\approx\alpha)?$, and therefore no need for $\neg$ and $\sim$ having two different interpretations. Hence, both negations shall evaluate (iii) in the usual way. In special, $(\neg\alpha)!$ and $(\neg\alpha)?$ will be true, respectively, iff α is false in all plausible worlds and iff α is false in at least one plausible world. This however is the same evaluation which, we have agreed above, should be given to $\neg(\alpha!)$ and $\neg(\alpha?)$ in order to account for the second reading of (I) and (II). Therefore, as far as our second negation is concerned, $\neg(\alpha!)$ is semantically equivalent to $(\neg\alpha)!$ and $(\neg\alpha)?$ is semantically equivalent to $\neg(\alpha?)$, or in symbols, $\neg(\alpha!) \leftrightarrow (\neg\alpha)!$ and $(\neg\alpha)? \leftrightarrow \neg(\alpha?)$. Needless to say, these are exactly the axioms (2) and (3) of Section 1.

Notice that since $\alpha!$ is true iff α is true in all worlds and $\neg(\alpha!)$ is true iff α is false in all worlds, it might happen that neither $\alpha!$ nor $\neg(\alpha!)$ are true, which of course means that regarding !-marked formulas the logic of skeptical and credulous plausibility has a *paracomplete* behavior. Similarly, $\alpha?$ is true iff α is true in at least one world. But since $\neg(\alpha?)$ is true iff α is false in at least one world, we might have a model that satisfies both $\alpha?$ and $\neg(\alpha?)$, making $\neg$ correspond to what I have called earlier a true paraconsistent negation. See however that since non-modal formulas have no connection whatsoever with our two approaches, along with such formulas $\neg$ shall behave classically. We have then that one of the negations has a plural behavior: in connection with ?-marked formulas it behaves paraconsistently, in connection with !-marked ones it behaves like a paracomplete negation, and along with non-modal formulas it behaves classically.

References

[1] J. Akker and Y. Tan. "Qml: a paraconsistent default logic". In: *Logique & Analyse* 143-144 (1993), pp. 311–328.

[2] D. Batens. "On the remarkable correspondence between paraconsistent logics, modal logics and ambiguity logics". In: *Paraconsistency: The Logical Way to the Inconsistency*. Ed. by M. Coniglio W. Carnielli and I. d'Ottaviano. New York: Marcel Dekker, 2002, pp. 445–454.

[3] J. Béziau. "The future of paraconsistent logic". In: *Logical Studies* 2 (1999), pp. 1–23.

[4] J. Béziau. "S5 is a paraconsistent logic and so is first-order classical logic". In: *Logical Studies* 9 (2002), pp. 301–309.

[5] J. Béziau. "Paraconsistent logic from a modal viewpoint". In: *Journal of Applied Logic* 3 (2005), pp. 7–14.

[6] P. Blackburn and M. de Rijke. "Why combine logics?" In: *Studia Logica* 59 (1997), pp. 5–27.

[7] A. Buchsbaum, T. Pequeno, and M. Pequeno. "A logical expression of reasoning". In: *Synthese* 154 (2007), pp. 431–466.

[8] R. Carnap. *Logical Foundations of Probability*. Chicago: University of Chicago Press, 1950.

[9] N. da Costa. "On the theory of inconsistent formal systems". In: *Notre Dame Journal of Formal Logic* 15 (1974), pp. 497–510.

[10] N. da Costa and W. Carnielli. "On paraconsistent deontic logic". In: *Philosophia* 16 (1986), pp. 293–305.

[11] M. Fitting. "Basic modal logic". In: *Handbook of Logic in Artificial Intelligence and Logic Programming, Vol. 1, Logical Foundations*. Ed. by D. Gabbay, D. Hogger, and J. Robinson. Oxford: Oxford University Press, 1993, pp. 368–448.

[12] A. Fuhrmann. "Models for relevant modal logics". In: *Studia Logica* 49 (1990), pp. 501–514.

[13] D. Gabbay and A. Hunter. "Making inconsistency respectable: A logical framework for inconsistency in reasoning". In: *Foundations of Artificial Intelligence Research (LNCS 535)*. Ed. by P. Jorrand and J. Kelemen. New York: Springer-Verlag, 1991, pp. 19–32.

[14] L. Goble. "Paraconsistent modal logic". In: *Logique et Analyse* 193 (2006), pp. 3–29.

[15] C. Hempel. "Studies in the logic of confirmation". In: *Mind* 54 (1945), pp. 1–26, 97–121.

[16] C. Hempel. "Inductive inconsistencies". In: *Synthese* 12 (1960), pp. 439–69.

[17] G. Hughes and M. Cresswell. *A New Introduction to Modal Logic*. New York: Routledge, 1996.

[18] D. Israel. "What's wrong with non-monotonic logic?" In: *Proceedings of the First Congress on Artificial Intelligence*. Elsevier Science, 1980, pp. 99–101.

[19] S. Jaśkowski. "Propositional Calculus for Contradictory Deductive Systems". In: *Studia Logica* 24 (1969), pp. 143–157.

[20] H. Kyburg. *Probability and the Logic of Rational Belief*. Middletown: Wesleyan University Press, 1961.

[21] A. Loparíc and N. C. A. da Costa. "Paraconsistency, paracompleteness, and valuations". In: *Logique et analyse* 27.106 (1984), pp. 119–131.

[22] A. Loparíc and L. Puga. "Two systems of deontic logic". In: *Bulletin of the Section of Logic* 15 (1986), pp. 137–144.

[23] D. Makinson. "General patterns in nonmonotonic reasoning". In: *Handbook of Logic in Artificial Intelligence and Logic Programming, Vol. 3, Nonmonotonic Reasoning and Uncertain Reasoning*. Ed. by Hogger Gabbay D. and Robinson D. Oxford: Oxford University Press, 1994, pp. 35–110.

[24] J. Marcos. "Nearly every normal modal logic is paranormal". In: *Logique et Analyse* 48 (2005), pp. 279–300.

[25] E. Mares and R. Meyer. "The semantics of R4". In: *Journal of Philosophical Logic* 22 (1993), pp. 95–110.

[26] A. Martins, L. Martins, and F. Morais. "Natural deduction and weak normalization for the paraconsistent logic of epistemic inconsistency". In: *Handbook of Paraconsistency: Studies in Logic andl Cognitive Systems*. Ed. by J. Béziau, W. Carnielli, and D. Gabbay. Amsterdam: North-Holland, 2007, pp. 355–382.

[27] A. Martins and T. Pequeno. "Proof-theoretical considerations about the logic of epistemic inconsistency". In: *Logique et Analyse* 143-4 (1996), pp. 245–260.

[28] A. Martins, T. Pequeno, and M. Pequeno. "A multiple worlds semantics to a paraconsistent nonmonotonic logic". In: *Paraconsistency: The Logical Way to the Inconsistency*. Ed. by W. Carnielli, M. Coniglio, and I. d'Ottaviano. New York: Marcel Dekker, 2002, pp. 187–211.

[29] D. McDermott. "Non-monotonic logic II". In: *Journal of the Association for Computing Machinery* 29 (1982), pp. 33–57.

[30] S. Odintsov and H. Wansing. "Constructive predicate logic and constructive modal logic. Formal duality versus semantical duality". In: *First-Order Logic Revisited*. Ed. by V. Hendricks. Berlin: Logos Verlag, 2004, pp. 269–286.

[31] T. H. C. Pequeno and A. R. V. Buchsbaum. "The logic of epistemic inconsistency". In: *Principles of Knowledge Representation and Reasoning: Proceedings of Second International Conference*. Ed. by J. Allen et al. San Mateo: Morgan Kaufmann Publishers Inc., 1991, pp. 453–460.

[32] D. Perlis. "On the consistency of commonsense reasoning". In: *Computational Intelligence* 2 (1987), pp. 180–190.

[33] R. Reiter. "A logic for default reasoning". In: *Artificial Intelligence* 13 (1980), pp. 81–132.

[34] T. A Seki. "Sahlqvist theorem for relevant modal logic". In: *Studia Logica* 73 (2003), pp. 383–411.

[35] R. S. Silvestre. "Modality, paraconsistency and paracompleteness". In: *Advances in Modal Logic. Volume 6*. Ed. by G. Governatori, I. Hodkinson, and Y. Venema. College Publications: Noosa, 2006, pp. 449–467.

[36] R. S. Silvestre. "Ambigüidades indutivas paraconsistência, paracompletude e as duas abordagens da indução". In: *Manuscrito* 30 (2007), pp. 101–134.

[37] R. S. Silvestre. *Induction and Plausibility. A Conceptual Analysis from the Standpoint of Nonmonotonicity, Paraconsistency and Modal Logic*. Saarbrucken: Lambert Academic Publishing, 2010.

[38] R. S. Silvestre. "Paranormal Modal Logic - Part I: The System K? and the Foundations of the Logic of Skeptical and Credulous Plausibility". In: *Logic and Logical Philosophy* 21 (2012), pp. 65–95.

[39] R. S. Silvestre. "Paranormal Modal Logic - Part II: K?, K and Classical Logic and other Paranormal Modal Systems". In: *Logic and Logical Philosophy* 22 (2013), pp. 89–130.

Paraconsistent Modalities as a Possible Way of Treating Epistemic-Doxastic Paradoxes

Itala M. Loffredo D'Ottaviano and Rogério J. de R. da Silva Júnior

1 Introduction

Modal epistemic logics pretend to deal formally with the notion of knowledge, whereas *doxastic modal logics* pretend to do so with the notion of belief. For most of the logical systems called *epistemic-doxastic logics* or simply *epistemic logics*, the two notions are interchangeable when they assume the principle that it is not possible to *know* a proposition without *believing* it. When the modal operators of belief and knowledge are not interchangeable, these logics can be treated as bi-modal systems where such operators are introduced into the language as primitives. These systems are particular types of modal logic with special interpretations of the modal operators: "$\Box\alpha$" is interpreted epistemically as "it is known that" (which is usually formalized by the operator "K"), and doxastically as "it is believed that" (formalized by the operator "B"). Although not so common, due to the interdefinability of modal operators, some authors also interpret the operator ($\Diamond$) epistemically. Thus "$\Diamond\alpha$" is epistemically interpreted as "is consistent with all that is known" (formalized as the operator "P"), and doxastically as "is compatible with all that is believed" (formalized as the operator "C").[1] There are several systems of epistemic logic, mostly based on normal systems of alethic logic. However, some are considered more adequate than others for representing agents of belief or knowledge, depending on the peculiarities of the axioms and rules adopted. For example, the system presented in [16] as a model for knowledge is an interpretation of the alethic system **KT4**, and the

---

Copyright © 2019 by Itala M. Loffredo D'Ottaviano and Rogério J. de R. da Silva Júnior. All rights reserved.

The second author is a PhD student.

[1] A fundamental reference for the study of epistemic-doxastic logics is [16]. We suggest in a complementary way [14] and [19].

system presented as a model for belief is an interpretation of the alethic system **KD4**.

The epistemic-doxastic paradoxes are related to the derivation of contradictions in epistemic-doxastic logical systems[2] and are referred to as, respectively, the *Paradox of Credibility* and the *Paradox of Knowability*[3]. The Paradox of Credibility is an undesirable contradictory result of certain alethic-doxastic modal logics, and its problem lies in the fact that it causes the collapse of the modal belief operator. In an analogous fashion, the Paradox of Knowability represents the collapse of the epistemic operator in the context of alethic-epistemic logics. We will describe these paradoxes in detail below.

2 Epistemic-doxastic paradoxes

2.1 The Paradox of Knowability

As already mentioned, the Paradox of Knowability is a result of the epistemic-alethic logic that leads to the collapse of the epistemic operator.

Let $\Sigma = \Sigma_1 \cup \Sigma_2$ be an alethic-epistemic signature, where $\Sigma_1 = \{\sim, \Box, K\}$ and $\Sigma_2 = \{\vee, \wedge, \supset\}$, and let Var be an enumerable set of propositional variables. Let $\mathcal{L}^{\Box K}$ be the language generated over Σ, and $For^{\Box K}$ be the set of well-formed formulas in the language $\mathcal{L}^{\Box K}$. The formulas of $\mathcal{L}^{\Box K}$ are defined recursively in the standard manner (for $p \in Var$):

$$\varphi ::= p \mid \psi \vee \gamma \mid \psi \wedge \gamma \mid \psi \supset \gamma \mid \sim\!\psi \mid \Box\psi \mid K\psi.$$

The operator ($\leftrightarrow$) can be defined from the primitive operators as usual. We will construct a system, denoted as $\mathbf{T}_K$, which is composed of the axioms

---

[2] Most of the known systems of epistemic logic have classical logic as their basis, due to their alethic heritage. These systems explode in the presence of contradictions, that is, they undergo deductive trivialization by virtue of the validity of the classical principle *Ex Falso Sequitur Quod Libet*: $\alpha \supset (\sim\!\alpha \supset \beta)$ (that is, from a contradiction everything follows; see [15]). This characteristic is also passed on to doxastic logics, where from a contradictory set of beliefs (or where contradictory consequences follow from these beliefs) the system undergoes deductive trivialization, that is, it follows that the agent would believe in anything, including other contradictions. The point is that agents of beliefs similar to real agents isolate contradictions when they encounter them, and from there it does not follow that they believe anything whatsoever. Contradiction in this case is put into a kind of quarantine until there appears new information that allows for the revision of belief, or until there is as much evidence in favor of a proposition as there is in favor of its contradiction, a situation which does not make the belief system trivial (as, for example, simultaneously believing in the assumptions of quantum mechanics and in those of the physics of relativity; these principles to date are irreconcilable).

[3] Also known as Fitch's Paradox. See [12].

(**PC**): all instances of valid formulas of the propositional calculus

(**K**): $\Box(\varphi \supset \psi) \supset (\Box\varphi \supset \Box\psi)$

(**K$_K$**): $K(\varphi \supset \psi) \supset (K\varphi \supset K\psi)$

(**T$_K$**): $K\varphi \supset \varphi$,

and closed by the rules

Modus Ponens (**MP**): $\alpha \supset \beta, \alpha \vdash \beta$

Necessity (**Nec**): $\vdash \alpha$ implies $\vdash \Box\alpha$

Epistemic necessity (**Nec$_K$**): $\vdash \alpha$ implies $\vdash K\alpha$.

Because it contains the minimal modal system **K**, the system **T$_K$** also has the exchange between the modal operators ($\Box$) and ($\Diamond$), which follows from

$$\text{Def}_\Diamond: \Diamond\alpha \stackrel{\text{def}}{=} {\sim}\Box{\sim}\alpha$$

and to some demonstrable theorems in **K**[4]. For simplicity, in this article we shall denote the possibilities of interchanging these alethic operators by the notation (**I$\Box\Diamond$**).

The Paradox is usually obtained whenever we try to reconcile the following two hypotheses in this system:

Fitch-Moore Thesis (**FMT**): ${\sim}\Diamond K(\varphi \land {\sim}K\varphi)$

Verificationist Thesis[5] (**VT**): $\varphi \supset \Diamond K\varphi$.

The (**FMT**) states that it is impossible to know (verify) that there is a true proposition and that this proposition is not known[6], whereas (**VT**) states that if a proposition is true, it is possible to know (or verify) it.

For its part, the acquisition of (**FMT**) depends on the introduction, as a hypothesis, of the *Thesis of the Logical Non-Omniscience* (**NO**)[7]:

$$(\mathbf{NO}): \varphi \land {\sim}K\varphi.$$

---

[4] These are the formulas ($\Box\varphi \leftrightarrow {\sim}\Diamond{\sim}\varphi$), demonstrated in [17], and ($\Box{\sim}\varphi \leftrightarrow {\sim}\Diamond\varphi$), whose proof may be found in [20].

[5] Also known as the *Principle of Knowability*.

[6] For a more extensive discussion of the so-called Fitch-Moore Thesis, see [1] and [10].

[7] It is, at a minimum, interesting to note that this formula, which directly contradicts epistemic necessity, is ultimately essential for obtaining the paradox.

The (**NO**) states that there is at least one true proposition that is not known.

Finally, we need the following theorem of epistemic logic:

$$K(\varphi \wedge \psi) \supset (K\varphi \wedge K\psi)^8.$$

The following is a version of the paradox based on [1] and [10]. First we obtain the Fitch-Moore Thesis:

1. $K(\varphi \wedge \sim K\varphi)$ [hypothesis]
2. $K\varphi \wedge K\sim K\varphi$ [1, Theorem of $\mathbf{T}_K$]
3. $K\varphi$ [2, (**PC**), (**MP**)]
4. $K\sim K\varphi$ [2, (**PC**), (**MP**)]
5. $\sim K\varphi$ [4, ($\mathbf{T}_K$) e (**MP**)]
6. $K\varphi \wedge \sim K\varphi$ [3, 5, (**PC**)]
7. $\sim K(\varphi \wedge \sim K\varphi)$ [1-6, Reduction to the Absurd]
8. $\Box \sim K(\varphi \wedge \sim K\varphi)$ [7, (**Nec**)]
9. $\sim \Diamond K(\varphi \wedge \sim K\varphi)$ [8, (I$\Box\Diamond$) – Fitch-Moore Thesis]

$\Box$

We will now see how there occurs, for example in [10], the collapse of the epistemic operator, adding (**VT**):

1. $\varphi \wedge \sim K\varphi$ [hypothesis (**NO**)]
2. $\Diamond K(\varphi \wedge \sim K\varphi)$ [1, (**VT**) and (**MP**)]
3. $\sim \Diamond K(\varphi \wedge \sim K\varphi)$ [(**FMT**)]
4. $\Diamond K(\varphi \wedge \sim K\varphi) \wedge \sim \Diamond K(\varphi \wedge \sim K\varphi)$ [2, 3, (**PC**)]
5. $\sim (\varphi \wedge \sim K\varphi)$ [1-4, Reduction to the Absurd]

$\Box$

Of course, this result is problematic, because:

---

[8]The proof of this theorem is analogous to that of its alethic counterpart $\Box(\varphi \wedge \psi) \supset (\Box\varphi \wedge \Box\psi)$, which is easily obtained in any system based on **K**. See [17].

1. $\sim(\varphi \wedge \sim K\varphi)$ [earlier result]
2. $\sim\varphi \vee \sim\sim K\varphi$ [1, (**PC**), (**MP**)]
3. $\varphi \supset \sim\sim K\varphi$ [2, (**PC**), (**MP**)]
4. $\sim\sim K\varphi \supset K\varphi$ [2, (**PC**)]
5. $\varphi \supset K\varphi$ [3, 4, (**PC**)]

The problem lies in the fact that the system in which this result is obtained contains the axiom (**T**$_K$): $K\varphi \supset \varphi$. Therefore, we have the formula ($K\varphi \leftrightarrow \varphi$), causing the collapse of the knowledge operator.

There is a strictly doxastic version of the Paradox of Cognoscibility, known as the *Paradox of Credibility*. From the modal point of view, it consists in weakening the above system using the Axiom (**D**) (written only with the strong operator), in place of Axiom (**T**). The novelty here is the addition of Axiom (**4**).

2.2 The Paradox of Credibility

The *Paradox of Credibility*[9] corresponds to the doxastic version of the *Paradox of Knowability*. It deals with the difficulty in reconciling the so-called *Principle of Credibility* (**PCD**), which may be stated as "if something is true then it is possible to believe it", and the doxastic version of the *Fitch-More Thesis*, here denoted as (**TFM**$_B$), according to which "it is unbelievable that something is true and one does not believe it". The problem here is that in a system where both are in force, there occurs the collapse of the belief operator, as happens with its epistemic counterpart. We will now briefly present a system in which this paradox occurs, the system **D4**$_B$.

The signature that gives rise to the language of **D4**$_B$ is similar to the signature Σ presented in the previous subsection, with the distinction that where the epistemic operator K appeared earlier, we now have the doxastic operator B, generating the language $\mathcal{L}^{\Box B}$. The clauses of formation for the set $For^{\Box B}$ of the language $\mathcal{L}^{\Box B}$ are defined recursively in the same manner, except that instead $\varphi ::= K\psi$, we now have:

---

[9] A synthetic approach can also be found in [1]. Although the Paradox of Knowability is more commonly known and treated in the literature, we argue, as does [1], that its doxastic counterpart presents characteristics that deserve a distinct analysis. Philosophically, the so-called "*Principle of Credibility*" can not be discarded with the same ease with which we can attempt to reject the "Principle of Knowability" when we try to avoid the epistemic paradox. Also, from the point of view of logical treatment, a possible paraconsistent treatment of the two paradoxes presents significant differences.

$$\varphi ::= B\psi.$$

As in the system $\mathbf{T}_K$ presented above, the system $\mathbf{D4}_B$ adds new axioms and rules to the minimal system:

(**K**$_B$) $B(\varphi \supset \psi) \supset (B\varphi \supset B\psi)$

(**D**$_B$) $B\varphi \supset {\sim}B{\sim}\varphi$

(**4**$_B$) $B\varphi \supset BB\varphi$

Doxastic Necessity (**Nec**$_B$): $\vdash \alpha$ implies $\vdash B\alpha$.

In $\mathbf{D4}_B$ we also have the possibility of the interchange (I$\Box\Diamond$), given that its alethic basis is the system $\mathbf{K}$. We will also need the following hypotheses:

Thesis of the Logical Non-Omnicredence (**NOC**): $\varphi \wedge {\sim}B\varphi$

Principle of Credibility (**PCD**): $\varphi \supset \Diamond B\varphi$.

As in the epistemic case, we will need a theorem of doxastic logic:

$$B(\varphi \wedge \psi) \supset (B\varphi \wedge B\psi)^{10}.$$

In $\mathbf{D4}_B$ it is possible to obtain the doxastic Fitch-Moore Thesis (**FMT**$_B$),

$${\sim}\Diamond B(\varphi \wedge {\sim}B\varphi):$$

---

[10] The proof of which is similar to that of the theorem $(\Box(\varphi \wedge \psi) \supset (\Box\varphi \wedge \Box\psi))$ in **K**. See [17].

1. $B(\varphi \wedge \sim B\varphi)$ [hypothesis]
2. $B\varphi \wedge B{\sim}B\varphi$ [1, Theorem of $\mathbf{D4}_B$ and $(\mathbf{MP})$]
3. $B\varphi \supset BB\varphi$ [$(\mathbf{4}_B)$]
4. $B{\sim}B\varphi \supset {\sim}B{\sim}{\sim}B\varphi$ [$(\mathbf{D}_B)$]
5. $B\varphi$ [2, $(\mathbf{PC})$, $(\mathbf{MP})$]
6. $B{\sim}B\varphi$ [2, $(\mathbf{PC})$, $(\mathbf{MP})$]
7. $BB\varphi$ [3, 5, $(\mathbf{MP})$]
8. ${\sim}B{\sim}{\sim}B\varphi$ [4, 6, $(\mathbf{MP})$]
9. $B\varphi \supset {\sim}{\sim}B\varphi$ [$(\mathbf{PC})$]
10. $BB\varphi \supset B{\sim}{\sim}B\varphi$ [9, $(\mathbf{Nec}_B)$, $(\mathbf{K}_B)$ and $(\mathbf{MP})$]
11. ${\sim}B{\sim}{\sim}B\varphi \supset {\sim}BB\varphi$ [10, $(\mathbf{PC})$, $(\mathbf{MP})$]
12. ${\sim}BB\varphi$ [8, 11, $(\mathbf{MP})$]
13. $BB\varphi \wedge {\sim}BB\varphi$ [7, 11, $(\mathbf{PC})$ and $(\mathbf{MP})$]
14. ${\sim}B(\varphi \wedge {\sim}B\varphi)$ [1-13, Reduction to the Absurd]
15. $\Box{\sim}B(\varphi \wedge {\sim}B\varphi)$ [14, $(\mathbf{Nec})$]
16. ${\sim}\Diamond B(\varphi \wedge {\sim}B\varphi)$ [15, $(\mathbf{I}\Box\Diamond)$ and $(\mathbf{MP})$ - doxastic Fitch-Moore Thesis]

$\Box$

Once we obtain the doxastic Fitch-Moore Thesis, and given that we accept the Principle of Credibility, we obtain the collapse of the belief operator, $(\varphi \leftrightarrow B\varphi)$, as follows:

$(\Rightarrow) \vdash \varphi \supset B\varphi$

Proof:

1. $(\varphi \wedge {\sim}B\varphi) \supset \Diamond B(\varphi \wedge {\sim}B\varphi)$ [$(\mathbf{PCD})$]
2. ${\sim}\Diamond B(\varphi \wedge {\sim}B\varphi)$ [$(\mathbf{FMT}_B)$]
3. ${\sim}(\varphi \wedge {\sim}B\varphi)$ [1,2, $(\mathbf{PC})$, $(\mathbf{MP})$]
4. ${\sim}\varphi \vee {\sim}{\sim}B\varphi$ [3, $(\mathbf{PC})$, $(\mathbf{MP})$]
5. $\varphi \supset {\sim}{\sim}B\varphi$ [4, $(\mathbf{PC})$, $(\mathbf{MP})$]
6. ${\sim}{\sim}B\varphi \supset B\varphi$ [$(\mathbf{PC})$]
7. $\varphi \supset B\varphi$ [5, 6, $(\mathbf{PC})$, $(\mathbf{MP})$]

$\Box$

Although the Axiom (**T**) is not included and interpreted in doxastic logics in general because it is too strong (it is hard to accept that if one believes something then that something is true), this formula ends up being derived anyway by ($\mathbf{D_B}$) and from the above result, which for its part resulted from combination of (**PCD**) and ($\mathbf{FMT_B}$):

($\Leftarrow$) $\vdash B\varphi \supset \varphi$
Proof:

1. $B\varphi \supset {\sim}B{\sim}\varphi$ [($\mathbf{D_B}$)]
2. ${\sim}\varphi \supset B{\sim}\varphi$ [instance of ($\varphi \supset B\varphi$)]
3. ${\sim}B{\sim}\varphi \supset {\sim}{\sim}\varphi$ [2, (**PC**), (**MP**)]
4. ${\sim}{\sim}\varphi \supset \varphi$ [(**PC**)]
5. ${\sim}B{\sim}\varphi \supset \varphi$ [3, 4, (**PC**), (**MP**)]
6. $B\varphi \supset \varphi$ [1, 5, (**PC**), (**MP**)]

□

The collapse of the doxastic operator, ($\varphi \leftrightarrow B\varphi$), follows directly from the two results above.

2.3 Alternatives to the Paradoxes

Some alternatives and solutions to the paradoxes have been proposed[11], and we will now exemplify some of these.

One of them is the question of whether the formalization used to capture the main theses and notions of the paradoxes is satisfactory or even adequate. It is possible in this context that the Epistemic-Doxastic Paradoxes are not paradoxes in a strict sense, but rather are undesired consequences of trying to express, in the language of a modal logic, concepts that are not fully expressible in this language.

Another proposal is to reject some of the hypotheses involved in the arguments. For example, the acceptance of the verificationist thesis (**VT**)[12], is controversial among philosophers, the thesis being rejected in some contexts because it is considered very strong. However, the demonstration below shows that the rejection of this assumption is not sufficient to avoid collapse in a system like $\mathbf{T_K}$, as we need at least the Thesis (**NO**):

---

[11] Some detailed examples of the treatment of the paradoxes can be found in [1] and [10].

[12] Less controversial, however, is its doxastic version, the Principle of Credibility, which is generally not refuted. See [1].

1. $\varphi \wedge \sim K\varphi$ [hypothesis (**NO**)]
2. $K(\varphi \wedge \sim K\varphi)$ [1, Nec_K]
3. $K\varphi \wedge K\sim K\varphi$ [2, Theorem of $\mathbf{T}_K$, (**PC**) and (**MP**)]
4. $K\varphi$ [3, (**PC**) and (**MP**)]
5. $K\sim K\varphi$ [3, (**PC**) and (**MP**)]
6. $\sim K\varphi$ [5, ($\mathbf{T}_K$) and (**MP**)
7. $K\varphi \wedge \sim K\varphi$ [4, 6, (**PC**), (**MP**)][contradiction]
8. $\sim(\varphi \wedge \sim K\varphi)$ [1-7, Reduction to the Absurd]
9. $\sim\varphi \vee \sim\sim K\varphi$ [8, (**PC**), (**MP**)]
10. $\varphi \supset \sim\sim K\varphi$ [9, (**PC**), (**MP**)]
11. $\sim\sim K\varphi \supset K\varphi$ [(**PC**)]
12. $\varphi \supset K\varphi$ [10, 11 (**PC**), (**MP**)]

□

We have verified that through the introduction of the thesis (**NO**), and by means of the procedure of Reduction to the Absurd, we directly obtain the formula that, associated to the axiom ($\mathbf{T}_K$), causes the collapse of the epistemic operator[13].

The path we explore in this paper, however, follows the trend developed in [10], which seeks to provide a paraconsistent basis for the system that gives origin to the Paradox of Knowability. The reason for this is that the core of the statements that give origin to the Paradox are applications of the classic procedure of Reduction to the Absurd.

We will explore some specific paraconsistent systems, namely, the systems **PI**, **mbC**, **bC** and **Ci**, the last three of these being classified as *Logics of Formal Inconsistency* or simply (**LFIs**)[14]. Among the characteristics that define an **LFI**, is the internalization of the notions of *consistency* and *inconsistency* in its object language, and the possibility of defining a negation with classical characteristics in its language. The **LFIs** are thus endowed with a more expressive object language, and their deductive power is greater than systems that do not allow the reconstruction of classical reasoning on the basis of their systems. We have adopted here the sequence of **LFIs** systems used by [3], which seeks to investigate

---

[13] We can verify that the same occurs in the doxastic case: from the introduction of the thesis (**NOC**) in the system $\mathbf{D4}_B$, we obtain the collapse of the doxastic operator without using the doxastic Fitch-Moore Thesis or the Principle of Credibility.

[14] *Logics of Formal Inconsistency*. Introduced in [5] and developed in [7]. See also [4].

modal systems with different paraconsistent negations, based primarily on a *minimal* or fundamental **LFI**, namely **mbC**, and later extending to its gradually more powerful extensions. The system **PI** does not fit the definition of **LFI** and is here initially investigated as the basis for the logic **mbC** and its extensions, but it acquires an extremely important role when extended to a class of cathodic systems, especially in the treatment of the epistemic paradoxes. The cathodic systems, for their part, have been proposed in [3][15], and can be understood as modal extensions of paraconsistent systems. In order to deal with the Paradoxes, we intend to endow these systems with epistemic-doxastic interpretations.

3   Paraconsistent logics and LFIs

Paraconsistent logics are those that do not trivialize in the presence of contradictory propositions[16]. In particular, Logics of Formal Inconsistency **LFI**s are types of paraconsistent logics that internalize the notion of consistency (denoted by the primitive connective ($\circ$)) in their language, and thus allow the recovery of the classical behavior of some propositions in their systems through the so-called *Principle of Gentle Explosion*. This principle states that in order for trivialization to occur, in addition to a contradiction, there must also be consistency of the formulas involved[17]:

$$\circ\alpha, \alpha, \neg\alpha \Vdash \beta.$$

3.1   The system PI

The system **PI**, introduced in [2] is a *strong* paraconsistent logic, that is, there is no formula in its language that makes it *partially explosive*[18] in relation to that formula. Moreover, it is also non-gently explosive. Following the axiomatization presented in [3], we will construct this system by adding its characteristic axiom to a positive propositional basis,

---

[15]In his thesis, Bueno-Soler also introduces the so-called *anodic systems*, a positive modal basis on which one can measure the logical effects of gradually adding subclassical negations, introducing them on the basis of weaker negations until finally a classical negation is defined which represents the limiting case. These negations, inserted in a gradual and controlled way, are called *cathodic elements*, and generate the so-called cathodic systems. For more information on the origin of the terms "anodic" and "cathodic" applied to logical systems, see [3].

[16]For a discussion of the origin and history of paraconsistency, we recommend [11] and [15].

[17]For more details on the history and development of **LFI**s, we recommend [5], [7], and [4].

[18]A logic is "partially explosive" if, when subjected to a contradiction, it explodes when containing a specific type of formula.

namely, the implicit-conjunctive fragment of the classical propositional calculus **PC**$^{\supset\wedge}$.

Let $\Sigma=\{\supset, \wedge\}$ be a propositional signature and Var be an enumerable set of propositional variables. Let $\mathcal{L}$ be a language generated on Σ and let For be the set of well-formed formulas of $\mathcal{L}$ defined recursively, for $p \in Var$:

$$\varphi ::= p \mid \psi \supset \gamma \mid \psi \wedge \gamma.$$

The operators ($\leftrightarrow$) and ($\vee$) are introduced by definition:

$$(\alpha \leftrightarrow \beta) \stackrel{def}{=} (\alpha \supset \beta) \wedge (\beta \supset \alpha)$$
$$(\alpha \vee \beta) \stackrel{def}{=} (\alpha \supset \beta) \supset \beta.$$

The system **PC**$^{\supset\wedge}$ has the following axioms,

(A1) $p \supset (q \supset p)$

(A2) $(p \supset q) \supset ((p \supset (q \supset r)) \supset (p \supset r))$

(A3) $(p \supset r) \supset (((p \supset q) \supset r) \supset r)$

(A4) $p \supset (q \supset (p \wedge q))$

(A5) $(p \wedge q) \supset p$

(A6) $(p \wedge q) \supset q$,

and is closed under the rules:

Modus Ponens (**MP**): α, $\alpha \supset \beta$ implies β

Uniform Substitution (**SU**): $\vdash \alpha$ implies $\vdash \alpha[p/\beta]$.

In [3], it is shown that the *Deduction Theorem* (**TD**) is valid for this axiomatization, and also that the following theorems are demonstrable in **PC**$^{\supset\wedge}$[19]:

(i) $p \supset (p \vee q)$

(ii) $q \supset (p \vee q)$

(iii) $(p \supset q) \supset ((q \supset r) \supset (p \supset r))$

---

[19] Bueno-Soler obtains proof of correction and completeness for **PC**$^{\supset\wedge}$, with respect to the semantics of valuations, these being defined in the usual way.

(iv) $(p \supset r) \supset ((q \supset r) \supset ((p \vee q) \supset r))$

(v) $p \vee (p \supset q)$

(vi) $(p \vee q) \supset (q \vee p)$

(vii) $(p \vee (q \vee r)) \supset ((p \vee q) \vee r)$

(viii) $((p \vee q) \vee r) \supset (p \vee (q \vee r))$

(ix) $(p \supset q) \supset ((p \supset r) \supset (p \supset (q \wedge r)))$

(x) $p \wedge (q \vee r) \supset (p \wedge q) \vee (p \wedge r)$.

To obtain the system **PI** from **PC$^{\supset\wedge}$**, we extend the signature Σ, adding as primitive connective a symbol for negation ($\neg$). Let $\mathcal{L}^\neg$ be the language generated by that signature, and let $For^\neg$ be the set of well-formed formulas with a new formula definition:

$$\varphi ::= \neg\psi.$$

The system **PI** includes the axioms and rules of **PC$^{\supset\wedge}$**, adding its characteristic axiom:

$$(\mathbf{PI}): p \vee \neg p.$$

The *Theorem of Deduction* (**TD**) is valid in **PI**, as no new rules of inference have been added. The valuations of **PI** include the usual valuations of the connectives of **PC$^{\supset\wedge}$**, with a clause for negation introduced:

$$\text{if } v(\alpha) = 0, \text{ then } v(\neg\alpha) = 1.$$

The negation of the system **PI** can be considered "weak" or "subclassical", and in fact is classified as *complementary*[20].

Some characteristics stand out in **PI**, namely:

(i) it is not possible to define a *supplementary negation*[21] in **PI**, nor a bottom particle;

---

[20] A formula is a *Verum* if it is implied by all of the formulas of the system, and in this case can be represented by a top particle ($\top$). A negation is considered "complementary" if it is not a *Verum* and satisfies the schema known as *Consequentia Mirabilis*: $(\neg\alpha \supset \alpha) \supset \alpha$ (see [7] and [3]).

[21] A formula is a *Falsum* if it implies all the formulas of the system, and in this case can be represented by a bottom particle ($\bot$). A negation is considered "supplementary" if it is not a *Falsum* and satisfies the *Reductio ad Absurdum* schema: $\alpha \supset (\neg\alpha \supset \beta)(Idem)$.

(ii) **PI** is not *finitely trivializable*[22], from which it follows that it is also *non-explosive*[23];

(iii) **PI** is not an **LFI**;[24]

(iv) For any extension of **PI** that is strongly paraconsistent, procedures of *reductio ad absurdum* are not valid, such as:

   (a) $(\Delta, \beta \vdash \alpha)$ and $(\Pi, \beta \vdash \neg\alpha)$ implies $(\Delta, \Pi \vdash \neg\beta)$;
   (b) $(\Delta, \neg\beta \vdash \alpha)$ and $(\Pi, \neg\beta \vdash \neg\alpha)$ implies $(\Delta, \Pi \vdash \beta)$.

   This fact is evident if we consider $\Delta = \Pi = \{\alpha, \neg\alpha\}$. If we apply *reductio ad absurdum*, the logic becomes partially explosive.

(v) Since $(\supset)$ is a *deductive implication*[25], the following rules of contraposition are *not valid* for any strongly paraconsistent extension of **PI**:

   (a) $\Gamma, \alpha \supset \beta \vdash \neg\beta \supset \neg\alpha$;
   (b) $\Gamma, \alpha \supset \neg\beta \vdash \beta \supset \neg\alpha$;
   (c) $\Gamma, \neg\alpha \supset \beta \vdash \neg\beta \supset \alpha$;
   (d) $\Gamma, \neg\alpha \supset \neg\beta \vdash \beta \supset \alpha$.

(vi) In general, the rule of *Substitution for Demonstrable Equivalents* in **PI** is not valid, a result that also fails in its extensions.

   As shown in [2], it is easy to verify that the formulas $(\neg p \leftrightarrow \neg p)$ and $(p \leftrightarrow (p \wedge p))$ are valid in **PI**, whereas the formula $\neg p \leftrightarrow \neg(p \wedge p)$ is not[26].

---

[22] A logic is finitely trivializable if there is a formula of its language, or a set of formulas taken in conjunction, which, when added to the logic, lead to its trivialization.

[23] A logic is explosive if the presence of a contradiction or of a bottom particle causes deductive trivialization.

[24] **PI** does not allow defining a negation with classical characteristics (a classical negation is simultaneously complementary and supplementary, and in **PI** it is not possible to define the supplementary), and it does not have a consistency connective in its object language. In this way the logic also does not comply with the Principle of Gentle Explosion, which states that a contradiction only entails trivialization if there is also consistency.

[25] A logic has a deductive implication if it is not equivalent to a top or a bottom particle and satisfies: (i) $\alpha \supset \beta$ implies $\alpha \Vdash \beta$ and (ii) $\alpha \Vdash \beta$ implies $\alpha \supset \beta$.

[26] However, such substitution remains valid for formulas in **PI** that are not in the scope of the operator $(\neg)$, that is, the formulas of $\mathbf{PC}^{\supset \wedge}$.

3.2 The system mbC

Adding the primitive connective ($\circ$), interpreted as an operator of consistency, to the signature that gives rise to the language of **PI**, we obtain a signature denoted by Σ°, generating the language $\mathcal{L}^\circ$. We obtain the set For° of well-formed formulas in this language by adding the clause:

$$\varphi ::= \circ\psi.$$

The system **mbC** is constructed on the signature Σ° and obtained from **PI** by the addition of its characteristic axiom:

$$(\mathbf{mbC}): \circ p \supset (p \supset (\neg p \supset q)).[27]$$

We add to clauses for valuations in **PI**, one for the consistency operator ($\circ$):

$$\text{if } v(\circ\alpha)=1, \text{ then } v(\alpha)=0 \text{ or } v(\neg\alpha)=0.$$

As a result of the axiom (**mbC**), and of the properties of deductive implication, we can obtain in **mbC** the following derived rule:

$$(\mathbf{RD})_{mbC}: \circ\alpha, \alpha, \neg\alpha \vdash_{mbC} \beta.$$

This rule intuitively states that deductive explosion occurs when we have both consistency and contradiction.

In **mbC** the following result can be obtained as a theorem which will be useful in later sections:

$$(\alpha \wedge \neg\alpha) \supset \neg\circ\alpha.[28]$$

Unlike the system **PI**, it can be seen that in **mbC** it is possible to define a bottom particle:

$$\bot_p \stackrel{def}{=} p \wedge (\neg p \wedge \circ p).$$

Thus, we can define a new operator of negation, the connective ($\sim$):

$$\sim\alpha \stackrel{def}{=} \alpha \supset \bot_p,$$

---

[27] Note that this is the same formula known as the *Principle of Gentle Explosion*, and it is presented in [7] as (**bc1**) in their re-presentation of the calculus C_1 of da Costa. The difference is that in that approach, ($\circ$) was an abbreviation, whereas here we have it as a primitive connective of language.

[28] That is, from a contradiction one obtains non-consistency. See the proof in [3].

for some propositional variable p.

The negation ($\sim$) can be shown as complementary and supplementary, thus fulfilling the requirements of a classical negation. Note that as for each formula α of **mbC** we have a distinct bottom particle, each defined classical negation will also be distinct. Furthermore, although these negations are equivalent, they are not freely interchangeable, as the property of Substitution by Demonstrable Equivalents is already not valid in the system **PI**.

Thus, to utilize a classical negation in **mbC**, we can designate any formula whatever, for example, a propositional variable p_0, which is useful for defining a bottom particle and thus also the negation that will be used in the system.

In virtue of the introduction of the connective ($\circ$) as a primitive, and of the Principle of Gentle Explosion expressed in the rule $(\mathbf{RD})_{mbC}$, some classical behaviors can be recovered, even without the prior definition of classical negation, provided that consistency assumptions are made. We then have in **mbC** new rules for reduction to the absurd and counterposition (see [7]).

3.3 The system bC

The paraconsistent system **bC** (which also qualifies as an **LFI**), shares the language of **mbC**, but adding the characteristic axiom:

$$(\mathbf{bC}) \quad \neg\neg p \supset p.$$

The valuations of **bC** include a specific additional clause:

$$\text{if } v(\neg\neg\alpha)=1, \text{ then } v(\alpha)=1.$$

3.4 The system Ci

The system $\mathbf{Ci}^{29}$ extends **bC** with the additional axiom:

$$(\mathbf{Ci}) \quad \neg \circ p \supset (p \wedge \neg p).^{30}$$

As the formula $(p \wedge \neg p) \supset \neg \circ p$ is a theorem of **mbC**, in **Ci** we can obtain an equivalence between the notions of *contradiction* and *non-consistency*, by means of which we can, using basic negation, define the operator of *inconsistency*:

$$\bullet\alpha \stackrel{\text{def}}{=} \neg\circ\alpha.$$

---

[29] There are two equivalent versions of **Ci**. The one we present here is the simplest, as noted in [3].

[30] That is, from a non-consistency one obtains a contradiction.

The valuations of **Ci** are the same as those of **bC** with the following clause added:[31]

if $v(\bullet\alpha)=1$ then $v(\alpha)=1$ and $v(\neg\alpha)=1$.

4   Anodic and cathodic systems

The "anodic" and "cathodic" nomenclatures, as applied to logical systems, were introduced in [3] and refer to special types of modalities. Specifically, *anodic systems* can be understood as modal extensions of positive propositional systems or as modal systems that do not have any kind of negation. *Cathodic systems* can be understood as modal systems that have subclassical negations in their language. Cathodic systems can be obtained in two ways:

- *extending anodic systems by adding axioms that have subclassical negations in their language – for example, axioms of paraconsistent logics*;

- *modally extending logics with weak negations, such as paraconsistent logics.*

The first way corresponds to the definition of cathodic systems given by [3] (the one adapted here), which simultaneously clarifies a construction mechanism for these systems: a *cathodic system* is an extension of an anodic system by the addition of the connective of negation ($\neg$), or by the simultaneous addition of ($\neg$) and the connective ($\circ$) as primitives in the language of the anodic system in question.[32]

The second way is to obtain paraconsistent modal systems by simply substituting the classical basis of the logic in question with a paraconsistent basis[33] (a resource that has already been used in the literature)[34], and it would be quite natural for us to do this here (we have already introduced the systems **PI**, **mbC**, **bC** and **Ci**, and thus it would suffice to extend them modally according to convenience).

In this work, we will follow the methodology presented in [3], for we agree that the effects of the addition of the paraconsistent axioms to

---

[31]The completeness of **Ci** has been proved in [5] by means of a semantics of valuations and also by means of a semantics of possible translations.

[32]Bueno-Soler originally defined this notion by stating as a requirement the simultaneous addition of the connectives ($\neg$) and ($\circ$); in this case, however, **PI**-based cathodic systems would not fit the definition because **PI** is not an **LFI** and does not admit the connective ($\circ$) in its language.

[33]This would be a case of the *fusion* of logics, and specifically a case of the so-called *algebraic fibrillations*. See [13], [6].

[34]See [10].

anodic systems allow for a better evaluation of the gradual effect that the different negations provoke in the modal systems. In any case, we are interested in the application of the resulting modal systems to the epistemic paradoxes. As Bueno-Soler tells us in [3]:

> On the one hand we are establishing a methodology for obtaining modal logics of different types, and on the other hand we will show that this apparatus allows us to identify, for example, the importance of negation for the solution of some deontic and epistemic paradoxes.

4.1 Anodic systems

Anodic systems can be understood as modal extensions of positive systems, that is, as systems without connectives of negation.

As in [3], we will construct the anodic system $\mathbf{K}^{\supset\wedge}$, extending the system $\mathbf{PC}^{\supset\wedge}$ to a minimal modal extension, by the axiom

$$(\mathbf{K}): \Box(\alpha \supset \beta) \supset (\Box\alpha \supset \Box\beta)$$

and by the rule

$$(Nec): \vdash\alpha \text{ implies } \vdash\Box\alpha.$$

Adding the operator ($\Diamond$) to the language of $\mathbf{K}^{\supset\wedge}$, which here is not definable on the basis of ($\Box$) because we do not have an operator for negation, we construct the bi-modal system $\mathbf{K}^{\supset\wedge\Diamond}$. The system $\mathbf{K}^{\supset\wedge\Diamond}$ extends $\mathbf{K}^{\supset\wedge}$ by adding the following axioms:

(K1) $\Box(p \supset q) \supset (\Diamond p \supset \Diamond q)$

(K2) $\Diamond(p \vee q) \supset (\Diamond p \vee \Diamond q)$

(K3) $(\Diamond p \supset \Box q) \supset \Box(p \supset q)$.

We can extend $\mathbf{K}^{\supset\wedge\Diamond}$ to a class of anodic systems by the addition of the schema:

$$\mathbf{G}^{k,l,m,n} := \Diamond^k\Box^l\alpha \supset \Box^m\Diamond^n\alpha^{35},$$

where $i \in \{k, l, m, n\}$ corresponds to the number of iterations of modal operators. This schema is capable of instantiating most of the known modal axioms[36] and satisfies the following semantic property:

---

[35] Introduced by Lemmon and Scott in [18].
[36] Exceptions to this are mentioned in [3].

$$\mathbf{P}^{k,l,m,n} := \forall w_1 \forall w_2 \forall w_3((w_1 R^k w_2 \wedge w_1 R^m w_3) \supset \exists w_4(w_2 R^l w_4 \wedge w_3 R^n w_4)),\text{[37]}$$

where $r \in \{k, l, m, n\}$ denotes the number of steps by which a state w_i is accessible on the basis of a state w_j. Because we have a bi-modal system, for the accessibility relation of $\mathbf{K}^{\supset \wedge \Diamond}$ to satisfy the property $\mathbf{P}^{k,l,m,n}$, we also introduce the dual schema $\mathbf{G}^{k,l,m,n}$:

$$\mathbf{G}^{m,n,k,l} := \Diamond^m \Box^n \alpha \supset \Box^k \Diamond^l \alpha.$$

A class of anodic systems can thus be obtained on the basis of the system $\mathbf{K}^{\supset \wedge \Diamond}$, as follows:

$$\mathbf{K}^{\supset \wedge \Diamond} + \mathbf{G}^{k,l,m,n} + \mathbf{G}^{m,n,k,l}.\text{ [38]}$$

4.2 Cathodic systems

Cathodic systems can be seen as paraconsistent modal logics, and one of the ways in which they are obtained is by extending anodic systems through the addition of characteristic axioms. Consider the bi-modal anodic system $\mathbf{K}^{\supset \wedge \Diamond}$, and its extensions obtained by adding the schemas $\mathbf{G}$. Entire classes of cathodic systems can be constructed in the following manner:

- $\mathbf{PI}^{k,l,m,n}$: $\mathbf{K}^{\supset \wedge \Diamond} + \mathbf{G}^{k,l,m,n} + \mathbf{G}^{m,n,k,l} + (\mathbf{PI})$;
- $\mathbf{mbC}^{k,l,m,n}$: $\mathbf{PI}^{k,l,m,n} + (\mathbf{mbC})$;
- $\mathbf{bC}^{k,l,m,n}$: $\mathbf{mbC}^{k,l,m,n} + (\mathbf{bC})$;
- $\mathbf{Ci}^{k,l,m,n}$: $\mathbf{bC}^{k,l,m,n} + \mathbf{Ci}$.

As in $\mathbf{mbC}$ it is possible to define a classic negation ($\sim$), some formulas that represent classical behaviors can be recovered as properties valid in $\mathbf{mbC}$, including rules such as classical contraposition and the classical procedure of reduction to the absurd (relative to the negation ($\sim$)).

The restoration of classical properties in $\mathbf{mbC}$ has a fundamental consequence for its modal extensions and for the class $\mathbf{mbC}^{k,l,m,n}$ in general, namely the recovery of the interdefinibility of the modal operators ($\Box$) and ($\Diamond$), which was not possible earlier when using only a subclassical negation.

---

[37] Referred to in [8] as the *diamond-property*, and in [9] as the *incestual property*.

[38] The correction and completeness of these classes of anodic systems are demonstrated in [3].

The possibility of recovering the interdefinibility of the modal operators, formerly dual, endows any modal system based on **mbC** with the usual properties of a modal system. In order to effect the recovery of this property in the class **mbC**^{k,l,m,n}, [3] indicates that it is necessary to add two *axioms of connection*[39], namely:

(AC1) $\Box \sim p \supset \sim \Diamond p$

(AC2) $\sim \Diamond p \supset \Box \sim p$.

With the help of these axioms, it is possible to demonstrate the following formulas in **mbC**^{k,l,m,n}, which affirm the relationship between the operators ($\Box$) and ($\Diamond$):

(i) $\Diamond \alpha \supset \sim \Box \sim \alpha$;

(ii) $\sim \Box \sim \alpha \supset \Diamond \alpha$;

(iii) $\Box \alpha \supset \sim \Diamond \sim \alpha$;

(iv) $\sim \Diamond \sim \alpha \supset \Box \alpha$.

Thus, it is easy to see that the systems belonging to the class **mbC**^{k,l,m,n}, as well as their paraconsistent extensions, can behave as monomodal systems. The axioms referred to here as (K1), (K2) and (K3), which are present on the basis of the anodic system $\mathbf{K}^{\supset \wedge \Diamond}$, can be derived as theorems of **mbC**^{k,l,m,n}; this is easily done on the basis of the axiom (**K**) and the classical properties recovered in **mbC** (as demonstrated in [3]).

Finally, because of the fact that in the presence of a classical negation the system loses its bi-modal behavior[40], it can be seen that the dual schemas $\mathbf{G}^{k,l,m,n}$ and $\mathbf{G}^{m,n,k,l}$, formerly required in the anodic bimodal system $\mathbf{K}^{\supset \wedge \Diamond}$, are equivalent in **mbC**^{k,l,m,n}. Clearly, its proof consists in showing that $\mathbf{G}^{k,l,m,n}$ implies $\mathbf{G}^{m,n,k,l}$, which also holds for its reciprocal.

All the results and properties discussed here also hold true for extensions of **mbC**^{k,l,m,n}, that is, for **bC**^{k,l,m,n} and **Ci**^{k,l,m,n}, given that these systems only produce new theorems related to their respective paraconsistent axioms.

The models for the anodic systems $\mathbf{K}^{\supset \wedge \Diamond} + \mathbf{G}^{k,l,m,n} + \mathbf{G}^{m,n,k,l}$ satisfy the property $\mathbf{P}^{k,l,m,n}$, mentioned earlier. In turn, the models for the cathodic systems will satisfy this property and add the characteristic valuations of the paraconsistent axioms.

---

[39] The *Bridge Principle*; see [6].
[40] In fact, the presence of two operators as primitives becomes superfluous, so that the class can actually be written in a monomodal fashion.

5 Revisiting the Paradoxes of Knowability and Credibility

As noted above, the Credibility Paradox is a specific contradictory result of alethic-doxastic modal logics that causes the collapse of the modal belief operator. Similarly, the Knowability Paradox corresponds to the collapse of the epistemic operator in the context of alethic-epistemic logics. Although similar in many respects, the doxastic version of Axiom (**T**), which is very strong for a belief logic, is not used in the alethic-doxastic system presented here. A version of Axiom (**D**), is used instead; this version contains classical negations and, according to the literature, causes collapse.[41] Another peculiarity that distinguishes the doxastic case from the epistemic one is the presence of Axiom (**4**) in logics of belief; this axiom is usually included to strengthen the deductive power of these logics, but here it becomes one of the formulas that directly contribute to the collapse. For these reasons, the doxastic paradox requires a differentiated treatment from that of its epistemic counterpart. We will first examine the epistemic case.

5.1 A cathodic system T_K based on mbC

We will construct an alethic-epistemic paraconsistent system T_K, based on **mbC**, with the objective of verifying if the Paradox of Knowability is still maintained in its usual form.

Let $\Sigma = \Sigma_1 \cup \Sigma_2$ be a signature for T_K and Var be an enumerable set of propositional variables, where $\Sigma_1 = \{\neg, \circ, \Box, \Diamond, K\}$ and $\Sigma_2 = \{\supset, \wedge\}$. The set For of well-formed formulas of language $\mathcal{L}$, generated over Σ, presents the following recursive definitions of formula for $p \in Var$:

$$\varphi ::= p \mid \psi \supset \gamma \mid \psi \wedge \gamma \mid \neg\psi \mid \circ\psi \mid \Box\psi \mid \Diamond\psi \mid K\psi.$$

The operator ($\vee$) is introduced by definition:

$$(\alpha \vee \beta) \stackrel{def}{=} (\alpha \supset \beta) \supset \beta.$$

The paraconsistent system T_K contains the following axioms:

($\mathbf{K}^{\supset \wedge \Diamond}$): All of the axioms and theorems of $\mathbf{K}^{\supset \wedge \Diamond}$

(**PI**): $p \vee \neg p$

(**mbC**): $\circ p \supset (p \supset (\neg p \supset q))$

($\mathbf{K}_K$): $K(p \supset q) \supset (Kp \supset Kq)$

---
[41] See [1].

($\mathbf{T}_K$): $Kp \supset p$[42].

The rules of inference of $\mathbf{T}_K$ are the rules of $\mathbf{K}^{\supset \wedge \Diamond}$, and the rule of Epistemic Necessity:

$$(\mathbf{Nec}_K): \vdash \alpha \text{ implies } \vdash K\alpha.$$

The attempt to obtain, in the usual way, the epistemic Fitch-Moore Thesis in a system $\mathbf{T}_K$ based on $\mathbf{mbC}$ is as follows:

1. $K(\varphi \wedge \neg K\varphi)$ [hypothesis]
2. $K\varphi \wedge K\neg K\varphi$ [1, Theorem of $\mathbf{T}_K$]
3. $K\varphi$ [2, (**PC**), (**MP**)]
4. $K\neg K\varphi$ [2, (**PC**), (**MP**)]
5. $\neg K\varphi$ [4, ($\mathbf{T}_K$) and (**MP**)]
6. $K\varphi \wedge \neg K\varphi$ [3, 5, (**PC**)][contradiction]
7. $\neg \circ K\varphi$ [6, Theorem of $\mathbf{mbC}$]

Considering only the signature presented here, based on the contradiction present in step (6), we obtain that the formula $K\varphi$ is not consistent according to the rules of the component logic $\mathbf{mbC}$. However, in the environment of $\mathbf{mbC}$, we can define a classical negation ($\sim$) for each propositional variable. We will then stipulate a propositional variable p_0 with the objective of defining a classical negation specific to the cathodic system $\mathbf{T}_K$ based on $\mathbf{mbC}$:

$$\sim\alpha \stackrel{\text{def}}{=} \alpha \supset \bot_{p_0}.$$

Due to the classical negation defined in the system $\mathbf{T}_K$, we can recover not only the usual classical properties (such as the Reduction to the Absurd procedure) but also the interdefinability of the modal operators ($\mathbf{I}\Box\Diamond$). This also makes it possible to fully recover the Paradox of Knowability in this logic[43], as well as in the classes $\mathbf{bC}^{k,l,m,n}$ and $\mathbf{Ci}^{k,l,m,n}$, relative, however, to the defined negation ($\sim$).

---

[42] The axiom $\mathbf{T}$, independent of its interpretation, can be instantiated by the schema $\mathbf{G}^{k,l,m,n}$ (as $\mathbf{G}^{0,1,0,0}$), which makes this system one of the cathodic systems introduced in [3].

[43] In a certain sense it is debatable whether we will have in this case the "same" Paradox. What the sentence ($\sim\alpha$) expresses with classical negation as a primitive operator is distinct from what its equivalent ($\alpha \supset (p_0 \wedge (\neg p_0 \wedge \circ p_0))$) expresses with the primitive paraconsistent primitive negation ($\neg$).

The case of the Credibility Paradox in an alethic-doxastic system based on **mbC**, has significant differences in relation to the Paradox of Knowability, and these will be discussed at the end of this work. We will now see how a system behaves in relation to the Paradox of Knowability in cases where we cannot define a classic negation.

5.2 A cathodic system T_K based on PI

The cathodic system constituted by the addition of Axiom (**PI**) to the anodic system $\mathbf{K}^{\supset \wedge \Diamond}$ does not allow the definition of classical negations in its language, and thus we do not have at hand the properties deriving from the introduction of this negation into the system. Its component logic **PI** does not rely on the consistency operator ($\circ$), and it is not an **LFI** because its basic negation is only complementary.

Its modal extensions therefore remain with a bi-modal basis, as we cannot interdefine the operators ($\Box$) and ($\Diamond$) and $\mathbf{K}^{\supset \wedge \Diamond}$, where they coexist as primitives. For this reason, the class of cathodic systems $\mathbf{PI}^{k,l,m,n}$, obtained by adding the schema $\mathbf{G}^{k,l,m,n}$, to the cathodic system in question, will also have, for each instance of this schema, an instance of its dual axiom: $\mathbf{G}^{m,n,k,l}$.

The language T_K, based on **PI**, is introduced in a way similar to that of the case based on **mbC**, but we do not have here the consistency operator ($\circ$). The axiomatization is also similar, except for the fact that we do not now have the axiom (**mbC**). The attempt in the usual way to obtain the epistemic Fitch-Moore Thesis in a system T_K based on **PI** is as follows:

1. $K(\varphi \wedge \neg K\varphi)$ [hypothesis]
2. $K\varphi \wedge K\neg K\varphi$ [1, Theorem de T_K]
3. $K\varphi$ [2, (**PC**), (**MP**)]
4. $K\neg K\varphi$ [2, (**PC**), (**MP**)]
5. $\neg K\varphi$ [4, (T_K) and (**MP**)]
6. $K\varphi \wedge \neg K\varphi$ [3, 5, (**PC**)][contradiction]

We thus verify that a non-finitely trivializable system such as **PI** meets the requirements for containing the advance of the paradox, in accord with the way that the paradox has been traditionally obtained and formulated.

The same occurs with the alternative way of obtaining the collapse of the epistemic operator, as presented in Subsection 2.3, since the formula $\varphi \supset K\varphi$, which together with (T_K) causes the collapse, is also obtained through the procedure of Reduction to the Absurd.

5.3 The Paradox of Credibility in paraconsistent systems

When we consider the possibility of constructing an alethic-doxastic system on a paraconsistent basis, the question that immediately arises is whether it is still possible to introduce the doxastically interpreted Axiom (**D**): $(B\varphi \supset \sim B\sim\varphi)$ (which in the version of the paradox that we are considering is essential to deriving the Fitch-Moore Thesis). This is because the notion conveyed here is that of classical consistency applied to the notion of belief[44], even if we rewrite it with a weak or paraconsistent negation[45]. The point is that, in general, what is wanted here is a system in which the situation represented by the formula $(B\varphi \land B\neg\varphi)$, can occur, and also that in this situation we can suspend judgment or maintain the contradiction in quarantine without summarily trivializing the system of beliefs.

The version of this axiom that uses the *compatibility* "*C*", namely $(B\varphi \supset C\varphi)$, is not immune to the same observation, as it is affirmed here that the information that is believed must be compatible with already established beliefs – a requirement that we would like to avoid in our system[46]. In any case, the syntactic structure of the Paradox of Credibility depends on the fact that this axiom contains negations.

6 Final remarks

In regard to the Paradox of Knowability, we have shown that a system not finitely trivializable like **PI** presents conditions for containing a certain way of obtaining the Paradox, that is, the usual way according to the literature. In possible future work, we intend to construct a counter-model that demonstrates the impossibility of the hypotheses used to generate the Paradox in a logic based on **PI**.

In relation to the Paradox of Credibility, the scenario is the following:

(i) without the presence of the doxastic Axiom (**D**) which is inadequate in a paraconsistent system, we will not have the Paradox of Credibility as well.

(ii) if we preserve the problematic axiom (written with subclassical negations), the resulting paraconsistent subsystems will no longer

---

[44] See [14].

[45] Intuitively, the reading of this formula suggests that believing that φ implies not believing in the negation of φ.

[46] In fact, the doxastic interpretation of the operator $\lozenge$, namely, "*C*", has up to the present been classical (see [16] and [14]). However, instead of the notion of compatibility, we could interpret it in the paraconsistent context as a "non-triviality" operator. We plan to explore this idea in future work.

be among the cathodic systems presented in this work, as this version of (**D**) is not an instance of the schema $\mathbf{G}^{k,l,m,n}$. In order to verify to what extent these new systems would contain the advance of the paradox in terms of whether or not to define a classic negation in their basic logics, it would be necessary to obtain their respective Theorems of Completeness.

References

[1] D. C. P. de Almeida. "A Persistência do Paradoxo da Cognoscibilidade". MA thesis. Campinas, SP, Brazil: Institute of Philosophy and Human Sciences, University of Campinas, 2011, pp. 1–74.

[2] D. Batens. "Paraconsistent extensional propositional logics". In: *Logique et Analyse* 90–91 (1980), pp. 195–234.

[3] J. Bueno-Soler. "Multimodalidades Anódicas e Catódicas: a negação controlada em lógicas multimodais e seu poder expressivo". PhD thesis. Campinas, Brazil: Institute of Philosophy and Human Sciences, University of Campinas, 2009, pp. 1–135.

[4] W. A. Carnielli and M. E. Coniglio. *Paraconsistent logic: Consistency, contradiction and negation.* Vol. 40. Springer, 2016.

[5] W. A. Carnielli and J. Marcos. "A taxonomy of **C**-systems". In: *Paraconsistency — the Logical Way to the Inconsistent.* Ed. by W. A. Carnielli, M. E. Coniglio, and I. M. L. D'Ottaviano. Vol. 228. Lectures Notes in Pure and Applied Mathematics. New York: CRC Press, 2002, pp. 1–94.

[6] W. Carnielli and M. E. Coniglio. "Combining Logics". In: *The Stanford Encyclopedia of Philosophy.* Ed. by Edward N. Zalta. Winter 2016. Metaphysics Research Lab, Stanford University, 2016.

[7] W. Carnielli, M. E. Coniglio, and Joao Marcos. "Logics of formal inconsistency". In: *Handbook of philosophical logic.* Ed. by F. Guenthner D. Gabbay. Springer, 2007, pp. 1–93.

[8] W. Carnielli and C. Pizzi. *Modalities and multimodalities.* Vol. 12. Springer Science & Business Media, 2008.

[9] B. F. Chellas. *Modal logic: an introduction.* Cambridge university press, 1980.

[10] A. Costa-Leite. "Paraconsistência, Modalidades e Cognoscibilidade". MA thesis. Campinas, SP, Brazil: Institute of Philosophy and Human Sciences, University of Campinas, 2003, pp. 1–98.

[11] I. M. L. D'Ottaviano. "On the development of paraconsistent logic and da Costa's work". In: *Journal (The) of Non-classical Logic* 7.1-2 (1990), pp. 89–152.

[12] F. B. Fitch. "A logical analysis of some value concepts". In: *The journal of symbolic logic* 28.2 (1963), pp. 135–142.

[13] D. M. Gabbay. *Fibring logics*. Vol. 38. Clarendon Press, 1999.

[14] R. Girle. *Modal logics and philosophy*. Montreal, Kingston, London, Ithaca: McGill-Queen's University Press, 2000.

[15] I. M. L. Gomes E. L.; D'Ottaviano. *Para além das Colunas de Hércules, uma História da Paraconsistência: de Heráclito a Newton da Costa*. Série Unicamp ano 50, vol. 50/Coleção CLE, vol. 80. Editora Unicamp/Centro de Lógica, Epistemologia e História da Ciência, 2017.

[16] J. Hintikka. *Knowledge and Belief: an introduction to the logic of the two notions*. Cornell University Press, 1962.

[17] G. E. Hughes and M. J. Cresswell. *A new introduction to modal logic*. London: Routledge, 1996.

[18] E. J. Lemmon and D. Scott. "An introduction to modal logic". In: *The Lemmon Notes*. Ed. by K. Segerberg. Oxford: Blackwell, 1977.

[19] C. A. Mortari. "Lógicas epistêmicas". In: *Nos Limites da Epistemologia Analítica*. Florianópolis: Federal University of Santa Catarina, 1999, pp. 17–68.

[20] R. J. R. Silva Júnior. "Modalidades Paraconsistentes como Base para Lógicas Epistêmico-Doxásticas". MA thesis. Campinas, SP, Brazil: Institute of Philosophy and Human Sciences, University of Campinas, 2018.

An Investigation into Reduction and Direct Approaches to the Computation of Argumentation Semantics

ODINALDO RODRIGUES

1 Introduction

Argumentation is a field of study involving several interdisciplinary areas, including logic, philosophy, linguistics and computer science, amongst others. In the late 1990s, Abstract Argumentation Frameworks were proposed as a means to formalise the relationships between arguments and evaluate the statuses of arguments under particular *semantics*. There are currently two leading approaches for the computation of these semantics: the so-called *reduction* and *direct* approaches.

A reduction approach translates a particular argumentation semantics and the structure of an argumentation framework into a problem specification in a related area for which a technique for finding solutions exists, then applies that technique and translates the results back into the argumentation context. An example of such an approach is to translate an argumentation problem to a propositional logic theory, then to employ a SAT solver to find models of the theory, and finally to translate back each model found by the solver into a possible solution to the original argumentation problem.

A direct approach to the computation of argumentation semantics, on the other hand, devises specific algorithms to solve the argumentation problem directly by operating on the argumentation graph itself.

Currently, SAT reduction-based solvers lead the performance tables [6]. However, recent advances in direct approaches [7, 14] have reduced the performance gap. The development of direct approach solvers is especially important when considering argumentation problems that cannot be translated to propositional logic (as is the case in some numerical argumentation frameworks). In [14], we proposed a new forward propagation algorithm for the computation of several semantics of argumentation frameworks, offering significant performance improvements over

Copyright © 2019 by Odinaldo Rodrigues. All rights reserved.

the well-known Modgil-Caminada's labelling algorithms [12]. This new algorithm was employed in the solver EqArgSolver, submitted to the 2nd International Competition of Computational Models of Argumentation (ICCMA 2017, http://argumentationcompetition.org).

Although the new algorithm offers improvements over some of the alternatives, it is not expected to always fare better against a SAT-based solution. In particular, previous experiments [14, 15] indicated that the size and geometry of the argumentation graph may have a strong effect on its performance. This paper aims to shed some light into the employment of the algorithm in the computation of preferred extensions and how it compares in terms of performance with SAT-based reduction approaches. For the comparison, we implemented two new SAT-based argumentation solvers and analysed the performance of all solvers on a relatively large set of argumentation graphs of varied structure.

The first new SAT-based solver is a naive no-frills implementation: it simply translates the geometry of the whole argumentation graph and the conditions characterising a complete extension into a propositional logic theory, submits the theory to the SAT solver Glucose (http://www.labri.fr/perso/lsimon/glucose/) and then translates models of the theory back as complete extensions. Of these, the maximal ones (with respect to set inclusion) are selected as the preferred extensions. The second SAT-based solver provides a more direct comparison with EqArgSolver because it uses the same decomposition of the original argumentation framework into strongly connected components (SCCs). These are again translated into propositional logic theories. However, the theories are obviously smaller than they would otherwise be for the entire graph. SCC solutions are then combined in an appropriate way to provide solutions for the argumentation framework as a whole.

Our objective with this analysis was to answer a number of open questions: 1) How does the forward propagation algorithm compare with a SAT-based approach? 2) Does the density of attacks influence the performance of the forward propagation algorithm? 3) Does the decomposition of the argumentation framework affect the performance of the SAT-based solvers? 4) Are there circumstances in which one approach is better than the other? These answers can hopefully pave the way for future improvements in the development of direct and reduction-based solvers alike.

The rest of the paper is organised as follows. In Section 2, we introduce some important preliminary concepts. In Section 4, we describe EqArgSolver and the forward propagation algorithm. We then describe different ways in which argumentation semantics can be captured as propositional logic theories in Section 5. This is followed by the empirical evaluation in Section 6 and we finish the paper with some conclusions and future work in Section 7.

2 Background

An *abstract argumentation framework* is a system for reasoning about arguments proposed by Dung [8] and defined in terms of a directed graph $\langle \mathcal{A}, \mathcal{R} \rangle$, where $\mathcal{A}$ is a *finite* non-empty set of arguments and $\mathcal{R}$ is a binary relation on $\mathcal{A}$, called the *attack relation*. If $(X, Y) \in \mathcal{R}$, we say that X attacks Y and depict it with an arrow from X to Y. In what follows, $X^- = \{Y \in \mathcal{A} | (Y, X) \in \mathcal{R}\}$, i.e., the set of arguments attacking X; and $X^+ = \{Y \in \mathcal{A} | (X, Y) \in \mathcal{R}\}$, the set of arguments that X attacks. If $X^- = \emptyset$, then we say that X is a *source argument*. For sets $E \subseteq \mathcal{A}$, $E^- = \cup_{X \in E} X^-$ and $E^+ = \cup_{X \in E} X^+$. We write $E \to X$ as a shorthand for $X \in E^+$. The *path-equivalence relation* $\sim_{\mathcal{R}} \subseteq \mathcal{A}^2$ is defined as $X \sim_{\mathcal{R}} Y$ iff $X = Y$ or there is a path from X to Y and a path from Y to X in $\mathcal{R}$. A *strongly connected component* (SCC) is an equivalence class of arguments under $\sim_{\mathcal{R}}$.

One of the main purposes of an argumentation framework is to provide a way of reasoning about the *status* of its arguments, i.e., whether an argument is accepted or is defeated by other arguments. Source arguments, having no attacks, always persist. Arguments attacked by attackers that persist are defeated. An attacked argument may still persist, provided all of its attackers are defeated. Thus, the statuses of arguments are determined systematically. In Dung's original formulation, this leads to the concept of *extensions* – subsets of $\mathcal{A}$ with special properties described in terms of the concepts that follow.

A set $E \subseteq \mathcal{A}$ is said to be *conflict-free* if for all elements $X, Y \in E$, we have that $(X, Y) \notin \mathcal{R}$. Moreover, for an argument $X \in \mathcal{A}$ to be *acceptable with respect to a set* E, we want E to "protect" it, i.e., for all $Y \in X^-$, $E \cap Y^- \neq \emptyset$. A set E is then said to be *admissible* if it is conflict-free and all of its elements are acceptable with respect to itself. An admissible set E is a *complete extension* iff E contains all arguments which are acceptable with respect to itself. The $\subseteq$-minimal complete extension is called the *grounded extension*. E is called a *preferred extension* iff E is a $\subseteq$-maximal complete extension; and E is a *stable extension* if E is preferred and $E \cup E^+ = \mathcal{A}$.

Dung's semantics can alternatively be presented in terms of a Caminada *labelling function* of the form $\lambda : \mathcal{A} \longrightarrow \{\mathbf{in}, \mathbf{out}, \mathbf{und}\}$ satisfying certain conditions [2, 3, 16]. Let dom denote the domain of a function and λ a labelling function, we define $in(\lambda) = \{X \in \text{dom } \lambda | \lambda(X) = \mathbf{in}\}$; $und(\lambda) = \{X \in \text{dom } \lambda | \lambda(X) = \mathbf{und}\}$; and $out(\lambda) = \{X \in \text{dom } \lambda | \lambda(X) = \mathbf{out}\}$. We say that an argument X is *illegally labelled* **in** by λ, if

$X^- \not\subseteq out(\lambda)$; X is *illegally labelled* **out** by λ, if $X^- \cap in(\lambda) = \emptyset$; and X is *illegally labelled* **und** by λ, if either $X^- \subseteq out(\lambda)$ or $X^- \cap in(\lambda) \neq \emptyset$. A labelling function is *legal* if it does not illegally label any arguments. A complete extension E_λ can be recovered from a legal labelling function λ by setting $E_\lambda = in(\lambda)$. Conversely, a labelling function λ_E can be defined from an extension E by setting $in(\lambda_E) = E$; $out(\lambda_E) = E^+$; and $und(\lambda_E) = \mathcal{A}\backslash(E \cup E^+)$.

3 Computing Extensions via the Strongly Connected Components of an Argumentation Framework

In [1], Baroni *et. al* proposed a general recursive schema for argumentation semantics based on the strongly connected components (SCCs) of an abstract argumentation framework. The schema can be used to obtain Dung's admissibility-based semantics with the advantage that the original problem is reduced into components of smaller complexity. Based on the recursive schema, many researchers showed how to compute the extensions of argumentation frameworks under several semantics. Liao's presentation in [10] is particularly easy to understand.

The overall process can be summarised as follows. Firstly, the framework is divided into SCCs. The SCCs form a partition of the original framework and can be organised into successive layers using the attack relation. Each layer can be computed independently with input from the solutions obtained for previous layers and according to a particular semantics. The solutions to the layers thus obtained are combined in an appropriate way. This process is illustrated with an example.

Consider the argumentation framework $\mathcal{N}$ in Fig. 1 with the SCCs $SCC_1 = \{X\}$, $SCC_2 = \{W, Y\}$, $SCC_3 = \{A, B, C\}$ and $SCC_4 = \{D, E\}$. These SCCs can be arranged into three layers following the attack relation: the first one containing SCC_1 and SCC_2; the second one containing SCC_3; and the last one containing SCC_4. Within a given layer, solutions for SCCs are independent from each other, but the attacks between arguments of different layers create dependencies of the solutions of an SCC on the solutions of the SCCs attacking it. In this case, we say that the solutions of the preceding layers *condition* the solutions of the SCC. For example, the computation of the solutions of SCC_3 depends on the labels assigned to X and W, and thus on the solutions for SCC_1 and SCC_2 found in the computation of layer 0. The solutions for SCC_1 and SCC_2 do not depend on any previous computation because they have no external attackers. The only solution for SCC_1 is to label X **in**. However, SCC_2 (which is independent of SCC_1) has three solutions corresponding to legal assignments: one in which both W and Y are labelled **und** and the other two in which one of them is labelled in and the other is labelled **out**. The solutions for layer 0 are obtained by combining the solutions for its two SCCs: f_1: $X = Y =$ **in**, $W =$ **out**, f_2: $X = W =$ **in**, $Y =$ **out**, and f_3: $X =$ **in**, $W = Y =$ **und**.[1]

Now let us turn to SCC_3, whose solutions are *conditioned* by the labels of the external attackers X and W in the solutions f_1, f_2 and f_3 for layer 0. In any such solution, $X =$ **in**, but the label of W could be either **in**, **out** or **und**. In order to generate, say all *complete* extensions for $\mathcal{N}$, each partial solution f_1, f_2 and f_3 needs to be *expanded* with the solutions for SCC_3 under the constraints that they impose. The following definition captures these constraints.

DEFINITION 1 (Initial Conditioned Solution for an SCC). Let f be a conditioning solution for an SCC S. The initial solution for S conditioned by f, in symbols $\lambda_S^f : S \mapsto \{$**out**, **und**, **in**$\}$, is the legal labelling function whose set $in(\lambda_S^f)$ is $\subseteq$-minimal with respect to all legal functions conditioned by f.

λ_S^f is the "minimal" (grounded) solution for S under the constraints imposed by f. For example, we know that in all partial solutions to layer 0, the label of the argument X must be **in**. Therefore, in any initial conditioned solution for SCC_3 the label of A must be **out**. As for the labels of B and C, they will depend on what the conditioning solution assigns to W. For example, if the conditioning solution is f_2, then the label of B must be **out** as well; f_1 allows it to be **in**, **out** or **und**; whereas f_3 only allows it to be **out** or **und**. In the case of f_1 and f_3, the initial conditioned solution for SCC_3 leaves the label of B undecided (**und**), which is the minimal commitment any legal assignment must satisfy (this minimises the set $in(\lambda)$ for any legal assignment λ). However, f_3 will prevent any assignment that gives the label **in** to B, because $f_3(W) =$ **und**. The initial conditioned solution can be seen as the result of propagating the labels of the external attackers through an SCC whose nodes are all initally labelled **und** (we call this the all-**und** labelling). The Discrete Gabbay-Rodrigues Iteration Schema [9] is an example of a method that can perform this propagation efficiently.

Once an SCC is conditioned by a solution, the computation of all solutions to the SCC under that solution can be thought of as the search of all possible ways to "expand" the initial conditioned solution by legally swapping labels from **und** to **in** or **out**.

More generally speaking, the whole process can be thought of as follows: given an SCC S, a conditioning solution f, and a partial labelling function λ_S^f, compute the set Λ of all expansions of λ_S^f satisfying some

---

[1] As this combination of SCC solutions is done within the same layer, it is called the *horizontal combination* of solutions of the layer [10].

constraints. In [14], we put forward a new algorithm to perform this computation. Obviously this can be done in different ways. For example, we can translate the constraints into propositional logic and use an external SAT solver to compute the models of the resulting theory (if any). Each model will correspond to a legal assignment satisfying the conditioning solution. Since the fastest argumentation solvers are currently all SAT-based, this paper aims to compare the SAT and our own direct approach under similar conditions to help understand the strengths and weaknesses of each.

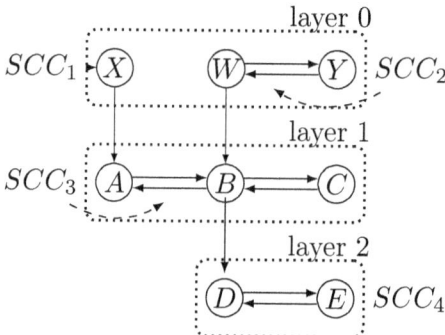

Figure 1. A complex argumentation framework and its decomposition into layers.

We now describe how the solver EqArgSolver and the forward propagation algorithm work.

4 EqArgSolver and the Forward Propagation Algorithm

EqArgSolver is a computer application that can be used to solve the following *enumeration* and *decision* problems in argumentation theory: *i)* Given an argumentation network $\langle \mathcal{A}, \mathcal{R} \rangle$, to produce one or all of the extensions of the network under the grounded, complete, preferred or stable semantics; and *ii)* Given an argument $X \in S$, to decide whether X is accepted in some extension or in all extensions under one of those semantics. EqArgSolver follows the general process of computation described in the previous section, using the decomposition of the argumentation framework into SCCs.

Algorithm 1 gives a high-level overview of the computation performed by EqArgSolver, which we now briefly describe. The argumentation framework is first divided into SCCs and arranged into layers. All arguments are initially labelled **und** and then each layer is conditioned by each partial solution found for the previous layers using the discrete version of the *Gabbay-Rodrigues Iteration Schema* [9]. In Algorithm 1, this conditioning process is performed by the call to the procedure GR-ground in line 8 generating initial solution λ. The process will not be described in detail here (details are available in [9]), but it is worth explaining that both the trivial and non-trivial SCC blocks may contain 0 or more SCCs and the conditioning will produce one single solution that represents the *minimal* constraints imposed by the labels of the attackers in the corresponding partial solution to the previous layer. $\lambda \downarrow TSB$ (Algorithm 1, line 9) is the restriction of λ to the nodes in the trivial SCC block, i.e., the "grounded" solution to the trivial SCC block. Since the first layer of an argumentation framework is not constrained by any solutions, the iteration schema (and any other propagation process) will simply propagate the **in** labels of the source arguments (if any). These source arguments have no attacks and therefore must be labelled **in**.

All top-level non-trivial SCCs will be left with all of their arguments undecided. Some of these could potentially be labelled **in** in a larger extension (line 12). The newly proposed forward propagation algorithm [14] described below, ensures that all such arguments are systematically tried for inclusion generating all possible partial solutions for the SCC starting from the minimal solution imposed by the conditioning of the SCC by the corresponding solution to the previous layer (line 13). The partial solutions thus obtained[2] are then combined using what Liao calls the *horizontal and vertical combinations of partial solutions* [11] (lines 14 and 16, respectively). All newly generated solutions are then applied to the next layer and this process is repeated until all layers are processed.

For example, in the argumentation framework of Fig. 1, the discrete iteration schema will produce the solution $X = $ **in** for SCC_1, and the solution $W = Y = $ **und** for SCC_2. The forward propagation algorithm would then operate on SCC_2 to try to attempt to label as **in** any viable argument left undecided in SCC_2. We have seen that we can either leave both W and Y undecided or to assign the label **in** to one and the label **out** to the other. The horizontal combination of the solutions for SCC_1 and SCC_2 (line 14) would then generate all solutions to layer 0, i.e., f_1, f_2 and f_3. This process repeats for the next layer as explained in the previous section. We now briefly describe the forward propagation algorithm itself in a bit more detail.

The Forward Propagation Algorithm.

---

[2] Some filtering to eliminate solutions not leading to maximal extensions in preferred/stable semantics problems is also done, although this is not shown in Algorithm 1. For full details, refer to [15].

```
1  EqArgSolver
2  |  Read and validate graph G
3  |  Decompose G into SCCs and arrange them into layers
   |  $\mathcal{L} = \{L_0, \ldots, L_{k-1}\}$
4  |  $Sols \leftarrow \{$all-und$\}$
5  |  for $i \leftarrow 0$ to $k-1$ do          /* Iterate through layers */
6  |  |  $newSols \leftarrow \varnothing$
7  |  |  foreach $f \in Sols$ do
8  |  |  |  $\lambda \leftarrow$ GR-ground$(L_i, f)$; $TSB \leftarrow$ trivial SCC block of $L_i$
9  |  |  |  $LayerSols \leftarrow \{\lambda \downarrow TSB\}$
10 |  |  |  $\mathcal{S} \leftarrow$ non-trivial SCCs in $L_i$
11 |  |  |  foreach $S \in \mathcal{S}$ do
12 |  |  |  |  $possIns \leftarrow$ candidate in-nodes of $S$ according to $\lambda$
13 |  |  |  |  $SCC$-sols$\leftarrow$findExtsFromArgs$(possIns, S, f, \lambda \downarrow S)$
14 |  |  |  |  Horizontally combine $SCC$-sols with solutions in
   |  |  |  |  $LayerSols$
15 |  |  |  end foreach
16 |  |  |  Add vertical combination of $f$ with each $\gamma \in LayerSols$ to
   |  |  |  $newSols$
17 |  |  end foreach
18 |  |  $Sols \leftarrow newSols$
19 |  end for
20 end
```

Algorithm 1: EqArgSolver's overall processing sequence.

The forward propagation algorithm is a new algorithm proposed in [14] allowing the generation of all complete extensions of an argumentation framework. It takes advantage of the constraints imposed by what it means to legally label an argument in order to prune the search space of feasible solutions. The algorithm is invoked by the procedure findExtsFromArgs at line 13 of Algorithm 1 and provides a huge performance improvement over Modgil-Caminada's algorithm (see [14] for a comparison). Essentially, each choice made for an argument being labelled **in** restricts the set of candidate labelling functions further. The more attacks there are in the argumentation graph the higher the number of such constraints to be satisfied. In [14], we suggested that the performance of the algorithm should therefore increase proportionally to the density of attacks in the graph. The empirical evaluation in Section 6 clearly confirms this conjecture.

As an example of execution, suppose we had at layer 0 the SCC shown in Fig. 2 (L). Initially, all of its arguments would be undecided. Some of

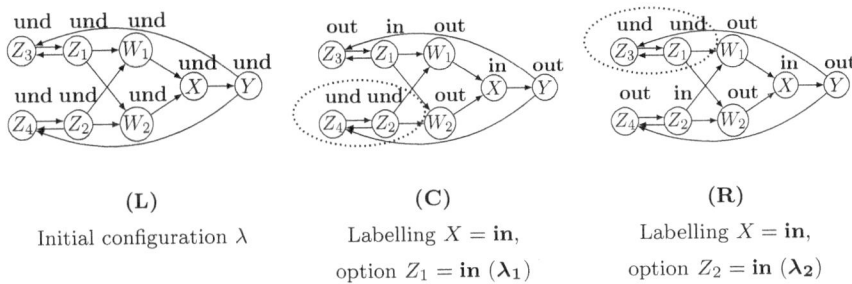

Figure 2. A sample SCC illustrating the forward propagation algorithm.

these arguments could potentially belong to an extension. The forward propagation algorithm would attempt to label **in** all viable arguments. Suppose we start with argument X (the order is not important). By labelling X **in** we would be forced to label **out** the argument Y that X attacks as well as X's attackers W_1 and W_2. In order to legally label the latter two arguments **out** we would also be required to label either Z_1 or Z_2 **in**. Again for Z_1 to be legally labelled **in**, Z_3 must be labelled **out** (λ_1), and for Z_2 to be labelled **in**, Z_4 must be labelled **out** (λ_2). This process is repeated until all options are considered. In the case of $Z_1 =$ **in** (λ_1), we would still have the two sub-options $Z_2 =$ **in**, $Z_4 =$ **out** (λ_{11}) and $Z_2 =$ **out**, $Z_4 =$ **in** (λ_{12}). This would result in all legal labellings in which X is labelled **in** (see Table 1).[3]

For *all* complete extensions to be generated, the algorithm will attempt to label **in** all viable nodes in the SCC. If the SCC of Fig. 2 is at layer 0, this will be all of its nodes. In the case of the SCC_3 of Fig. 1 with conditioning solution f_2, there would be nothing to be done, because the iteration schema would force A and B to be labelled **out** and this would force C to be labelled **in**, leaving no argument left undecided. In general, the algorithm would only attempt to label **in** the suitable candidate arguments. There are further subtle requirements to be satisfied, and sometimes the forward propagation may fail, leaving just the initial solution which will be legal because the iteration schema always provides a legal assignment when it starts from a legal assignment. The full details can be found in [14].

At the point where the forward propagation algorithm is invoked (Algorithm 1, line 13), we can place any suitable procedure that computes all legal label assignments to the SCC taking into account a particular conditioning solution. In Section 5.2 we will show how to use a SAT

---

[3] Note that Table 1 does not list *all* complete labellings for the SCC in Figure 2, it simply lists all complete labellings in which the label of X is **in**.

	X	Y	W_1	W_2	Z_1	Z_2	Z_3	Z_4
λ_1	in	out	out	out	in	und	out	und
λ_2	in	out	out	out	und	in	und	out
λ_{11}	in	out	out	out	in	in	out	out
λ_{12}	in	out	out	out	in	out	out	in
λ_{21}	in	out	out	out	in	in	out	out
λ_{22}	in	out	out	out	out	in	in	out

Table 1. All possible complete labelling functions in which the label of argument X of the SCC in Fig. 2 is **in**.

solver to perform this computation. The replacement of the SAT solver procedure for the forward propagation algorithm yielded a new SAT-based solver that we called `glucsccsolver`, after its use of `Glucose` and the decomposition into SCCs.

5 Computing the Semantics via a SAT Solver

It is pretty straightforward to search for complete extensions of an argumentation framework using an existing SAT solver. The main idea is to characterise the legal labelling conditions as a theory in propositional logic and search for models of the theory using the solver. A model can then be translated back as an extension via an appropriate mapping. In this Section, we show one of many possible characterisation and discuss possible optimisations for the stable semantics. In addition, we show how to restrict the labelling conditions to individual SCCs conditioned by solutions, allowing the combination of the technique described in Section 3 with a SAT solver. The formalisations given here were used to implement the two SAT-based solvers `glucsolver` and `glucsccsolver` discussed in Section 6.

We focus our presentation on the complete semantics, because it can be used in the computation of the grounded, preferred and stable semantics as well (see Section 2). We must also stress that a more efficient SAT-based implementation would take advantage of the intrinsic characteristics of the underlying SAT solver to optimise the search for solutions. The characterisation below is merely illustrative, but will serve to compare implementations that differ only in the approach used in the computation of the solutions for individual SCCs (see Section 6).

5.1 Translating the Complete Semantics Conditions to Propositional Logic

Let $\langle \mathcal{A}, \mathcal{R} \rangle$ be an argumentation framework. In order to encode the semantics as a logical theory, we need a propositional logic language with the usual logical connectives and $3 * |\mathcal{A}|$ symbols where for each $X \in \mathcal{A}$, the symbols $in(X)$, $out(X)$ and $und(X)$ denote that the argument X is either labelled **in**, **out**, or **und**, respectively. The formulae below, which we refer to as the COMPLETE conditions, are then defined for all arguments $X \in \mathcal{A}$.

We start with the uniqueness labelling conditions implicit in the definition of a function.

(UNIQUE) $(in(X) \vee out(X) \vee und(X)) \wedge \neg(in(X) \wedge out(X)) \wedge$
$\neg(in(X) \wedge und(X)) \wedge \neg(out(X) \wedge und(X))$

We then need to capture the conditions under which arguments are legally labelled **in** and **out**:

(IN-C1) $\bigwedge_{Y \in X^-} out(Y) \rightarrow in(X)$

(IN-C2) $in(X) \rightarrow \bigwedge_{Y \in X^+} out(Y)$

(OUT-C1) $\bigvee_{Y \in X^-} in(Y) \rightarrow out(X)$

(OUT-C2) $out(X) \rightarrow \bigvee_{Y \in X^-} in(Y)$

(IN-C1) says that if all attackers of an argument are labelled **out**, the argument must be labelled **in**. (IN-C2) says that if an argument is labelled **in**, then all of the arguments that it attacks must be labelled **out**. (OUT-C1) says that if any attacker of an argument is labelled **in**, the argument must be labelled **out**. Finally, (IN-C2) says that if and argument is labelled **out**, then at least one of its attackers must be labelled **in**.

It is easy to see that all source arguments, having no attackers, will be labelled **in** in any model that satisfies (IN-C1). This is because the empty conjunction of **out**'s is equivalent to $\top$, which means those arguments must be labelled **in**. This will force any arguments attacked by those arguments to be labelled **out** by (IN-C2) and so forth.

In [5], Cerutti et al. showed that a propositional logic formalisation similar to (UNIQUE), (IN-C1), (IN-C2), (OUT-C1) and (OUT-C2) precisely captures the notion of a complete extension. The authors divided the set $\mathcal{A}$ into the two parts $C, \mathcal{A} \backslash C$ (for some $C \subseteq \mathcal{A}$). The perceptive reader will note that formulae (1)–(5) of [5, Definition 14] correspond to (UNIQUE), (IN-C1), (IN-C2), (OUT-C1) and (OUT-C2) and that NC_i^x of [5, Proposition 5] is $\top$ when $C = \mathcal{A}$. Therefore, $(1) \wedge (2) \wedge (3) \wedge (4) \wedge (5)$ is sufficient to capture the notion of a complete extension semantically in propositional logic.

A SAT solver such as Glucose accepts a formula in *conjunctive normal form* (CNF) which are conjunctions of disjunctions of literals. Each clause is a disjunction of positive and negative literals l_k of the form $\bigvee_m l_m \vee \bigvee_n \neg l_n$. Let $X^- = \{Y_1, \ldots, Y_{x_i}\}$ and $X^+ = \{W_1, \ldots, W_{x_j}\}$. The conditions above can be re-written as disjunctive clauses as follows:

(UNIQUE) $\text{in}(X) \vee \text{out}(X) \vee \text{und}(X)$
$\neg \text{in}(X) \vee \neg \text{out}(X)$
$\neg \text{in}(X) \vee \neg \text{und}(X)$
$\neg \text{out}(X) \vee \neg \text{und}(X)$

(IN-C1) $\neg \text{out}(Y_1) \vee \neg \text{out}(Y_2) \vee \ldots \vee \neg \text{out}(Y_{x_i}) \vee \text{in}(X)$

(IN-C2) $\neg \text{in}(X) \vee \text{out}(W_1)$
$\neg \text{in}(X) \vee \text{out}(W_2)$
$\vdots$
$\neg \text{in}(X) \vee \text{out}(W_{x_j})$

(OUT-C1) $\neg \text{in}(Y_1) \vee \text{out}(X)$
$\neg \text{in}(Y_2) \vee \text{out}(X)$
$\vdots$
$\neg \text{in}(Y_{x_i}) \vee \text{out}(X)$

(OUT-C2) $\neg \text{out}(X) \vee \text{in}(Y_1) \vee \text{in}(Y_2) \vee \ldots \vee \text{in}(Y_{x_i})$

EXAMPLE 2. Consider the argumentation framework below:

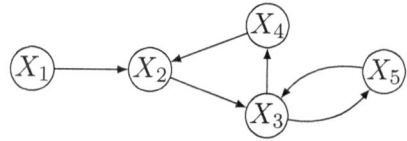

The framework would yield the following clauses for the node X_1:

(UNIQUE) $\text{in}(X_1) \vee \text{out}(X_1) \vee \text{und}(X_1)$
$\neg \text{in}(X_1) \vee \neg \text{out}(X_1), \neg \text{in}(X_1) \vee \neg \text{und}(X_1)$
$\neg \text{out}(X_1) \vee \neg \text{und}(X_1)$

(IN-C1) $\text{in}(X_1)$

(IN-C2) $\neg \text{in}(X_1) \vee \text{out}(X_2)$

Since X_1 has no attackers, the attack disjunction in (IN-C1) is false, leaving just the positive literal $\text{in}(X_1)$; in (IN-C2), we only have one clause for the node that X_1 attacks, i.e., X_2; finally, (OUT-C1) $\equiv \top$, and (OUT-C2) $\equiv \text{out}(X_1) \to \bot \equiv \neg \text{out}(X_1)$, since X_1 has no attacks and hence the disjunction of its attackers is false. Since (IN-C1) + (UNIQUE) $\vdash \neg \text{out}(X_1)$, this is not needed.

Any model of a theory with the clauses above would satisfy $\text{in}(X_1)$ and $\text{out}(X_2)$ (because of (IN-C1) and (IN-C2)). As for X_3–X_5, one of the possibilities would be to make und(X_3) = und(X_4) = und(X_5) = *true*. An extension would simply be the set of nodes X for which the proposition in(X) = *true*. In this case, the model described would correspond to the complete (but not preferred) extension $\{X_1\}$.

Glucose would provide a model (if any) according to some internal computation strategy. In order to obtain further models, one can re-submit the original set of clauses *plus* a clause negating the previous solution. As the solution is presented as the set of literals that are true (to be understood as a conjunction of literals describing the model), all we need to do is to add the negation of these literals as a new clause and re-submit the call to Glucose. Eventually, no more models will be found and Glucose will return the word "UNSAT". This simply means that no more models (i.e., complete extensions) can be found.

We saw above that each model will correspond to a complete extension that is not necessarily preferred. At the *meta-level*, we will need to filter out these extensions keeping only those that are maximal w.r.t. set inclusion in order to obtain the preferred extensions.

5.2 Conditions for SCCs Conditioned by Solutions

We can employ the same principles presented in Section 3 and write a modified theory that can be used to find all solutions for a non-trivial SCC.[4] These solutions can then be combined horizontally and vertically as before to yield solutions for the argumentation framework as a whole. The advantage of doing this is that the theorem prover would have to deal with a smaller theory and this may speed up the computation of solutions of larger argumentation frameworks. The drawback is that more external calls to the solver will be needed. Obviously there is a trade-off between the two approaches. The results in Section 6 indicate that there is an advantage of doing so when the size of the network is relatively large. For now, let us describe what is needed.

Consider the argumentation framework of Fig. 1 again depicted on the left of Fig. 3 and suppose we wished to find the solutions to SCC_3 alone, which is in layer 1. In Section 3 we saw that the solutions to SCC_3 are conditioned by the solutions f_1, f_2 and f_3 to layer 0, where $f_1(X) = $ **in**, $f_1(W) = f_1(Y) = $ **und**; $f_2(X) = f_2(W) = $ **in**, $f_2(Y) = $ **out**; and $f_3(X) = f_3(Y) = $ **in**, $f_3(W) = $ **out**.

The part of the argumentation framework relevant to the computation of SCC_3's solutions is shown in Fig. 3 (R).

---

[4] Trivial SCCs are dealt with directly by the discrete Gabbay-Rodrigues Iteration Schema (see [15] for details).

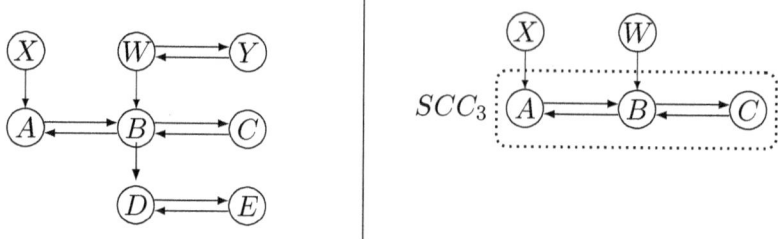

Figure 3. The argumentation framework of Fig. 1 and its relevant part to the computation of the solutions to SCC_3

The only modification we need to do is to write special formulae for the nodes attacking the SCC (which have fixed values) and write the usual formulae for the nodes in the SCC itself, eliminating those associated with attacks going out of the SCC. Let us start with the former. In this example, the formulae need to describe what the labels of X and W are in the corresponding solution.

Axioms for the External Attackers of the SCC

The values of the incoming attacks are determined by the particular conditioning solution. This means one formula for each attacking argument (each being written as three clauses). For example, for solution f_1, we would need the two formulae:

$$\text{in}(X) \land \neg\text{out}(X) \land \neg\text{und}(X), \text{und}(W) \land \neg\text{in}(W) \land \neg\text{out}(W)$$

Alternatively, one could write the actual label of the argument, e.g., in(X) and then instantiate (UNIQUE) with X, but this overcomplicates matters somewhat. Similarly, although (IN-C2) could be applied to X, forcing A to be **out**, this can be taken care of by (OUT-C1). So for all external attackers Y_i of an SCC $\mathcal{S}$, we only need to write one of

(ATT-IN) $\text{in}(Y_i) \land \neg\text{out}(Y_i) \land \neg\text{und}(Y_i)$
(ATT-OUT) $\neg\text{in}(Y_i) \land \text{out}(Y_i) \land \neg\text{und}(Y_i)$
(ATT-UND) $\neg\text{in}(Y_i) \land \neg\text{out}(Y_i) \land \text{und}(Y_i)$

depending on whether the attacker Y_i is labelled **in**, **out**, or **und**, respectively, by the corresponding conditioning solution.

Axioms for the Nodes in the SCC

For all nodes in the SCC itself, we need the axioms (UNIQUE), (IN-C1), (OUT-C1) and (OUT-C2). As for (IN-C2), we only need the axioms where the attacked node belongs to the SCC. For example, for the full argumentation framework, we would normally write the axioms

$$\text{in}(B) \to \text{out}(A) \land \text{out}(C) \land \text{out}(D)$$

which in clausal form would yield the three clauses

$$\neg \text{in}(B) \vee \text{out}(A), \neg \text{in}(B) \vee \text{out}(C), \neg \text{in}(B) \vee \text{out}(D)$$

However, D is outside of the SCC, so its value is not relevant in the solution to SCC_3, so instead, we simply write

$$\neg \text{in}(B) \vee \text{out}(A), \neg \text{in}(B) \vee \text{out}(C)$$

because the nodes A and C that B attacks are within the same SCC as B.

5.3 Putting it All Together

We built two SAT-based argumentation solvers based on the translations provided in Sections 5.1 and 5.2 above. Both employ Glucose as an external SAT solver. The first solver uses a unified approach, i.e., it translates the whole argumentation framework as a set of clauses and then invokes Glucose successively to compute all models of that set. Each model is then translated back as a single extension, with those that are maximal being the preferred ones. We called this solver glucsolver.

The second solver makes use of EqArgSolver's decomposition of the argumentation framework into SCCs and their arrangement into layers. As glucsolver, it also invokes Glucose but only to find models of the logical translations of the non-trivial SCCs. This means the theories it has to solve are smaller, although a higher number of external calls to Glucose need to be made. We called this solver glucsccsolver.

These two solvers allowed us to compare the impact of the decomposition into SCCs in the SAT reduction-based approach and their relationship with our direct approach employing the forward propagation algorithm.

Before we move on to the evaluation of the benchmark results in Section 6, it is worth emphasising that for the special case of the stable semantics, the logical encoding can be greatly (and trivially) simplified as shown next.

5.4 Optimising for the Stable Semantics

We had to utilise three propositional symbols for each argument because classical logic is two-valued whereas Dung's semantics is three-valued. The proposition in(X) can only encode, say, whether or not the argument X is labelled **in**. If it is not labelled **in**, it could still be labelled **out** or **und**, so we need more symbols to distinguish between these two states. However, in the case of the *stable semantics*, extensions correspond to assignments in which arguments can only be in one of two states: **in** or out (as no arguments can be labelled **und**). In this case, one propositional symbol per argument will suffice. Accordingly, we could choose the propositional symbol "X" to represent the fact that the label of the argument X is **in**, and then infer from the formula $\neg X$ that the label of the argument X is **out**. Under this assumption, the COMPLETE conditions could then be simplified as the conditions that follow (which we refer to, collectively, as STABLE):

(ST IN-C1) $\quad \bigwedge_{Y \in X^-} \neg Y \to X$

(ST IN-C2) $\quad X \to \bigwedge_{Y \in X^+} \neg Y$

(ST OUT-C1) $\quad \bigvee_{Y \in X^-} Y \to \neg X$

(ST OUT-C2) $\quad \neg X \to \bigvee_{Y \in X^-} Y$

PROPOSITION 3. *If a model w satisfies conditions STABLE, then the set of arguments $\{X | w \models X\}$ corresponds to a stable extension.*

Proof. *A model of COMPLETE corresponds to a complete extension. When a complete extension is stable, then it attacks all arguments outside of it. Therefore, all arguments outside of the extension must be labelled* **out** *and hence no argument is labelled* **und**. *By (UNIQUE) and the fact that no argument is labelled* **und**, *any model w corresponding to a stable extension will satisfy in(X)$\leftrightarrow\neg$out(X). We can simply use the formula "X" for in(X) $\equiv \neg$out(X) and "$\neg X$" for $\neg$in(X) $\equiv$ out(X). From this, (ST IN-C1), (ST IN-C2), (ST OUT-C1) and (ST OUT-C2) will correspond to (IN-C1), (IN-C2), (OUT-C1) and (OUT-C2).* ∎

Since (ST IN-C1) $\equiv$ (ST OUT-C2) and (ST IN-C2) $\equiv$ (ST OUT-C1) (via the contraposition of the implication), we only really need an appropriate combination of these, for example (ST IN-C1) and (ST IN-C2).

6 Empirical Evaluation

We conducted three sets of experiments to evaluate different aspects of the performance of the forward propagation algorithm proposed in [14] in relation to a generic search for preferred extensions of an argumentation framework using the Glucose SAT solver (http://www.labri.fr/perso/lsimon/glucose/).

Our objective with the experiments was not to conduct an extensive empirical evaluation between solvers. Indeed this is being done by ICCMA 2017 (http://argumentationcompetition.org/). We merely proposed to draw some conclusions about the behaviour of the forward propagation algorithm with respect to a "generic" SAT-based approach.

Experimental Setup

All graphs in the experiments were generated using probo's SCC generator [4], which generates graphs with a bounded number of SCCs and a given probability of attacks between arguments. We were not too concerned about attacks between SCCs, because the search for solutions for an SCC lies at the base of the recursive problem, so our main focus was on the SCC cardinality and the density of attacks *within* SCCs. Thus, in all experiments we set the probability of attacks between arguments of *different* SCCs to 0.1, but varied the internal probability of attacks between arguments within an SCC between 0.1 and 1.0. This variation allowed us to confirm a conjecture we made in [14] suggesting that the performance of the forward propagation algorithm should increase proportionally to the probability of attacks between arguments. Intuitively, the higher this probability, the higher the density of attacks in the graph, and hence the more the number of constraints that are added to the labelling functions and, consequently, the smaller the search space of feasible solutions. Our experiments confirmed this conjecture (we will revisit this later).

The graphs thus generated were submitted to the three solvers: EqArgSolver [15] (which employs the forward propagation algorithm); glucsolver which employs Glucose on the translation of the whole argumentation graph as a SAT problem (as described in Section 5.1); and finally glucsccsolver, which employs a combination of the decomposition technique into SCCs and Glucose restricted to theories encoding single non-trivial SCCs (as described in Section 5.2). The solvers ran on a PC with an Intel i7 4690K processor and 32Gb RAM.

Fig. 4, Fig. 5 and Fig. 6 show the comparative average execution time per graph. We discuss the findings in three categories of graphs as follows.

Small cardinality, small number of SCCs

In the first set of experiments, we randomly generated 6,000 graphs with up to 50, 100, 120 140, 160 and 200 nodes. For each cardinality boundary, we generated 1,000 graphs with up to 2 SCCs each, 100 for each probability 0.1, 0.2, ..., 1.0. We then searched for preferred extensions using the three solvers. The performance of the solvers for this set of graphs is depicted in Fig. 4. From the results, we can conclude the following:

1. There was not much difference in performance between the SAT solvers with or without decomposition of the graphs into SCCs. This is due to the fact that the graphs are relatively small and

only contain up to 2 SCCs, so the performance gain obtained by searching for models in smaller theories could not be offset by the extra effort used in the decomposition into SCCs.

2. The SAT-based solvers outperformed the forward propagation algorithm when the probability of internal attack was roughly below 0.5. After a certain probability threshold (of around 0.5), the forward propagation algorithm outperformed the SAT-based solvers in all cardinalities.

3. The performance of the forward propagation algorithm *did* increase proportionally with the increase in the probability of attacks between arguments.

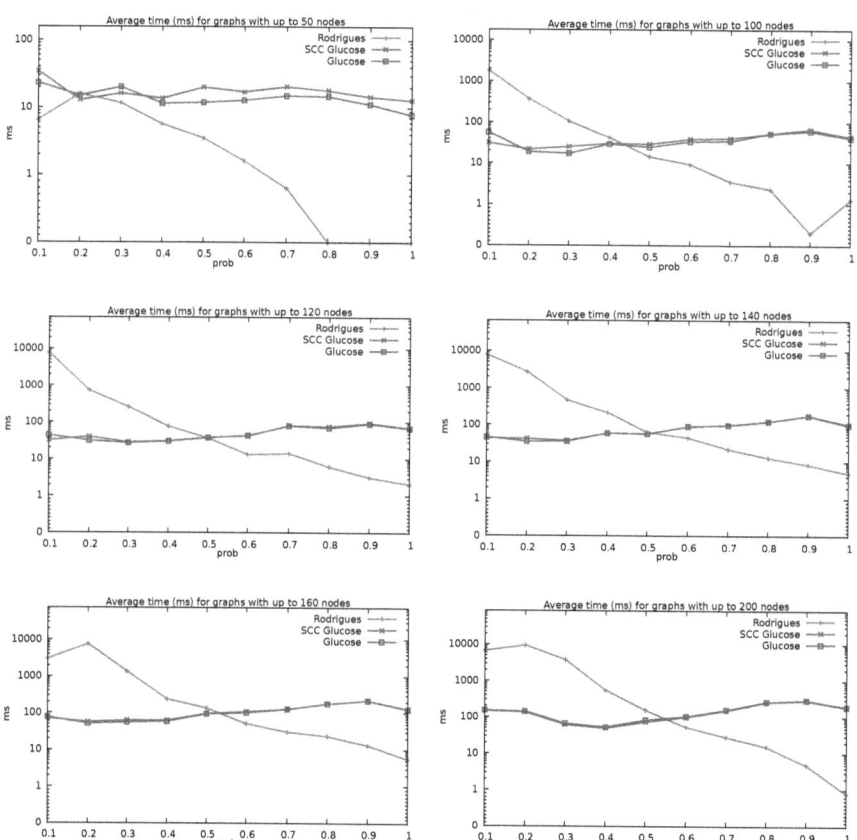

Figure 4. Average computation time for graphs with up to 50–200 nodes and up to 2 SCCs.

Medium cardinality, small number of SCCs

We then generated a second set of graphs with a larger upper bound on the number of arguments varying from 300 to 600 in increments of 100 and a slightly larger upper bound on the number of SCCs of 5. Again we set the probability of internal attacks from 0.1 to 1.0 in 0.1 increments. The results are depicted in Fig. 5. From this set we can conclude the following:

1. Again there was no significant difference in performance between the SAT solvers with or without decomposition of the graphs into SCCs. We again conclude that the relatively small number of SCCs was not sufficient to take advantage of the decomposition into SCCs.

2. The SAT-based solvers still outperform the forward propagation algorithm when the probability of internal attack is roughly below 0.5. After a certain threshold, the forward propagation algorithm then outperforms the SAT-based solvers.

3. There is more variance in performance between the SAT-based solvers and the one using the forward propagation algorithm and the performance gap is somewhat reduced between the three. This may suggest that there may be a graph cardinality threshold after which the performance of the SAT-based solvers degrade.

4. Again, the performance of the forward propagation algorithm *did* increase proportionally with the increase in the probability of attacks between arguments.

Large cardinality, medium number of SCCs

Subsequently, we randomly generated 100 graphs with up to 2,500 arguments each and a number of SCCs between 35 and 50 each, again with internal probability of attacks between arguments from 0.1 to 1.0 in 0.1 increments. The results of this set of experiments are depicted in Fig. 6, from which we can conclude:

1. The SAT solver using the decomposition outperformed the one without it independently of the internal probability of attacks. This confirms the result ($I2$) given in [5], which asserted that there would be a threshold after which the decomposition into SCCs provides performance benefits for SAT-reduction solvers.

2. The forward propagation algorithm outperformed both SAT solvers independently of the internal probabilities of attacks.

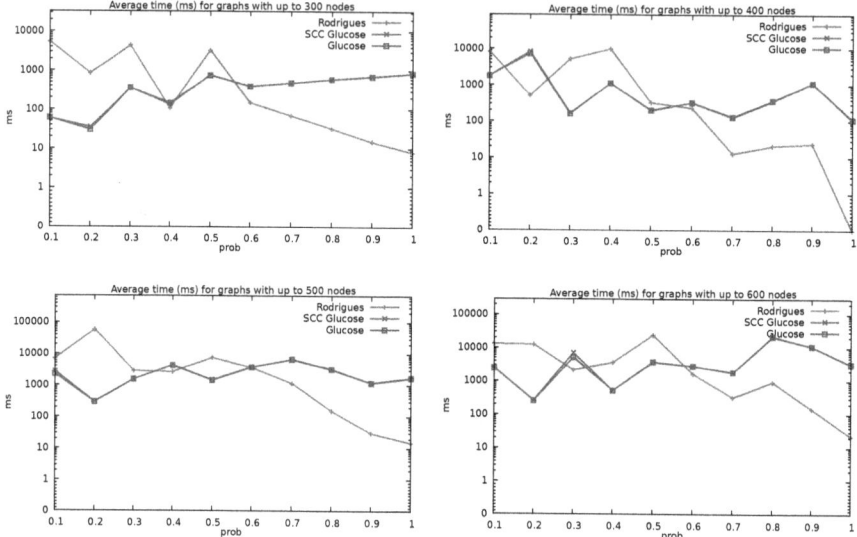

Figure 5. Average computation time for graphs with 300–600 nodes and up to 5 SCCs.

Although the graph in Fig. 6 still shows some improvement in EqArg-Solver's performance with the increase in the probability of attacks, this was not very noticeable. There may be many reasons for this. Firstly, the sample was relatively small. Secondly, the large number of SCCs means that the relative size of each individual SCC may not be large enough to show a more marked variation. This investigation will continue in future work.

7 Conclusions and Future Work

The motivation for the development of the forward propagation algorithm proposed in [14] came from the need to replace Modgil-Caminada's algorithm in the solver GRIS [13]. Modgil-Caminada's algorithm proved very inefficient for all but the simplest graphs and can only compute the preferred extensions. The forward propagation algorithm can compute all complete extensions and is able to decide on argument acceptability without necessarily having to generate all extensions. This new algorithm was used in the solver EqArgSolver [15], which was submitted to ICCMA 2017 (see http://argumentationcompetition.org/).

Solvers using SAT-reduction approaches took the top spots in the 1st ICCMA and we would expect them to continue to outperform solvers using direct approaches for some time, but there are application areas where solvers using a direct approach are the only alternatives, e.g., in certain argumentation frameworks where attacks and arguments receive values in the [0, 1] interval. Therefore continued research in direct algo-

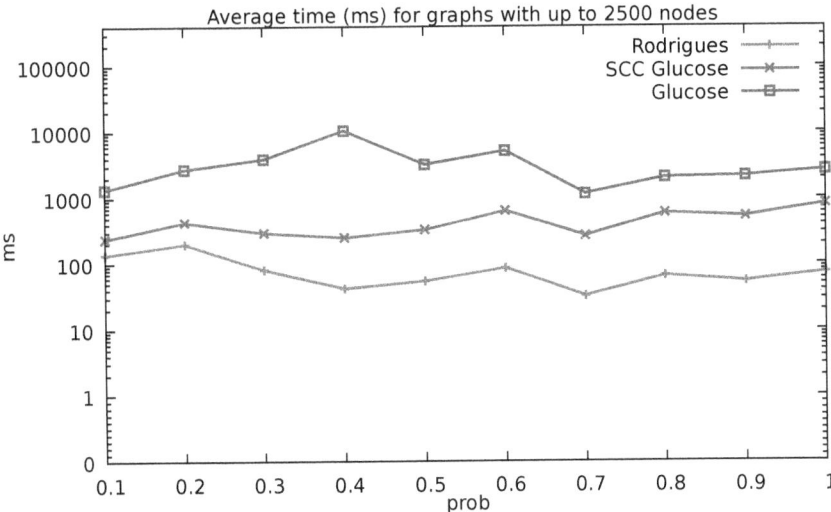

Figure 6. Average computation time for graphs with up to $2,500$ nodes and in between 35 and 50 SCCs.

rithms remains important for the development of the field. The objective of this paper was to shed some light into the performance differences between the SAT approaches and the forward propagation algorithm, paving the way for future performance enhancements arising from this analysis.

We showed how to build a rudimentary meta argumentation solver based on the Glucose SAT solver, which we called glucsolver. It simply does a naive translation of the entire argumentation graph and the COMPLETE labelling conditions as a logical theory and uses the models found by Glucose's for the theory to provide all complete extensions of the argumentation framework. The maximal extensions of these will correspond to the argumentation framework's preferred extensions. We then provided two optimisation techniques: in problems involving the *stable* semantics, we showed how to simplify the translation of the argumentation problem into propositional logic by minimising the number of required propositional variables. Secondly, we showed how to restrict the logical conditions for complete labellings of an SCC when they are conditioned by a solution. This means that the SAT solver can be invoked to find solutions restricted to SCCs and hence work on smaller theories. We called the second argumentation solver employing this technique glucsccsolver. It simply replaces the forward propagation algorithm in EqArgSolver with a call to Glucose for each theory representing an individual non-trivial SCC and hence provides a pathway for a more direct comparison between the two approaches.

We ran three sets of experiments with each of the above solvers. Our experiments confirmed that the density of attacks within SCCs is inversely proportional to the execution time of the forward propagation algorithm. This means that this algorithm is particularly useful in dense graphs. The results also indicate that the decomposition into SCCs is only advantageous when the number of SCCs is relatively large (confirming an observation made earlier by [5]). For larger graphs, EqArgSolver using the forward propagation algorithm outperformed the SAT-based solvers. This could arise from several factors. `Glucose` may suffer some performance degradation when the underlying theory is large. This would explain why `glucsccsolver` outperforms `glucsolver`, since each call to `Glucose` invoked by `glucsccsolver` is done on a smaller logical theory. The performance gain is however offset by the extra overhead incurred by the additional external system calls and would explain why EqArgSolver outperformed both SAT-based solvers. In a more robust SAT-based implementation, the SAT solver would be embedded in the meta-solver to eliminate the extra effort in extenal system calls and its invocation would be carefully fine-tuned for the problem at hand.

Our next step is to try to incorporate into the forward propagation algorithm techniques that in the logical context allow a faster search for models of the underlying logical theory. If we can translate these into algorithmic constraints or heuristics, they may help us prune the search space of solutions further, improving the overall performance of the propagation algorithm.

References

[1] P. Baroni, M. Giacomin, and G. Guida. "SCC-recursiveness: a general schema for argumentation semantics". In: *Artificial Intelligence* 168.1 (2005), pp. 162–210. ISSN: 0004-3702. DOI: http://dx.doi.org/10.1016/j.artint.2005.05.006.

[2] M. Caminada. "A labelling approach for ideal and stage semantics". In: *Argument and Computation* 2.1 (2011), pp. 1–21.

[3] M. Caminada and D. M. Gabbay. "A Logical Account of Formal Argumentation". In: *Studia Logica* 93.2-3 (2009), pp. 109–145.

[4] F. Cerutti et al. "A Benchmark Framework for a Computational Argumentation Competition (Demo paper)". In: *Proceedings of the Fifth International Conference on Computational Models of Argumentation (COMMA'14)*. Sept. 2014.

[5] F. Cerutti et al. "A SCC Recursive Meta-Algorithm for Computing Preferred Labellings in Abstract Argumentation". In: *14th Intl. Conference on Principles of Knowledge Representation and Reasoning (KR)*. 2014.

[6] F. Cerutti et al. "Summary Report of the First International Competition on Computational Models of Argumentation". In: *AI Magazine* 37.1 (2016), p. 102.

[7] G. Charwat et al. "Methods for solving reasoning problems in abstract argumentation – A survey". In: *Artificial Intelligence* 220 (2015), pp. 28–63. DOI: http://doi.org/10.1016/j.artint.2014.11.008.

[8] P. M. Dung. "On the acceptability of arguments and its fundamental role in nonmonotonic reasoning, logic programming and n-person games". In: *Artificial Intelligence* 77 (1995), pp. 321–357.

[9] D. M. Gabbay and O. Rodrigues. "Further Applications of the Gabbay-Rodrigues Iteration Schema". In: *Computational Models of Rationality*. Ed. by C. Beierle, G. Brewka, and M. Thimm. Vol. 29. College Publications, 2016, pp. 392–407.

[10] B. Liao. "Toward incremental computation of argumentation semantics: A decomposition-based approach". In: *Annals of Mathematics and Artificial Intelligence* 67.3 (2013), pp. 319–358. ISSN: 1573-7470. DOI: 10.1007/s10472-013-9364-8. URL: http://dx.doi.org/10.1007/s10472-013-9364-8.

[11] B. Liao. *Efficient Computation of Argumentation Semantics*. Elsevier, 2014, p. 148. ISBN: 978-0-12-410406-8.

[12] S. Modgil and M. Caminada. "Proof Theories and Algorithms for Abstract Argumentation Frameworks". In: *Argumentation in Artificial Intelligence*. Ed. by Guillermo Simari and Iyad Rahwan. Boston, MA: Springer US, 2009, pp. 105–129. ISBN: 978-0-387-98197-0. DOI: 10.1007/978-0-387-98197-0_6.

[13] O. Rodrigues. "GRIS System Description". In: *System Descriptions of the 1st International Competition on Computational Models of Argumentation*. Ed. by M. Thimm and S. Villata. Cornell University Library, 2015, pp. 37–40.

[14] O. Rodrigues. "A Forward Propagation Algorithm for the Computation of the Semantics of Argumentation Frameworks". In: *Proceedings of the 4th Intl. Conference on Theory and Applications of Formal Argumentation*. TAFA'17. To appear. 2017.

[15] O. Rodrigues. "EqArgSolver – System Description". In: *Proceedings of the 4th Intl. Conference on Theory and Applications of Formal Argumentation*. TAFA'17. To appear. 2017.

[16] Y. Wu and M. Caminada. "A labelling-based justification status of arguments". In: *Studies in Logic* 3.4 (2010), pp. 12–29.

A Semantic-Inferentialist Framework for Natural Language Understanding

Vládia Pinheiro, Tarcísio Pequeno, Vasco Furtado

1 Introduction

Natural Language systems depend on models for expressing semantic knowledge, which define the nature and characteristics of knowledge necessary to text understanding. Distributional Semantic models like Hyperspace Analog to Language (HAL) model [16] and Latent Semantic Analysis (LSA) [15] rely on the distributional hypothesis that words occurring in the same contexts tend to have similar meanings [14]. These models have been used in several tasks such as semantic priming, synonymy tests [13], and NLP tasks like Semantic Textual Similarity [17]. However, we share the claim of other researchers such as Cambria and White (2014) [5] that the information necessary for a deep understanding of texts is sometimes implicit and is insufficient to approximate the meaning of words with vectors summarizing their patterns of co-occurrence in corpora.

For instance, when we, human beings proficient in natural language, read the news "*Maria was murdered by her husband after an argument on Solon Pinheiro Street*", we are able to argue that the type of crime was "homicide" probably due to a "crime of passion." This is possible because people know the conditions in which the concepts "to murder" and "husband" can be used and the consequences of using these concepts in the aforementioned sentence. Such inferences do not come only from the individual content of the concepts, but are generated by a holistic thinking about the use of these concepts in a particular sentence structure. In this example, one conclusion of the text was "*Maria was murdered by someone who is capable of being jealous*" (generated by a consequence of use of the concept "husband"—"husband is **capable of** being jealous"). This inference matches with one of the causes of crime—"crime of passion"—whose related concepts are "*jealousy*", "*lover*", "*violence*", "*death*".

Copyright © 2019 by Vládia Pinheiro, Tarcísio Pequeno, and Vasco Furtado. All rights reserved.

This example shows that it is within the linguistic practice that the circumstances to use a word and the consequences thereof can be grasped, and—by disregarding them—much of what could be inferred is lost.

Philosophers such as Sellars (1980) [22], Dummett (1973, 1978) [8, 9] and Brandom (1994, 2000)[3, 4] proposed semantic-inferentialist theories, which present a different approach to define the content of concepts. According to these theories, the expression of the semantic value of concepts should favor the role that such concepts play in reasoning, as premises and conclusions, rather than their referents and representational characteristics (Brandom, 2000, pg 1) [4]. Following this inferentialist view, we propose a computational model to treat the semantic-pragmatic level of natural languages—the Semantic-Inferentialist Model (SIM).

The Semantic-Inferentialist Model (SIM) follows the pragmatic view of language, which enables a top layer of semantic reasoning in NLP systems, where semantic relations are considered according to their roles in inferences, as premises or conclusions. SIM defines new requirements for expressing and reasoning about semantic knowledge, in a way as to enable computer systems to have better quality in handling linguistic signs in order to make inferences. In SIM, what needs to be expressed in a concept must be manipulated and qualified in terms of preconditions or postconditions of the use of the concepts in linguistic situations, that is to say, the inferential content. NLP systems that follow SIM will thus have a framework to represent the inferential relation between concepts and the knowledge of how to combine these relations into sentence structures, in order to make inferences considering the pragmatic aspect of natural language.

This article initially presents SIM, its foundations and its components, especially the reasoner that explores the semantic bases of the model and how it uses an inferential relatedness measure between concepts in the reasoning process. Finally we detail how SIM was used in the development of the WikiCrimes$_{IE}$ system—a system to extract information that characterizes a criminal occurrence from news stories about crimes (e.g., type of crime, scene of the crime, cause of the crime, weapon used). The results obtained in the identification of implicit and explicit information in news articles demonstrate the feasibility and limitations of the model.

2 The Semantic-Inferentialist Model (SIM)

We argue that—in order to answer questions, extract information, refute arguments and provide explanations on a given text—one must grasp the content of concepts based on their roles in the inferences that participate. The various roles of the concepts and sentences in inferences are defined by the preconditions and postconditions associated with the use thereof, or as we call it, by its inferential content. In turn, the inferential content is responsible for generating new premises and conclusions for the sentence, which feed reasoning. Based on these prerequisites, we propose the Semantic-Inferentialist Model (SIM). SIM defines the main requirements for semantic expression and reasoning in order to enable NL systems to generate premises and conclusions from sentences and texts, which support various NLP tasks. It is based on the existence of two knowledge bases for NL systems: Conceptual Base and Pattern-Sentence Base.

The **Conceptual Base** contains the inferential content of concepts in a natural language—the set of preconditions and postconditions of usage of that concept, expressed through binary inferential relations between this concept and other concepts of the base. Hence, there is a network of concepts interlinked by inferential relations that express the situations of use of the concepts. This network follows the holistic view that to know a concept is to know its relations—in the form of premises or conclusions—with other concepts.

Formally, the Conceptual Base is represented in a directed graph $G_c(C, R_c)$, where C(vertices) is the set of concepts of a natural language, and R_c (edges) is the set of inferential relation between any two concepts $c_i, c_k \in C$. Each inferential relation $rc_j \in R_c$ is represented by a tuple $(rel\_name_j, c_i, c_k, type)$, where rel_name_j is the type of semantic relation between the concepts $c_i, c_k \in C$, and type = "Pre" or "Pos", respectively, indicates a precondition or a postcondition of concept c_i. Additionally, rc_j has its strength $\phi_{rc_j} = w_{rc_j} \beta_{rel}$, which is a function of an index of credibility w_{rc_j} and of a weight β_{rel}, assigned per parameter, for each type of semantic relation *(rel_ name)*. The concept **crime** has, for example, two inferential relations—(*capableOf*, **crime**, **victim**, **Pre**) and (*desirableEffectof*, **crime**, **judgment**, **Pos**).

The **Pattern-Sentence Base** contains the syntactic structure of pattern-sentences and their inferential contents, which are binary relations between one phrase (noun, verb or complement) of a pattern-sentence and a concept from the Conceptual Base. Hence, there is a network of inferential relations that express the situations of use of a sentence that follows the syntactic structure defined by the pattern-sentence.

Formally, this base is represented by a directed graph $Gp(V, Rp)$, where V (vertices) is formed by pattern-sentences p_j and concepts $c_i \in C$, and Rp (edges) is the set of inferential relation between a phrase of p_j and a concept c_i. Each inferential relation $rp_j \in Rp$ is represented by a tuple $(rel\_name_j, phrase(p_j), c_i, type)$, where rel_name_j is the type of semantic relation between a phrase of p_j (e.g. subject of p_j) and a concept c_i, and type = "Pre" or "Pos", respectively, indicates a precondition or a postcondition of p_j.

It is worth remarking that a challenge for the application of SIM in NL systems is the construction of large databases with semantic-inferentialist knowledge. We have constructed InferenceNet (www.inferencenet.org) —the first bilingual linguistic resource (in English and Portuguese) that contain common sense and inferentialist knowledge about concepts and pattern-sentences. A detailed description of this process of construction can be accessed in [20]. In addition to the conceptual and sentence-pattern knowledge bases, SIM includes an Inferential Relatedness Measure and a semantic parser—the Semantic-Inferentialist Analyzer (SIA). These two components will be described in Sections 3 and 4.

3 Inferential Relatedness Measure

The inferential relatedness measure calculates how similar two concepts are. From the inferential point of view, as much as the set of preconditions (or postconditions) of one concept is similar to the set of preconditions (or postconditions) of another concept, the more similar are the inferences in which the concepts can participate. This measure is used by the SIA (semantic parser) in order to define (1) which concept, among two concepts named homonymously, was used in the sentence and (2) the semantic contribution (preconditions and postconditions) of a concept c_i used in a sentence s_k—considering the preconditions and postconditions of c_i that match the preconditions and postconditions of other concepts used in the sentence s_k (see details in step 5 of the SIA algorithm, explained in section 4).

To calculate the inferential relatedness between two concepts c_1 and c_2, three forms of inferential proximity have been defined:

1. Proximity by Direct Relationship (F_1)—when a precondition (or postcondition) of c_1 expresses a direct relationship with c_2, or vice-versa. For example, c_1 = "crime" and c_2 = "robbery," and the inferential relation (isA, "robbery", "crime", "Pre"). In this case, the set X_1 is built, comprised of the inferential relations of c_1, which has concept c_2 as an image, and vice versa. Formally,

    - $X_1 = \{(rel\_name_j, c_i, c_k, type) | (c_i = c_1; c_k = c_2) \text{ or } (c_i = c_2; c_k = c_1)\}$.
    - $F_1 = \sum_{j=1}^{n} \phi_{rc_j}$, where ϕ_{rc_j} is the strength of the inferential relation $rc_j \in X_1$.

2. Proximity by Common Relationship (F_2)—this form of proximity occurs when c_1 and c_2 express the same type of semantic relation with

the same concept c_3. For example, c_1="crime" and c_2="robbery" and both have inferential relations

$$(capableOf, \text{``crime''}, \text{``have\_victim''}, \text{``Pre''})$$

and

$$(capableOf, \text{``robbery''}, \text{``have\_victim,''} \text{``Pre''}).$$

In this case, the set X_2 is built, comprised of the inferential relations of c_1 and c_2 that have the same type of inferential relation rel_name with the same image concept. Formally,

- $X_2 = (rel\_name, c_1, c_3, type), (rel\_name, c_2, c_3, type).$
- $F_2 = \sum_{j=1}^{m} \phi_{rc_j}$, where ϕ_{rc_j} is the strength of the inferential relation $rc_j \in X_2$.

3. Proximity by Relationship of the Same Nature (F_3)—this form of proximity occurs when c_1 and c_2 express inferential relations of the same nature (functional, causal, event-based, etc.) with the same concept. For example, c_1 ="shot" and c_2 ="finger", both have inferential relations (usedFor, "shot," "wound," "Pre") and (capableOfReceiveAction, "finger," "wound," "Pos"), and "usedFor" and "capableOfReceiveAction" are of the same nature. In this case, the set X_3 is built, comprised of inferential relations of c_1 and c_2 that have relations of the same nature with the same image concept, and express the same type of inferential relation. Formally,

- $X_3 = \{(rel\_name_1, c_1, c_3, type), (rel\_name_2, c_2, c_3, type) \mid nature(rel\_name_1) = nature(rel\_name_2)\}.$
- $F_3 = \sum_{j=1}^{m} \phi_{rc_j}$, where ϕ_{rc_j} is the strength of the inferential relation $rc_j \in X_3$.

The measure of inferential relation $\theta(c_1, c_2)$ between two concepts c_1 and c_2 is calculated by the formula (1).

$$\theta_{c_1,c_2} = (F_1 \omega_1 + F_2 \omega_2 + F_3 \omega_3) \mu_{c_1,c_2} \tag{1}$$

Where,

- F_1, F_2, F_3 are the summations of the strengths of inferential relations of c_1 and c_2 that satisfy the three forms of inferential proximities, as defined above;
- $\omega_1, \omega_2, \omega_3$ are the weights, attributed per parameter, of the three forms of inferential proximity defined above; and

- μ_{c_1,c_2} is the normalization factor between concepts c_1 and c_2, calculated by the following formula (2).

$$\mu_{c_1,c_2} = \frac{n+m+p}{|R_{c_2}|} \qquad (2)$$

Where,

- $(n+m+p)$ is the total of elements of sets X_1, X_2 e X_3; and
- $|R_{c_2}|$ is the number of inferential relations of c_2

The normalization factor serves to avoid a concept c_1 from being considered more inferentially related to c_2 than to c_3, just because c_2 has a greater number of inferential relations and, therefore, the values of F_1, F_2, F_3, will probably be higher, calculated between c_1 and c_2.

4 The SIA Algorithm

The SIA algorithm defines a semantic-inferentialist reasoning process that—based on an input text in natural language and on the content expressed in the semantic bases of the SIM—generates the premises and conclusions of the sentences in the text [21]. These premises and conclusions are sentences related to the input text, forming an inferential network among them. Therefore, the result of the semantic parser is a directed graph $Gs(S, Rs)$, where S (vertices) is a set of the sentences of the input text, and Rs (edges) is the set of inferential relations between any two sentences $s_i, s_k \in S$. Each $rs_j \in Rs$ is represented by the tuple $(type, s_i, s_k)$, where type indicates whether s_k is a precondition (*Pre*) or a postcondition (*Pos*) of s_i.

In general, the steps of the algorithm are: (i) analyze the syntactic dependency tree of the input text, breaking down the clauses of the text into simple sentences; (ii) associate, for each sentence, one or more pattern-sentences that dictate the syntactic structure of the sentence; (iii) define the concepts used in sentences and their contribution to the meaning of the sentence through the Inferential Measure; (iv) manipulate and combine the inferential content of the concepts and the pattern-sentences (through the inference rules described in Section 4.1) in order to generate the network of premises and conclusions of the sentences. Below we present a simplified execution of the SIA for the following text: "*A man identified as Marcos Santana was executed by gunshot on Dutra Street in the neighborhood of Maraponga. The crime occurred Saturday afternoon*".

SIA Algorithm

Step 1 - Receive syntactic dependency tree from the input text;
The tree is generated by a morphosyntactic parser (e.g. PALAVRAS [2]).

Step 2 - Break down the text into sentences s_i;
/* the steps of the algorithm are exemplified only for sentence s_3.*/
s_3="a man identified as Marcos_ Santana was executed on Dutra_-street."

Step 3 - Search pattern-sentences p_j for each sentence s_i
/* the pattern-sentences that correspond to the grammatical structure of sentences of the text are selected. The graphs are instantiated with the preconditions and postconditions of pattern-sentence selected.*/
For s_3, the pattern-sentence is $p4336$ = "<X> <be execute> <on> <Y>"

Step 4 - Pre-select the candidate concepts c_i used in each sentence s_i
/* concepts that match the terms of each sentence are pre-selected.*/
In s_3, concepts were pre-selected for the terms w_1 = "man", w_2 = "be", and w_3 = "execute". For the latter, two concepts were pre-selected: c_{358} = "execute" (to do something) and c_{182167} = "execute" (to murder).

Step 5 - Define concepts c_i used in each sentence s_i
/* the concepts used in each s_i of the input text are defined using the highest value of the inferential relatedness measure between concept c_1 (pre-selected in step 4) and concept c_2 (each concept defined for each term related to the term in question). Then, the graphs are instantiated with the preconditions and postconditions of the concepts defined.*/
For w_3 = "execute" we have the following related terms: W_{r3} ={man, crime}. To define which of the two concepts pre-selected in step 4 was in fact used in s_3, $\theta(c_1, c_2)$ is calculated for each $c_1 \in C_{w2}$ = {c_{358}(execute), c_{182167}(execute)} (concepts pre-selected in step 4 for the term "execute") $c_2 \in C_{man} \cup C_{crime}$ = {c_{18428}(man), c_{25847}(crime)} (concepts of the terms related to the term "execute" (W_{r3}).
As the concept c_{182167} (execute (murder)) showed a higher value of $\theta(c_1, c_2)$, it was defined as the concept used in the sentence s_3. In this example, the weights $w_1 = 5$, $w_2 = 2$, $w_1 = 1$ were attributed per parameter for the three forms of inferential proximity.

Step 6 - Instantiate pattern-sentences p_j of each sentence s_i
/* the pattern-sentences selected in step 3 are instantiated with the concepts defined and with other terms of the original sentence.*/
The pattern-sentence p_{4336} = "<X> <be execute> <on> <Y>", was instantiated with the phrases of s_3 "<a man identified as Marcos_ San-

tana>" and "<Dutra_Street>" in place of <X> and <Y>, respectively, and with the concepts "man", "be" and "execute" (defined in step 5).

Step 7 - Generate inferential network for each sentence s_i
/* the inference rules (**I-c**), (**E$_1$-c**) and (**E$_2$-c**) are applied for each concept c_i of s_i, and the inference rules (**P-p**) and (**C-p**) are applied for each pattern-sentence p_j of s_i, generating an inferential network of each sentence s_i. */

The complete inferential network of s_3 was generated with 1005 premises and 30 conclusions. Below are some premises and conclusions generated for s_3 and the respective inference rule that generated it:
PREMISES:

- <a person> <was executed> <on> <Dutra=Street> (**P-p**)

- <something capable of thinking> <was executed> <on> <Dutra=Street> (**E$_1$-c**)

CONCLUSIONS:

- <a man as Marcos=Santana> <was affected by action that> <has effect of> <suffering> (**E$_2$-c**)

- <a victim> <was executed> <on> < Dutra=Street> (**C-p**)

- <a man as Marcos=Santana> <was executed> <on> <a place> (**P-p**)

4.1 Inference Rules of SIA

Before explaining the inferential rules used by SIA, it is necessary to formally define several elements that are used in the inference rules.

- $\mathbf{P}_{s_i}$ is the set of pattern-sentences of a sentence $s_i \in S : \mathbf{P}_{s_i} = \{p_j | s_i$ follows the syntactic structure of the pattern-sentence $p_j \in P\}$. P is the set of pattern-sentences.

- $sn(s)$, $sv(s)$, $sp(s)$ are functions that return, respectively, the noun-phrase, the verb-phrase, and complement-phrase of a sentence s or of a pattern-sentence p. These functions are generically called $phrase(s)$.

- $s\_replace(x, y, s)$ is a function that returns the sentence s with the element y used in place of x.

- $c\_usage(s, c_i)$ is a boolean function that returns **true** if the concept c_i is used in the sentence s, and returns **false** otherwise

The SIA algorithm defines two forms of semantic reasoning, in which inference rules systematize the generation of premises and conclusions of the sentences of the input text, based on the inferential content of concepts and pattern-sentences. These two forms are described as following.

Generation based on the inferential content of concepts

In this form of reasoning, premises and conclusions are generated from the sentences of the input text based on the inferential content of concepts used. We defined generic rules of introduction and elimination of concepts that can be instantiated for each concept. These rules follow the pattern of definition of logical connectives proposed by Gentzen (1935) [11]. Dummett (1973) [9] transposed the Gentzen's model to the concepts of a language: a concept is defined by specifying rules of introduction of the concept (preconditions for the use of the concept or sufficient conditions for the use of the concept) and rules for elimination of the concept (postconditions of use of the concept or the necessary consequences of use of the concept). Below we present the interpretation and the syntax[1] of the rules of introduction and rules of elimination of concepts in sentences, and how these rules are used by the SIA for generating premises and conclusions of the sentences of the input text.

1. The rule **(I-c)** defines that, if a precondition of a concept is met, which meets a precondition of a pattern-sentence, then the concept can be used in the phrase of the sentence that follows the structure of the pattern-sentence (phrase defined in the precondition of the pattern-sentence). Formally, if $(rel\_name, c_1, c_2, Pre)$ is expressed in the subgraph G_{c_1}, and $(rel_name, phrase(p_1), c_2, Pre)$ is expressed in the subgraph G_{p_1}, then the concept c_1 can be used in the place of $phrase(s)$, such that p_1 is the sentence-pattern.

$$\frac{(rel\_name, c_1, c_2, \text{"}Pre\text{"})(rel\_name, phrase(p_1), c_2, \text{"}Pre\text{"})}{s\_replace(phrase(s), c_1, s), p_1 \in P}(\mathbf{I-c})$$
(3)

It is noteworthy that, under this rule, the concept c_2, an image of the precondition of c_1, must be the same concept image of the precondition of the pattern-sentence p_1, and the type of the semantic relation $name\_relation$ must be equal in both preconditions. In this case, the concept c_1 can be used in the $phrase(s)$ that corresponds

---
[1] A formalization of the inference rules of the SIA follows the pattern of formalization of the inference rules of the logical system Natural Deduction proposed by Prawitz (1965).

syntactically to the $phrase(p_1)$. See below an example of the rule of inference (I-c).

EXAMPLE 1. Let

- $c_1 =$ "young" and precondition of c_1:

$$(isA, \text{"young"}, \text{"person"}, \text{"Pre"})$$

- $p_1 =$ "<X> <murder>" and precondition of p_1:

$$(isA, sn(p1), \text{"person"}, \text{"Pre"})$$

Therefore, by (I-c), sentence s can be generated:

<A young> <murder>.

2. The rules $(E_1 - c)$ and $(E_2 - c)$ respectively define that, if a concept is used in a sentence, then the preconditions [postconditions] of the concept can be used to generate preconditions [postconditions] of the sentence. Formally, given a sentence s in which concept c_1 is used, and let the precondition be $(rel\_name, c_1, c_2, Pre)$ [or postcondition $(rel\_name, c_1, c_2, Pos)$], expressed in the G_{c_1} subgraph, then a sentence s_j can be generated as a precondition [postcondition] of s_i, generated by replacing c_1 by the relation with c_2.

$$\frac{(rel\_name, c_1, c_2, \text{"Pre"}) c\_usage(s, c_1)}{(\text{"Pre"}, s, s\_replace(c_1, (rel\_name, c_2), s))}(\mathbf{E_1 - c}) \qquad (4)$$

$$\frac{(rel\_name, c_1, c_2, \text{"Pos"}) c\_usage(s, c_1)}{(\text{"Pos"}, s, s\_replace(c_1, (rel\_name, c_2), s))}(\mathbf{E_2 - c}) \qquad (5)$$

See below two examples of application of the inference rule $(E_1 - c)$.

EXAMPLE 2. Let

- $s_1 =$ "The crime occurred at 33 Titan Street",
- $c_1 =$ "crime" and precondition of c_1:

$$(isA, \text{"crime"}, \text{"violation of the law"}, \text{"Pre"})$$

Therefore, by $(E1 - c)$, sentence s_2 can be generated as a precondition of s_1: $s_2 =$ <A(an)> <violation of the law> <occurred> <at 33 Titan Street>

EXAMPLE 3. Let

- s_1 = "A young man was <u>executed</u> with several gunshots",
- c_1 = "execute" and precondition of c_1:

$$(usedFor, \text{"execute"}, \text{"revenge"}, \text{"Pre"})$$

Therefore, by $(E1-c)$, sentence s_2 can be generated as a precondition of s_1: s_2 = <A young man> <suffered an action that> <is used for> <revenge>

See below two examples of application of the inference rule $(E2-c)$.

EXAMPLE 4. Let

- s_1 = "The <u>crime</u> occurred at 33 Titan Street",
- c_1 = "crime" and poscondition of c_1:

$$(effectOf, \text{"crime"}, \text{"suffering"}, \text{"Pos"})$$

Therefore, by $(E2-c)$, sentence s_2 can be generated as a poscondition of s_1: s_2 =<Something that has the effect of> <suffering> <occurred> <at 33 Titan Street>

EXAMPLE 5. Let

- s_1 = "A young man was <u>executed</u> with several gunshots",
- c_1 = "execute" and postcondition of c_1:

$$(effectOf, \text{"execute"}, \text{"death"}, \text{"Pre"})$$

Therefore, by $(E2-c)$, sentence s_2 can be generated as a postcondition of s_1: s_2 = <a young man> <suffered an action that> <has the effect of> <death>

Generation based on the inferential content of the pattern-sentences

In this form of reasoning, premises and conclusions of the sentences of the input text are generated based on the inferential content of the corresponding pattern-sentences. We defined two generic rules for premises (P-p) and conclusions (C-p) of a pattern-sentence, which can be instantiated for each pattern-sentence used in the sentences of input text.

1. The rules (**P-p**) and (**C-p**) respectively define that if one sentence follows the structure of a pattern-sentence, then the preconditions [postconditions] of the pattern-sentence can be used

to generate preconditions [postconditions] of the sentence. Formally, given a sentence s_i with $p_1 \in P_{S_i}$, and let the precondition $(rel\_name, phrase(p_1), c_1, Pre)$ [or the postcondition $(rel\_name, phrase(p_1), c_1, Pos)$], expressed in the subgraph G_{p_1}, then a sentence s_j can be generated as a precondition [postcondition] of s_i of the form $rel\_name(phrase(s_i), c_1)$.

$$\frac{(rel\_name, phrase(p_1), c_1, , \text{``}Pre\text{''}), p_1 \in P_{s_i}}{(\text{``}Pre\text{''}, s_i, rel\_name(phrase(s_i), c_i))}(\mathbf{P-p}) \quad (6)$$

$$\frac{(rel\_name, phrase(p_1), c_1, , \text{``}Pos\text{''}), p_1 \in P_{s_i}}{(\text{``}Pos\text{''}, s_i, rel\_name(phrase(s_i), c_i))}(\mathbf{C-p}) \quad (7)$$

See below examples of application of the rules of inference (**P-p**) and (**C-p**).

EXAMPLE 6. Let

- s_1 = "Maria da Rocha was murdered by her lover",
- p_1 = "<X> <be murder> <by> <Y>", such that $p_1 \in P_{s_1}$
- a precondition of p_1: $(isA, sn(p_1), \text{``}person\text{''}, \text{``}Pre\text{''})$

Therefore, by $(P-p)$, sentence s_2 can be generated as a precondition of s_1: s_2 = "<Maria da Rocha> <is a(an)> <person>"

EXAMPLE 7. Let

- s_1 = "Maria da Rocha was murdered by her lover",
- p_1 = "<X> <be murder> <by> <Y>", such that $p_1 \in P_{s_1}$
- a postcondition of p_1: $(isA, sp(p_1), \text{``}murderer\text{''}, \text{``}Pos\text{''})$

Therefore, by $(C-p)$, sentence s_2 can be generated as a postcondition of s_1: s_2 = "<her lover> <is a(an)> <murderer>"

5 Use of SIM for Information Extraction

The WikiCrimes system [10] (www.wikicrimes.org) is a Web 2.0 application that offers a collaborative environment for registering and searching crime events, based on use and direct manipulation of geoprocessed maps. One of the needs of the WikiCrimes project is to provide its users with a tool that facilitates interaction and the registration of crimes through the automatic extraction of information from texts of crime-related articles, published in newspapers, on the web, or in the text of police reports, among other sources. Moreover, the nature of the information to be extracted for WikiCrimes is a motivating factor. Information about use of weapons and violence, involvement of gangs, profile of the criminal, cause of the crime, type of crime, and likely crime scenes, are for the most part not explicit in the text. Current approaches to information extraction that use semantic role annotation, named entity recognition, grammatical rules, among others, would even be able to extract explicit information in text, but encounter difficulties in handling implicit knowledge.

5.1 WikiCrimes$_{IE}$

Based on SIM, we have developed a Knowledge-Based System for crime information extraction, named as WikiCrimesIE [19]. It aims at extracting information—from news stories about crimes—that characterizes a criminal occurrence and to make an automatic mashup with a digital map in WikiCrimes.

WikiCrimes$_{IE}$ uses the SIA as a module for semantic reasoning and the broad base with inferentialist and common-sense content—InferenceNet. In short, the user selects a text from a web page and copy to the WikiCrimes$_{IE}$, which, in turn, sends the text to the syntactic parser (in this experiment, we use the parser PALAVRAS [2]), and to the SIA. The parameters necessary for using the Inferential Measure were defined as follows: $\beta_{rel} = 1$, for every rel_name, and $\omega_1 = 5$, $\omega_2 = 2$, $\omega_3 = 1$.

As seen in section 4, the SIA algorithm, because of its generality, is capable of generating all the premises and conclusions of the sentences in the input text, forming an inference network. For the case of an information extraction system, the architectural decision was to model information extraction objectives (IE Goals), and to design a component that selects, from among the inferences, which one best responds and justifies each one of the goals. It is noteworthy that this component was not developed for a specific IE system, and can be reused in various IE applications that use the SIM as the model of expression and semantic reasoning.

IE Goals are templates that contain information to be filled in by the system, based on the inferences network generated by the SIA. Each template represents a goal. For example, the template "The crime scene is _____" represents the "Extract crime scene" goal. Each goal is defined by:

- a description, for example: "Extract crime scene";

- a main subject, expressed by a concept, for example, "crime";

- a list of related concepts that define the information required about the main subject, for example, concepts related to geographical locations;

- the type of information to be extracted—(i) descriptive; (ii) single alternative; (iii) multiple options;

- a list of response options, including a description and a list of related concepts.

All concepts related to one IE goal must be expressed in the SIM conceptual base. These concepts are used to select the premises and conclusions of the sentence, which justifies and explains the answer. Table 1 presents the definition of two IE Goals: extract cause of the crime and type of weapon.

For example, considering the following text: *"Again, a crime with characteristics of summary execution occurred in Fortaleza. On the evening of Tuesday, the young Marcelo dos Santos Vasconcelos, 29, was shot on his doorstep. The crime occurred at Casimiro de Abreu Street, in Parangaba"*.

WikiCrimesIE system extracted the following information: the crime scene "Casimiro de Abreu Street, Parangaba" was extracted from the sentence "The crime occurred on Casimiro de Abreu Street, in Parangaba"; the type of crime "homicide" was extracted from the sentence "Marcelo dos Santos Vasconcelos, 29, was shot on his doorstep"; the type of weapon was "firearm", extracted from the sentence "Marcelo dos Santos Vasconcelos, 29, was shot on his doorstep (because of the inferential similarity between "shot" and "firearm"); and the cause of crime "contract killing" was extracted from the sentence "a crime with characteristics of summary execution occurred in Fortaleza" (because of the inferential similarity between "contract killing" and "summary execution").

5.2 SIM Evaluation in the Information Extraction Task

A Golden Collection (GC) was elaborated with 200 news texts, published on crime pages of Brazilian newspapers on the Internet. Two adult persons, proficient in the Portuguese language, who were then asked to annotate the GC with the information about the scene of the crime, type of crime, cause of the crime and weapon used in the crime. Special attention must be given to the accuracy of the cause of crime and the weapon used. These types of information are normally implicit in texts and require more complex inferences in order to extract them. Beforehand, both persons were asked to read the instruction for annotation

Table 1. Definition of IE Goals for WikiCrimes$_{IE}$

Description of Goald	Subject	Related Concepts	Response Options	Concepts ref. Response Options
Extract Cause of the Crime	Crime "crime"	"cause" "reason" "motivation"	Poor public lighting	"darkness", "poor lighting"
			Gang disputes	"gang"
			Use of alcohol	"alcohol", "inebriation", "drunk"
			Police violence	"police violence"
			Organized crime	"organized crime"
			Revenge	"revenge"
			High concentration of people	"crowd", "many people"
			Drug use/ trafficking	"drug", " crack", "cocaine", "drug traffic"
			Children/ teens in the streets	"children in the street" "teens in the street"
			Contract killing	"contract killing"
			Crime of passion	"jealousy", "lover" "violence", "death"
Extract Type of Weapon	Weapon "weapon"	"type" "category"	Firearm	"revolver", "pistol", "firearm"
			Cold weapon	"knife", "pocketknife", "glass"
			Other	

Table 2. Results of WikiCrimes$_{IE}$

Evaluation Measure	Crime Scene	Cause of Crime	Type of Crime	Weapon Used	Average
Precision	87%	76%	72%	85%	80%
Recall	71%	70%	68%	76%	71%
F-Measure	**78%**	**73%**	**70%**	**80%**	**75%**
Processing errors	3%	8%	8%	8%	8%
Parsing errors	2%	7%	7%	7%	7%

and a text describing all types of crimes, the causes of crime and the types of weapons that should be considered as response options. In the case of different responses between the two persons, a third person was consulted regarding which of the responses was the correct one. In the end, only one response for each question was recorded.

The responses annotated in the GC and inferred by WikiCrimes$_{IE}$ were compared in terms of the precision, recall, f-measure. Additionally, the results of the system were impacted by processing errors caused by problems related to poorly- formed sentences (no subject) and complexes clauses, and by errors caused by the syntactic parser.

In all, there were three iterations to test WikiCrimesIE vis-à-vis the GC. After each one, a set of improvements was incorporated into the SIA: optimizing the SIA algorithm related to the processing of graphs, the data structures, and the manipulation in memory, as well as evolution of the measure of inferential similarity and the inference mechanism detailed in sections 3 and 4. Table 2 presents the results of the evaluation measures of WikiCrimes$_{IE}$ in the task of extracting the crime scene, cause of the crime, type of crime, type of weapon, and the average result.

In a quantitative analysis, the results of WikiCrimesIE were compared with the state of the art in SRL and IE systems, which competed in the last NLP contest events (CoNLL, SemEval). WikiCrimes$_{IE}$, with F-measure = 75% (overall average), showed the best result among all the systems evaluated for Portuguese and English texts. For English language, it surpassed the system described by [7], which showed the best result (F-measure = 73.66%) in the SRL task for event CoNLL-2009. In the event SemEval-2007, according to Girju et al. (2007) [12], the better system showed F-measure = 72.40% in the task of semantic classification. For Portuguese language, the Priberam system[1] showed a better F-measure (57,11%) in NER task in the second HAREM [18]. Specifically for the "PLACE" category (like crime scene information), the REMBRANDT system [6], obtained an F-measure = 59.93%. It is important to point out that this comparison is still unfair, because the contest events commonly require information that is explicit in the text, and the current approaches to IE that use SRL, NER, grammatical rules, among others, are able to extract information explicit in the text, but have difficulties in handling implicit knowledge.

A qualitative analysis showed that the information extracted by WikiCrimes$_{IE}$ such as type of crime, type of weapon and causes of crime were not explicitly mentioned in 82% of the texts in the GC. The text used in the SIA description in Section 4 has an example that the crime was a "homicide with firearm," but this information was not explicit there. From the preconditions and postconditions of the concepts "execute" and "gunshot" used in the pattern-sentence "<X> <be execute> <on> <Y>" "<X> <be execute> <by> <Y/gunshot>", it was possible to infer and extract such information.

6 Conclusion

Most systems for natural language understanding use syntactic and statistical approaches or, in the best case—when they take advantage of semantic knowledge—they are limited to knowledge of the concepts, considered individually, and adopt an atomist approach to the semantic analysis of natural language. Thus the semantic reasoning process does not consider the use of a concept in sentences (the expression of the pragmatic aspect of a language) or how this concept, enunciated with other concepts, contributes to conclusions and inferences of the sentence in which it appears.

In contrast, the inferential approach proposed here enables Knowledge-based NLP systems to make inferences based on the linguistic practice (from preconditions and postconditions of use expressd on the Conceptual and Pattern-sentences bases). These inferences generate premises and conclusions, which, in turn, support the extraction of explicit and implicit information. In this work, we showed how the Semantic-Inferentialist Model (SIM) expresses situations of use of the concepts and how the semantic reasoner of the model manipulates the knowledge bases taking into account the context in which the concepts and the sentences are used. With this model, Knowledge-based NLP systems are provided with an inferential skeleton that allow them to, for instance, generate new sentences because they know about the preconditions of use of the concepts and pattern-sentences. The model also defines a relatedness measure based on the inferential relation between concepts. Such a measure may be used for solving conceptual anaphors and disambiguation as well as for defining the semantic contribution of a concept into a sentence.

A Knowledge-based NLP system for extracting information about crimes—WikiCrimes$_{IE}$—was developed using the proposed model. We evaluated the precision and recall of this system in the execution of the task to extract the crime scene, type of crime, weapon used, and cause of the crime from news reports about crimes available on the web. The results are comparable to the state of the art. The use of SIM in this context is very important, due to the necessity of taking into account implicit information, often present in journalistic texts, and more complex inferences-derived from the use of the concepts in linguistic practice-must be performed.

References

[1] C. Amaral et al. "Adaptação do sistema de reconhecimento de entidades mencionadas da Priberam ao HAREM". In: ed. by C. Mota and D. Santos. Linguateca, 2008, pp. 171–179.

[2] E. Bick. *The parsing system Palavras. Automatic Grammatical Analysis of Portuguese in a Constraint Grammar Framework*. University of Arhus, 2000.

[3] R. Brandom. *Making it explicit: Reasoning, representing, and discursive commitment*. Harvard university press, 1998.

[4] R. Brandom. *Articulating Reasons. An Introduction to Inferentialism*. Cambridge, MA, 2000.

[5] E. Cambria and B. White. "Jumping NLP curves: A review of natural language processing research". In: *IEEE Computational intelligence magazine* 9.2 (2014), pp. 48–57.

[6] N. Cardoso. "Rembrandt-reconhecimento de entidades mencionadas baseado em relaçoes e análise detalhada do texto". In: *Encontro do Segundo HAREM* (Sept. 2008).

[7] W. Che et al. "Multilingual dependency-based syntactic and semantic parsing". In: *Proceedings of the thirteenth conference on computational natural language learning: shared task*. Association for Computational Linguistics. 2009, pp. 49–54.

[8] M. Dummett. *Truth and other enigmas*. Harvard University Press, 1978.

[9] M. Dummett. *Frege: Philosophy of language*. Harvard University Press, 1981.

[10] V. Furtado et al. "Collective intelligence in law enforcement–The WikiCrimes system". In: *Information Sciences* 180.1 (2010), pp. 4–17.

[11] G Gentzen. "Untersuchungen über das logische Schließen, Mathematical Zeitschrift 39, 176–210, 405–431". In: *English translation in (Szabo, 1969)* (1934).

[12] R. Girju et al. "Semeval-2007 task 04: Classification of semantic relations between nominals". In: *Proceedings of the 4th International Workshop on Semantic Evaluations*. Association for Computational Linguistics. 2007, pp. 13–18.

[13] T. L. Griffiths, M. Steyvers, and J. B. Tenenbaum. "Topics in semantic representation." In: *Psychological review* 114.2 (2007), p. 211.

[14] Z. S. Harris. "Mathematical structures of language". In: (1968).

[15] T. K. Landauer and S. T. Dumais. "A solution to Plato's problem: The latent semantic analysis theory of acquisition, induction, and representation of knowledge." In: *Psychological review* 104.2 (1997), p. 211.

[16] K. Lund and C. Burgess. "Producing high-dimensional semantic spaces from lexical co-occurrence". In: *Behavior research methods, instruments, & computers* 28.2 (1996), pp. 203–208.

[17] J. Mitchell and M. Lapata. "Composition in distributional models of semantics". In: *Cognitive science* 34.8 (2010), pp. 1388–1429.

[18] C. Mota and D. Santos. *Desafios na avaliação conjunta do reconhecimento de entidades mencionadas: O Segundo HAREM*. 2008.

[19] V. Pinheiro et al. "Information extraction from text based on semantic inferentialism". In: *International Conference on Flexible Query Answering Systems*. Springer. 2009, pp. 333–344.

[20] V. Pinheiro et al. "InferenceNet. Br: expression of inferentialist semantic content of the Portuguese language". In: *International Conference on Computational Processing of the Portuguese Language*. Springer. 2010, pp. 90–99.

[21] V. Pinheiro et al. "Um Analisador Semântico Inferencialista de Sentenças em Linguagem Natural". In: *Linguamática* (2010), pp. 111–130.

[22] W. Sellars. *Pure Pragmatics and Possible Worlds: The Early Essays of W. Sellars*. Ed. by J. Sicha. Atascadero: Ridgeview, 1980.

A Short Note on the Formalism of Johann von Neumann

ABEL LASSALLE CASANAVE AND LUIZ CARLOS PEREIRA

That the so-called 'foundational crisis' decanted into the formulation of a mathematical problem is something that should never be forgotten. While recognizing that the question about mathematical reliability is clearly a philosophical question, von Neumann, in the very beginning of his paper (1931), "The Formalist Foundation of Mathematics" reminded us:

> "Noteworthy is the fact that this question, in and of itself philosophico-epistemological, is turning into a logico-mathematical one."[10, p. 50]

Von Neumann had been invited by the organizers of the Second Conference on Epistemology of Exact Sciences, organized by the Society for Empirical Philosophy (Berlin), as representative of the formalist position. The conference, which took place in Könnisberg from 5 to 7 September 1930, also had as invited speakers Rudolf Carnap, representing the logicist position, and Arend Heyting, representing the intuitionist position.[1] But how could such a clear philosophico-epistemological question have turned into a logico-mathematical one? The different contributions of the intuitionist (constructivist), logicist, and formalist positions to the transformation of the question of the reliability of classical mathematics into a mathematical problem (or a series of them) is recapitulated as follows by von Neumann: in the first place, Brouwer's intuitionist critique of the methods of classical mathematics; secondly, Russell's description of the methods of classical mathematics by means of a *symbolism*, although without distinguishing the good ones from the bad ones;

Copyright © 2019 by Abel Lassalle Casanave and Luiz Carlos Pereira. All rights reserved.

[1]On September 5, Carnap, Heyting and von Neumann gave an hour talk at the meeting. A fourth talk was given by Friedrich Waismann on Wittgenstein's Philosophy of Mathematics. The papers by the first three speakers (with slightly different titles) were published in the second volume of *Erkenntis*. On September 6, Gödel gave a twenty-minute talk on his completeness theorem. On September 7 the famous discussion took place.

and thirdly, Hilbert's mathematical-combinatorial investigations of these methods and their relationships. Von Neumann's version, essentially correct, was still that of Hilbert's formalism, which until then was, in a certain sense, the "winner" in the foundational arena. In the pages that follow, we will analyse the contribution of von Neumann to the mentioned conference from a perspective that considers more than a century of research on foundations of mathematics.

1 The question of the reliability of classical mathematics.

In fact, even though von Neumann formulated the problem in the very general terms of classical mathematics, the specific mathematical problem to which the so-called Hilbert school was at the time searching a solution was the consistency of Peano's arithmetic, although their *members* certainly had no doubt about the reliability of this theory. It should then be noted that the proof of the consistency of arithmetic was just the first step of a stratified strategy for solving the more ambitious problem of the consistency of classical mathematics: arithmetic, analysis, transfinite number theory.

Now, how do Brouwer's intuitionism, Russell's logicism, and Hilbert's formalism combine to reduce the philosophical question about the reliability of classical mathematics to the mathematical problem of a proof of consistency? Certainly, Brouwer's critique of the methods used in classical mathematics, particularly the unrestricted validity of the principle of excluded middle, is a relevant element, although criticisms of procedures that are generically termed infinitary or transfinite were already found, among others, in Kronecker and Poincaré. Hilbert had distinguished between reliable finitary methods and transfinite methods about which their reliability could be doubted, but in his paper, von Neumann simply identifies the former with intuitionistically acceptable methods. The question is not trivial, because what is involved is the determination of what methods, what demonstrative tools, are acceptable for the solution (negative or positive, partial or total) of the problem. For von Neumann it is a matter of proving the consistency of arithmetic by methods acceptable in intuitionistic or constructive arithmetic. Indeed, in general, it could be argued that there was no reason to use in the solution of a given problem methods deductively weaker than those that are accepted by the *adversaries*.

The role of Russell's logicism is of a completely different nature: von Neumann, as well as Hilbert, found in Russell's logicism an adequate symbolism for the complete formalization of mathematical theories, including the underlying classical logic. In particular, Hilbert introduced a notion of proof in a formal system F as a sequence of formulas of a given formal language such that each formula is either an axiom or follows from previous formulas in the sequence by an inference rule. Thus, the formal system in question and its proofs could now become mathematical objects themselves. The problem could then be formulated more precisely as follows: to prove the consistency of a formal system F consists in demonstrating by finitary methods that no sequence of formulas of the type that constitutes a proof has as final formula a contradiction.

The theory in which Hilbert's logical-combinatorial research, to which von Neumann refers, is carried out is the so-called 'Proof Theory'. The expression 'Proof Theory', in German, *Beweistheorie*, was introduced by Hilbert in 1922 to designate a new scientific enterprise whose conceptual significance can be expressed as follows: "As the physicist investigates his apparatuses, the astronomer his position and the philosopher exercises the criticism of reason, so, in my opinion, the mathematician only asserts his propositions through a critique of the proof and for this he needs a Proof Theory."[2] Undoubtedly, Hilbert's expectation with respect to his Proof Theory was not at all unambitious. Beyond the very problem of the consistency of mathematics, the horizon of questions and themes which such a theory was intended to deal with and whose solution it sought was already stated in 1917: "the problem of the solubility in principle of all mathematical questions, the problem of the subsequent verifiability (controllability) of the result of a mathematical investigation, even more, the question of a criterion of simplicity for mathematical demonstrations, of the relation between content and formalism (Inhaltlichkeit und Formalismus) in mathematics and logic, and finally, the problem of the decidability of a mathematical question by means of a finite number of operations."[4, pp. 412–413]

However, it may not be irrelevant to recall that there was in fact a philosophico-epistemological question, and that the idea was to reduce its solution to the solution of a mathematical problem. For the clarification of this previous philosophico-epistemological question, the reading of the work of von Neumann serves as invaluable support. In particular, von Neumann omitted that the reduction of mathematics to a formal system, the 'play with formulas' critically referred to by Hermann Weyl, is completely alien to Brouwer's ideas, though ultimately Heyting or Weyl himself concede in formalizing the construction procedures, which could already be too much a concession. Nor did logicism conceived symbolism as a regulated manipulation of meaningless symbols. To reduce

---

[2][5, p. 170]. A similar sentence, but without the expression "Proof Theory", can be found in [4, p. 415].

the philosophico-epistemological question to a mathematical problem it would be necessary to justify this merely symbolic conception of mathematics. In regard to Hilbert and his proof theory, von Neumann wrote:

> "The leading idea of Hilbert's theory of proof is that even if the statements of classical mathematics should turn out to be unreliable as to content, nevertheless, classical mathematics involves an internally closed procedure, which operates according to fixed rules known to all mathematicians, and which consists basically in constructing successively certain combinations of primitive symbols which are considered 'correct' or 'proved'."[10, p. 50]

The idea that mathematics involves a (symbolic) procedure according to rules is undoubtedly a hilbertian idea, but what Hilbert called the 'fundamental idea' of his proof theory, which is reproduced in several of his texts, is its precondition:

> "The fundamental idea of my proof theory is as follows: everything that makes up mathematics in the traditional sense is rigorously formalized, so that mathematics proper (or mathematics in the narrow sense) becomes a stock of formulae."[7, p. 489]

The reduction of mathematics to formal systems then depends on the thesis of complete formalization. But this thesis does not conclude in a substantive philosophical thesis about the nature of mathematics in general, but rather is a methodological thesis: mathematics is seen exclusively as a procedure (which can be represented symbolically), because that is the way to reduce the philosophico-epistemological problem to the mathematical problem of the demonstration of consistency. And while the procedures that are considered are not constructively acceptable from the point of view of content, what makes them 'unreliable', the construction procedure according to the formal notion of proof that represents them is reliable.

Von Neumann illustrates the intuitionist critique considering the proof of a proposition that affirms the existence of a real number x with a property E, but a proof from which a procedure to construct such x can not be derived. Without going into the details of the proof, some comments are pertinent. First, the so-called 'Goldbach Conjecture' is involved in the proof, namely, that any even number greater than 2 is the sum of two primes. This conjecture is easily expressed in the formal language of Peano's arithmetic, but the proof considered is developed in the analytic theory of numbers. The example chosen indirectly shows that the dispute with the intuitionists in relation to the usual theory of numbers was theoretical rather than practical, since the effective mathematical divergences related to the principle of excluded third or to non-constructive proofs of existence occurred in the scope of analysis. Now, what is proved is indeed a conditional: if Goldbach's conjecture is false, then a certain property holds for odd numbers, but not for even numbers, or vice versa, if it holds for even numbers, it does not hold for the odd numbers. In either case, there weren't then—and still today there are none—mathematical tools available to prove them.

Two remarks on the proof:

(1) First, it is not a formal proof; it has the usual form of proofs in mathematical practice (for a competent audience in the topic). But we must remember that the 'fundamental principle' according to which every proof is formalizable is in operation.

(2) Second, the proof uses the principle of excluded middle. For Hilbert, the principle excluded middle is a transfinite logical principle and therefore, a principle whose content is not reliable, but not without content, as evidenced by the following passage:

> "The tertium non datur occupies a distinguished position among the axioms and theorems of logic in general: for while all the other axioms and theorems can be immediately traced back without difficulty to definitions, the tertium non datur expresses a new, contentually meaningful fact that stands in need of proof."[6, pp. 124–125]

According to von Neumann, the principle of excluded middle is one of those procedures that all mathematicians know and use, a principle whose use must be proved reliable by means of a proof of consistency. Now, in formalized mathematics, if there were an error in the proof, it could be found by a finitary procedure. As von Neumann wrote:

> "In other words, although the content of a mathematical statement cannot always (i.e., generally) be finitely verified, the formal way in which we arrive at the sentence can be."[10, p. 51]

Thus, according to von Neumann, the reliability of classical mathematics must consist of an investigation of methods of proof, not of isolated statements:

"We must regard classical mathematics in a finitary combinatorial game played with the primitive symbols and we must determine in a finitary combinatorial way to which combination of primitive symbols the construction methods or "proofs" lead."[10, p. 51]

But to consider Mathematics as a combinatorial game, that we must conceive it so from the foundational perspective, is not the same thing as saying that mathematics is a combinatorial game of symbolic manipulations of formulas devoid of meaning. Mathematics is a game of this nature from the perspective of meta-mathematics or proof theory, but even this reduction is not arbitrary: leaving aside its mathematical value, this game of formulas, says Hilbert, expresses the technique of our thinking, which is eminently symbolic. Indeed: *In the beginning was the sign.*

2 The Proof Theory

In the second section, von Neumann listed the tasks to be performed by Hilbert's proof theory. The first three have to do with the formalization of logic and mathematics, which essentially was already carried out by Russell; the fourth is the difficult one: the demonstration of consistency. Von Neumann warns that the formalization of logic and mathematics can be carried out in several different ways. In "On the Hilbert theory of demonstration" [9], from 1927, von Neumann presents a Hilbert-style axiomatic formalism for a fragment of arithmetic, in which each group of axioms seeks to characterize the different concepts involved, including, but in this case, fundamentally, also the logical ones. Let us review, with the aid of the work of 1927, how the first three tasks that von Neumann enunciates were carried out in 1931.

The first is to list the symbols used in logic and mathematics. In "On the Hilbert's Proof Theory", von Neumann distinguishes five kinds of symbols: variables (both propositional and individual), constants (as '0'); operations (which includes '¬' and '·', but also addition '+' and product '×', the symbols for equality and abstractions (such as the existential and the universal quantifiers), and punctuation (such as commas, etc.).

The second task is to unambiguously characterize the class of 'meaningful' formulas of classical mathematics, basically by means of a recursive definition. That class, of course, includes not only the true formulas, but also the false ones. In the work of 1927, a rule of substitution played a fundamental role, since it was involved in the formulation of the axioms.

The last of these three tasks that are linked to the formalization of classical logic and mathematics consists in presenting a construction procedure for the provable formulas that correspond to the theorems of classical mathematics. In other words, for the present case, it is a question of selecting a set of formulas as axioms and offering a definition of proof as a sequence of formulas, i.e., the definition of formal proof that represents within the formalism the results obtained in the practice of classical mathematical (some of whose principles and methods of proof are unreliable).

The fourth and final task is the consistency proof, which von Neumann states in his 1931 paper in the following way:

> "(4) To shown (in a finitary combinatorial way) that those formulas that correspond to statements of classical mathematics which can be checked by finitary arithmetical methods can be proved (i.e., constructed) according to the process described in (3) if and only if the check of the corresponding statement shows it to be true."[10, p. 52]

The idea is to establish the reliability of classical mathematics as a shortcut to justify arithmetical statements whose verification by reliable means would be too tedious. This—von Neumann seems to say—would be sufficient to justify the reliability of mathematical practice, reliability that is seen as an empirical fact. Now, if the *check* of a numerical formula shows that this formula is false, then from it the equation '$p = q$' can be derived for two different numerals p and q. Given (3), then we could give a formal proof of '$p = q$', from which it would follow that '$1 = 2$'. Then, it is enough to prove consistency to show that there is no formal proof of '$1 = 2$'.

Thus, Hilbert's main goal was a result of 'conservativeness', consistent with the kind of interpretation that Prawitz calls 'moderate formalist'.[12, p. 253] Such a result presupposes, of course, the distinction between real and ideal sentences, which in this meta-mathematical context are those that involve reference to an infinite totality of objects conceived as 'rounding' the theory, that is, allowing results to be obtained 'more easily' on the real part, conceived according to the same context as verifiable. If one thinks that these ideal sentences introduce "ideal elements", then the problem would be to show that the use of those ideal elements in proofs of real statements, in the sense of verifiable, does not lead to incorrect results. Now, under what conditions is that conservativeness result equivalent to the problem of consistency?

According to Prawitz [12, p. 257], the formulation of the project, given the formalization of a mathematical theory T, takes the following form:

> For any proof P in T and for every real sentence A in T, if
> P is a proof of A in T, then A is true.

This conditional is only one side of the bi-conditional formulated by von Neumann. The structure of the proof of this conditional can be described as follows:

1. First, the contrapositive is proved.

2. Suppose a numerical statement E is shown to be false / incorrect by a finitary procedure.

3. The relation $p = q$ can then be derived from this formula for two numbers actually given.

4. Given (3), a formal proof of $p = q$ can be produced.

5. From this proof, one can trivially produce a formal proof of $1 = 2$.

6. The formal system does not prove $1 = 2$.

7. Thus, if the statement (verifiable by finitary methods) corresponding to a formula A in the formal system is verified false, then A cannot be proved in formal system.

8. Thus, by contrapositive, given a formula A corresponding to a statement of classical mathematics that can be 'checked' by finite means, if it can be proved in formal system, then the statement corresponding to A is verified as true.

3 Partial results

The way in which (3) is carried out by von Neumann is not explicit in his paper, but he highlighted its two fundamental elements that are related to the notion of proof:

3.1. First, the characterization in a univocal and finitary way of some formulas that are called 'axioms'.

3.2. Second, the characterization of the notion of provable formula.

In the paper of 1927, there are six groups of axioms. The so-called finitary axioms are: axioms for propositional logic (formulated exclusively with '.' and '¬'); the axioms for equality '='; the axioms for non negative numbers (0 is a number, if a is a number, the successor of a is also a number, 0 is not a successor of any number, if the successors of a and b are equal, then a and b are equal too). The axioms for the quantifiers are transfinite: they imply, in particular, the principle of the excluded middle. There are two groups of axioms for functions and definitions that are also transfinite. The reader will have noticed that the 1927 system does not explicitly contain the principle of induction, although the given axioms do contain a weaker version of it. It is therefore a fragment of classical arithmetic that could be completed.

Axioms are considered to be proved. In addition, if A and B are two formulas, and and $A \to B$ have been proved, then B has also been proved. A proof is a "checkable" sequence of formulas. Thus, Hilbert's way of seeing mathematics as a stock of formulas or a combinatorial game is complete: in it, the formulas differ from the usual formulas of mathematics insofar as they include the logical symbols; in addition, a proof is an intuitively presented figure consisting of inferences where each premiss is an axiom, the conclusion of an inference from previous formulas in the sequence, or results from a formula by substitution. In relation to inferences in the usual sense of mathematics he writes:

> "Instead of a contentual inference, in proof theory we have an external action according to rules, namely, the use of inference schemata and of substitution."[7, pp. 489–490]

A formula is formally provable if it is an axiom or the final formula of a proof, but the meta-mathematics or proof theory, where the consistency of the theory in question is proved, is mathematics with content in a very specific sense. Indeed, in meta-mathematics only concepts and methods with finitary (or intuitionistically acceptable, as von Neumann would put it) content can be used to prove the consistency of the formal axiomatic theory in question.

In the paper just quoted, Hilbert mentions the proofs of Ackermann and von Neumann of the consistency of weaker fragments of classical arithmetic. In particular, von Neumann had obtained the corresponding consistency proof for the fragment we characterized very schematically earlier. At the end of his paper of 1931, von Neumann enthusiastically stated that although the consistency of classical mathematics had not yet been proved, the consistency of a weaker system than classical mathematics (1927) had been demonstrated. And he then added:

> "Thus Hilbert's system has passed the first test of strength: the validity of a non-finitary, not purely constructive mathematical system has been established through finitary constructive means. Whether someone will succeed in extending this validation to the more difficult and important system of classical mathematics, only the future will tell."[10, p. 54]

In the perfect Greek tragedy, recognition coincides with the change of fortune: in the same conference, Kurt Gödel presented in the discussion session the first of his results on the incompleteness of arithmetic.[3] Von Neumann immediately saw the consequences of the theorem that Gödel would enunciate shortly after: under certain reasonable conditions, the consistency of arithmetic could not be proved by methods that could be represented within it, as the supposed finitary or constructive methods were considered to be. Von Neumann's somewhat disillusioned reaction was to abandon his foundational studies to concentrate on matters pertaining to the application of mathematics. For him, the philosophico-epistemological question of the reliability of classical mathematics could no longer be solved along the lines indicated by Hilbert.[4]

Hilbert's reaction was not that of von Neumann. Hilbert had the attitude of a mathematician, who, facing difficulties in the solution of a problem, difficulties that, in spite of being unexpected, are not unusual in the history of mathematics, considers that the means considered sufficient for the resolution of a problem are not so. Thus, more powerful methods must be used, but specially in this case, with well-justified philosophico-epistemological restrictions. As Paul Bernays will say, the notion of finitary acceptable methods must be extended. In this respect, the contributions of Gerhard Gentzen were undoubtedly fundamental.

Between 1934 and 1939, Gentzen produced four different proofs of the consistency of classical arithmetic (PA). The first proof (1934) was not published. After sending it to Gödel and to von Neumann, Gentzen gave up the idea of publishing it because of the criticisms he received: the proof would not be constructive / finitary. In 1968, at a conference on proof theory and intuitionism, Bernays presented a paper, later published in 1970, in which he recalls those criticisms about the first proof of Gentzen. In that paper Bernays wrote:

> "The first published Gentzen consistency proof for the formal system of first-order number theory, including standard logic, the Peano axioms and recursive definitions, was given in Gentzen's paper "Die Wiederspruchsfreiheit der reinen Zahlentheorie" [3]. It was however not his original proof but

---

[3]The discussion section took place on 7 September, and Gödel's fundamental and historical intervention was: "One can (assuming the consistency of classical mathematics) even give examples of propositions (and indeed, of such of the type of Goldbach or Fermat) which are really contentually true but are unprovable in the formal system of classical mathematics. Therefore if one adjoins the negation of such a proposition to the axioms of classical mathematics, one obtains a consistent system in which a contentually false proposition is provable." (see [2]).

[4]Cf. [8], sections 4 and 5.

a revised version of it. The revision was motivated by a criticism, in which I myself for some time concurred of the original proof, on the grounds that it implicitly included an application of the fan theorem. Gentzen did not expressly oppose this opinion; he took care of the criticism by modifying his consistency proof before it was published. Fortunately, the text of the original proof is preserved in galley proof."[1, p. 409]

It was only in the second proof, the first published proof (1936), and in the third proof, published in 1938, that the use of transfinite induction up to ϵ_0 appears.[5] The fourth demonstration, presented in 1942, as his *Habilitationsschrift*, and published in 1943, is in fact based on a result of 'unprovability': Gentzen formalized transfinite induction up to ϵ_0 in a sentence of the language of PA and proved that this sentence is not provable in PA, thus guaranteeing that at least one sentence is not provable in PA, i.e., that PA is consistent. (Gentzen also proved that all cases of transfinite induction to ordinals less than ϵ_0 are provable in PA.)[6]

A final observation concerning method is pertinent here. Von Neumann is right when he asserts in 1931 that the question of the reliability of classical mathematics had become, "in the last decades", a logico-mathematical problem. In fact, perhaps somewhat roughly said, two

---

[5] The reactions to Gentzen's consistency proofs that make use of transfinite induction up to ϵ_0 can be quite different. Takeuti, for example, asserts that "since I [Takeuti] am a logician and am very familiar with the magic of quantifiers Gentzen's consistency proof, which consists of the elimination of quantifiers and [are] accessibility proof for the ordinal less than ϵ_0, is greatly reassuring. It does add to my confidence in the consistency and truth of Peano Arithmetic" (see [13], emphasis ours). On the other side we have Tarski's famous statement: "Furthermore I should like to remark that there seems to be a tendency among mathematical logicians to overemphasize the importance of consistency problems, and that the philosophical value of the results obtained so far in this direction seems somewhat dubious. Gentzen's proof of the consistency of arithmetic is undoubtedly a very interesting meta-mathematical result, which may prove very stimulating and fruitful. I cannot say, however, that the consistency of arithmetic is now much more evident to me (at any rate, perhaps, to use the terminology of the differential calculus, more evident than by epsilon) than it was before the proof was given" [14].

[6] In fact, Gentzen had previously produced another consistency proof (a fifth proof), which is historically the first! In 1933 Gentzen published an interpretation of classical arithmetic into Heyting's intuitionistic arithmetic and showed that, as a result of that interpretation, if intuitionist arithmetic is consistent, then classical arithmetic is consistent too. That result is explicitly a reduction of the problem of the consistency of classical arithmetic to the problem of the consistency of intuitionist arithmetic! It is the same 'reductionist' spirit that presided over translation / interpretation / coding results of the late 1920s and early 1930s.

main (and traditional) foundational programs, namely, logicism and formalism, depended for their achievement on mathematical results. However, it is not clear how a philosophical inquiry may depend on a mathematical result (or, if applicable, an empirical result). Wittgenstein might say that if a philosophical question could be solved by a mathematical proof (or, for that matter, by an experiment) that would only show that the question was not a real philosophical question. We know the history: without considering the 'new X' so dear to certain revisionist positions, these programs were shipwrecked, at least in their original formulations, by the force of mathematical results.

Nevertheless, something of the wreck survived: on the firm ground of Gentzen's results proof theory was consolidated as one of the four great areas of logic, both in its reductive version, certainly closer to the Hilbert project, and in its general form, free from any foundational expectation, which seeks to investigate "proofs" without imposing restrictions on the methods used in the investigation.[7] Proof theory produced and continues to produce technical fruits of unequivocal importance and it has even presented an interesting semantical perspective of constructive nature, something that perhaps would be surprising for von Neumann and for formalism in general.

References

[1] P. Bernays. "On the original Gentzen consistency proof for number theory". In: *Studies in Logic and the Foundations of Mathematics* 60 (1970), pp. 409–417.

[2] J. W. Dawson Jr. "Discussion on the Foundations of Mathematics". In: *History and Philosophy of Logic* 5 (1984), pp. 111–29.

[3] G. Gentzen. "Die widerspruchsfreiheit der reinen zahlentheorie". In: *Mathematische Annalen* 112.1 (1936), pp. 493–565.

[4] D. Hilbert. "Axiomatisches Denken". In: *Mathematische Annalen* 78 (1917), pp. 405–415.

[5] D. Hilbert. "Neubegründung der Mathematik. Erste Mitteilung". In: *Abhandlungen aus dem Seminar der Hamburgischen Universität*. Vol. I. Reprint: D. Hilbert, Gesammelte Abhandlungen, Berlin 1935. 1922, pp. 157–177.

[6] D. Hilbert. "Beweis des Tertium non datur". In: *Nachrichten von der Gesellschaft der Wissenschaften zu Göttingen, Mathematisch-Physikalische Klasse* (1931), pp. 120–125.

---

[7]The distinction between a "general proof theory" and a "reductive proof theory" was introduced in [11].

[7] D. Hilbert. "Die grundlegung der elementaren zahlenlehre". In: *Mathematische Annalen* 104.1 (1931), pp. 485–494.

[8] R. Murawski. "John von Neumann and Hilbert's School of Foundations of Mathematics". In: *Studies in Logic, Grammar and Rhetoric* 7 (20 2004), pp. 37–55.

[9] J. von Neumann. "Zur hilbertschen beweistheorie". In: *Mathematische Zeitschrift* 26 (1927), pp. 1–46.

[10] J. von Neumann. "Die formalistische Grundlegung der Mathematik". In: *Erkenntnis* 2.1 (1931). (The quotations are taken from the English translation in P. Benacerraf and H. Putnam (eds): Philosophy of Mathematics – Selected Readings, New Jersey, Prentice-Hall, pp.50-54), pp. 116–121.

[11] D. Prawitz. "Ideas and results in proof theory". In: *Proceedings of the second Scandinavian Logic Symposium*. Ed. by J. E. Fenstad. Amsterdam: North-Holland.

[12] D. Prawitz. "Philosophical aspects of proof theory". In: *Contemporary philosophy: A new survey: Tome 1 Philosophie du langage, Logique philosophique/Volume 1 Philosophy of language, Philosophical logic*. Dordrecht, Boston, Lancaster: Martinus Nijhoff Publishers, 1981, pp. 235–277.

[13] G. Takeuti. "Consistency proofs and ordinals". In: *Proof Theory Symposium – Kiel 1974*. Ed. by A. Dold and B. Eckmann. Springer, 1975, pp. 365–369.

[14] A. Tarski. "Contribution to Theorie de la preuve formalisee: Discussion". In: *Revue Internationale de Philosophie* 8 (1954), pp. 15–21.

Context and Computation
ANDRÉ LECLERC

A lot of work has been done to develop a man-machine interaction in a way that simulates as much as possible a man-man interaction. To construct intelligent machines capable of interacting naturally with potential users, we need to develop a language with a very powerful syntax, enabling the machine to simulate what would be a usual interaction between competent speakers. The syntax should contain, for instance, 'illocutionary forces indicating devices' to allow the recognition of the illocutionary acts performed in the interaction (orders or commands, statements, declarations, etc.), rules of introduction and elimination for the terms used in inference chains, and to each term should be associated a lot of information about possible contexts of use and the most frequent uses of them. A huge data base concerning natural and social regularities, patterns of behavior, usual ways of doing things and to react to given situation, etc., should be useful. After all, we speak the way we do because we possess such a knowledge. This is a Herculean task, but the program is still at the beginning and some interesting results are already there to encourage its continuation.

Descartes in the fifth part of the *Discours de la méthode* presented a negative prognostic for the success of this program: it is 'morally impossible', he says, to construct a machine with a capacity for "genuine discourse". No machine can do what a stupid human being does: communicating thoughts in a *relevant* way in any *new* situation. More optimistic is Turing's position. It could be state, in a nutshell, just like this: 'if my machine can fool you long enough, be honest and admit that it can 'think' in its own way'. I know this can be seen as 'conservative', but in that debate I would take side with Descartes: the language used by the machine needs to be highly regimented, while ordinary language has an irreducible 'open texture'. All the tokens of the same sentence-type (in a regimented language) must be interpreted in the same way by all in order to be understood correctly, while the tokens of the same sentence-type (in ordinary language) must be interpreted differently according to

Copyright © 2019 by André Leclerc. All rights reserved.

the context, to be understood correctly and they can determine different truth-conditions (or conditions of satisfaction) in different contexts.

In the last decade there has been a hotly debated issue in the semantics of natural languages concerning the status of the content of our utterances, also known as 'what is said'. Is it a semantic or a pragmatic notion? Are the contextual influences on the content restricted to a small set of expressions (indexicals and demonstratives), or do they reach words of any category and whole sentences? Another related and important question is: Is spontaneous linguistic understanding really a kind of modular or computational process? These are important issues in the debate opposing minimalism and contextualism in recent philosophy of language.

Minimalism[1] remains closer to the tradition of logical or formal semantics. What is said by the literal utterance of a sentence is the direct output of a process consisting in the application of syntactic compositional rules to the meanings of lexical items. Thence what is said is a semantic notion. The distance between sentence meaning and what is said is 'minimal', because the only difference between the literal meaning of the sentence and what is said in a context of utterance comes from the 'saturation' of indexicals and demonstratives, a process by which their variable denotations are determined.

All expressions have a character and a content [4]. Minimalism says that most linguistic expression have a standing (constant) meaning, that is, their intension determines the extension in the same way in all contexts of use. Their understanding would be constant. There wouldn't be difference between standing meaning and occasion meaning. Only indexicals and demonstratives would have a different semantics: their literal meaning, which is their 'character' is a function from contexts of use to content, and their content is variable. The linguistic meaning or character of 'I' is something like *the agent of the context* (or *the person who is saying 'I' in the context*). The other classes of expressions in a language have a fixed content (their linguistic meaning always determines the same content). With the exception of indexicals and demonstratives, the semantic of an expression is *insensitive to context*. Any other contextual effects understood by the hearer is *implicated* by the speaker; then the hearer must engage in reasoning and follow pragmatic rules in order to identify the speaker meaning (in cases of ironies, metaphors, implicatures, indirect speech acts, etc.). The domain of pragmatics would be the domain of non-literality. Thus, for minimalists, once the content of indexicals or demonstratives is determined in a sentence, the literal

---

[1]The term 'minimalism' has been introduced by Recanati [8].

meaning is fully determined, and the literal meaning *is* what is said by a literal utterance of the sentence. So our ability to put together lexical items according to syntactic rules could be describe as a computational process, context sensitivity being limited to a very small set of expressions.

Contextualism is the thesis that different tokens of the same sentence-type can determine different truth-conditions according to the context of utterance, even when the sentence-type does not contain indexicals or demonstratives. In other words, the output of our semantic competence —lexical items combined according to syntactic rules—is not enough to determine truth-conditions, or does not yield a fully interpretable sentence. Contextualism stresses context-dependence as never before. A sentence like 'Jones had a walk' must be interpreted differently according to the context in order to be interpreted correctly: if Jones is a healthy adult, we understand that he just walked a few kilometers to stay in shape; if Jones is a toddler, we understand that he just took his first steps in his life; if Jones is an old man in a hospital trying to recover from a serious disease, we understand that he just walk slowly in the hallway or from his bed to the bathroom and back; if Jones is an athlete that just undergone a knee surgery, we understand that he is recovering well and will be back to his team soon, etc. The expression 'open texture' means that there is no principled way to close the list; the only limit is the limit of our own imagination. Lexical meanings are *modulated* in context, that is, they get through what Recanati calls Primary Pragmatics Processes.[2] So we are not leading here with ambiguity or polysemy.

Since Frege and the introduction of the idea of functional application in logic and philosophy of language, understanding (or determining the meaning of complex expressions) has been seen as a kind of calculation or computation. In extensional contexts for instance, a one-place predicate denotes a function whose value is a truth-value when applied to an object as its argument. And the truth-value of complex sentences is a function of the truth-value of the simpler sentences. To understand a sentence is to grasp its meaning, which is, most of the time [3], the same as

---

[2] The primary pragmatic processes alter the lexical meanings in different ways: *transfer* is a kind of metonymy ("the city is asleep" for "the inhabitants of the city are asleep"); *loosening* is a way of widening the conditions of application of a term; normally it produces metaphor ("The machine swallow my card"); and *enrichment*, when a term must be understood as meaning something more specific ("I had breakfast" for "I had breakfast *this morning*"). See [8].

[3] There might be exceptions. According to Frege, grasping a thought would be the same as grasping truth conditions, but in cognitive dynamics, thoughts are not preserved through the changes we have to do in the sentences to track them. 'Today is a nice day' and 'Yesterday was a nice day' cannot express the same thought,

grasping its truth conditions. Is understanding a kind of computation? If by 'computing' we mean something like following an algorithm that leads, in a finite sequence of steps, to a unique result, then contextualism can be seen as a serious opponent to the view that understanding is a kind of computation, which would undermine the minimalist view on the semantics of natural language. I will argue that this is the case. The main tenet of contextualism is 'that meaning underdetermines truth-conditions. What is expressed by the utterance of a sentence in a context goes beyond what is encoded in the sentence itself.' [2, p. 105] What we understand are *intuitive truth conditions* (Recanati's expression), not the mechanical, merely disquotational truth conditions. The intuitive truth-conditions are those actually processed and immediately understood by the agents of the context. We will see soon why it is so.

I shall first present the principles of classical philosophical semantics, criticizing them and showing that they are wanting. Then I shall present new principles, those of contextualism, more faithful to our linguistic practices.

1 Some Guiding Principles of Semantics and their Necessary Reformulation

The first noticeable semantic property according to traditional philosophical semantics could be formulated in this way:

> *Having a determined truth-conditional content, or expressing a truth-valuable proposition, is a semantic property of a sentence-type.*

Semantic, as Frege taught us, must be systematic. It is a theoretical representation of our ability to produce and understand a potentially infinite number of meaningful sentences. Semantic must show how our knowledge of the sense and reference of any complex expression depend uniquely on the knowledge of the sense and reference of simpler expressions and the way they are syntactically combined. Our semantic knowledge is associated to expression-types. The tokens inherits the semantic properties of their corresponding types. A competent speaker-hearer does not have to learn again the meaning of expressions at each new utterances. But that principle of inheritance does not hold for indexicals and demonstratives. The character alone is not enough to determine the content. There is no absolute Fregean mode of presentation conventionally associated with 'this', 'that', 'I', 'now', etc. The meaning (or

---

because 'today' and 'yesterday' have different modes of presentation, and modes of presentation are the constituents of thoughts. But the two sentences have the same truth conditions, or so it seems to me.

character) just tells us where to look, so to speak, in order to proceed to the saturation. There is no 'completing Fregean sense' to use Perry's expression. [7] The *Sinn* (the character of the indexical) is not sufficient to determine alone the content (*Bedeutung*).

But if context-sensitivity pervades the use of language as contextualists think it does, the principle must be reformulated like this:

> *Having a determined truth conditional content or expressing a full-fledged proposition is the property of an act-of-utterance-in-a-specific-context, not the property of a sentence-type.*

As we shall see, we always apply generic meanings in highly specific context of use. The truth conditional content is not fully determined before the modulation of meanings in a specific context.

The systematic character of our semantics relies heavily on one principle known as the compositionality principle:

> *The sense and reference of a complex expression is a function of the sense and reference of its constitutive parts and the way they are combined syntactically.*

This principle is extremely important, but it needs a reformulation to be adequate to natural language semantics. Suppose that $I(\alpha) = m$ is a function of interpretation associating to an arbitrary expression its semantic value. In the simple, direct version of the compositionality principle, we have that, for two expressions α and β, $I(\alpha \star \beta) = f(I(\alpha), I(\beta))$, where '$\star$' is any mode of composition whatsoever, and 'f' is a function taking as argument the interpretation of each lexical items. The rule says that the interpretation of the complex expression '$\alpha \star \beta$' is the value of the function f. But that will not do for contextualists, because lexical items, most of the time, get through Primary Pragmatic Processes, which somehow change (or enrich) the lexical meaning. These processes are 'pre-propositional'. They take place before the application of compositional rules. In 'Jones had a walk', 'walk' must be interpreted differently in different context. The meaning of 'walk' is *modulated* according to the context of use. So the interpretation function yields a different value according to the context. We have something like $f(mod(walk, c_1))$, for context c_1, but the value of function f for context c_2 will be something different, etc. The contextualist reformulation for the rule of compositionality would be:

$$I(\alpha \star \beta)_c = f(mod(\alpha, c_1)(I(\alpha)_{c_1}), mod(\beta, c_2)(I(\beta)_{c_2})$$
$$= f(g_1(I(\alpha)_{c_1}), g_2(I(\beta)_{c_2})),$$

where 'g_1' and 'g_2' are different primary pragmatic processes, and 'c_1' and 'c_2' are subparts of context c in which '$\alpha \star \beta$' is used. [10, 9] In this way, contextualism is compatible with compositionality, but in a rather indirect way. The content of a complex expression is a function of the modulated meanings of its parts, and not a function of the content *simpliciter*, as understood in the minimalist way.

The principle of context is another one needing reformulation. That famous principle is in fact a prescription: never ask for the sense or the reference of an expression outside the context of a full sentence. Initially, Frege means the *context* of a sentence (*Satz*). As a matter of fact, the adjective 'light' must be interpreted differently according to the name it modifies; thus, a light meal is *light* in a way that is very different from that of a *light* piece of luggage, for instance. But the context that interest us is something much broader than that. It includes much more than a speaker, a hearer, a place, and a time, which are necessary for the semantic process of saturation. We must also consider how things are and could be in the context of utterance, the intention or plan of the speaker, encyclopedic knowledge mutually shared (especially knowledge of natural and social regularities), expectations of the agents of the context, patterns or ways of doing things, and salient objects or events in the context. All these elements can contribute to determine *what has been said* in a context of use.

If I say, out of the blue, 'I have five fingers in my right hand', the reaction will be perplexity: 'That's obvious! Why does he call attention to something standard?' But in front of two policemen searching for the murderer with four fingers in his right hand, my intention or plan is clear for everyone: to free myself of a possible accusation. Utterances are acts and acts are performed for some reasons; when the reason is obscure, the speech act cannot be fully understood. The identification of speaker's intention (or plan, expectations) is an essential step for the uptake. We also presuppose all the time encyclopedic knowledge about a lot of regularities, natural and social. There are standard ways of doing things. If I say to my son 'Please, cut the cake!', and he do it with an ax, that would not count as a way to satisfy my request.

So by 'context', we need to understand something that goes much beyond what Frege had in mind to include anything that could determine what is said, or modifies the lexical meanings applied in a specific situation.

Finally, the meanings of lexical items are always generic in character. Here John Austin's distinction between descriptive and demonstrative conventions can be very useful.[1] Descriptive conventions are what we find in dictionaries. For instance, in a dictionary, we find that definition for the word 'table': 'a flat surface, usually supported by four legs, used for putting things on'. No specific table is mentioned; the lexical meaning is of course always *generic*. A dictionary entry only mentions generic things, generic events, generic situations, etc. But we always speak in highly specific contexts. The demonstrative conventions are what anchor our language in specific situations of use. When at home I say something like 'There is a cat on the table in the dining room', I am speaking of a very specific material table, a specific cat, etc. The table is the one salient in the context of use. The principle could be state simply like this:

> *We always apply generic meaning in specific situation. So, what is said is determined only in part by the lexical meaning; a decisive contribution comes from the processes of modulation, the intention or plan of the speaker-hearer, and their interests and expectations when they assess the meanings of words as used in a specific context.*

So, the primary pragmatic processes do not come as a surprise. We have to adjust constantly our meanings (those of our idiolect) to the most diverse contexts of use.

2 Reasonableness

Two philosophers, Charles Travis and Hillary Putnam, make explicit some dramatic consequences of the contextualism they embrace. First, Travis stresses the strong connection between the intentions and expectations of the speaker-hearer and what is said:

> We see words as taking responsibility for serving certain purposes, in that we will count them as having said what is correct, so true, only where we count these purposes as (adequately) served. [11, p. 463]

Ten years later, and relying on Wittgenstein, Travis continue his crusade:

> Content is inseparable from point. What is communicated in *our* words lies, inseparably, in what we would expect of them. [12, p. 33]

> ...for Wittgenstein, our representation in language is essentially situated. My saying what I did is *not* detachable from the circumstances of my saying it. [12, ibidem]

We use words in specific contexts of use because these contexts are similar enough to previous contexts in which the same words have been used. So, it is reasonable to use the words in similar contexts. Putnam describes very well this process:

> The meanings of the words does (*sic*) restrict what can be said using them; but what can be said using them, consistently with the meaning of the words, depends on our ability to figure out how it is reasonable to use those words, given those meanings (given a certain history of prior uses), in novel circumstances. And, pace Chomsky, the idea that reasonableness itself can be reduced to an algorithm is a scientific fantasy. [6, p. 125]

There are no universal criteria of reasonableness. Reasonableness is itself determined contextually. So, as Putnam noticed, it cannot be reduced to an algorithm.

3 Linguistic Understanding

Undoubtedly, we have the ability to understand differently different tokens of the same type according to the context. So, *our understanding of the situation contributes in a decisive way to the correct understanding of a token produce in a specific context of use.* What gives us access to the intuitive truth conditions, those immediately grasped by the agents of the context, is a previous understanding of the whole situation, *in particular a grasping of their expectations and plans.*

But there is more. We understand easily ungrammatical sentences, nonsense, malapropisms, incomplete sentences. Consider the following sentences:

a) You are very much Paris.

b) Give peace a chance!

c) Pride has a city!

d) The Cardinal Mazarin has sent here his hemispheres.

e) I am ready!

f) John speaks fluently English.

The first three are nonsenses, but they are easily understood; d) is a malapropism (just switch 'hemispheres' for 'emissaries'); e) is incomplete (ready for what?) but understood easily on each occasion of use; and the last one is perceived as ungrammatical (the adverb should go to the end of the sentence).

These cases are marginal, of course, but they reveal something interesting for my purpose. Spontaneous linguistic understanding is understanding of utterances, of actions performed for such and such a reason, and its first and primary object is occasion meaning. Occasion meaning is the result of a pre-propositional 'modulation' of standing meaning. The common use of language is a rule-governed rational activity, and rationality is precisely an unlimited capacity to realize intentional adjustments in a huge diversity of context. Linguistic understanding goes much beyond the mere mechanical parsing of sentences. It presupposes much more. Here is my working hypothesis:

> Actions are performed for different reasons, and are parts of bigger plans. *Words and sentences are seen and understood as instruments used in the performance of actions.* **My suggestion is that we should consider plans and not only actions, as the unit of investigation.** I also suggest that the expectations of agents (speakers-hearers), and the correct identification and understanding of these expectations, are especially important for the determination of the sense of the words used in an occasion. In other words: **The root of any distinction in thought and in the sense of linguistic expressions is found in its sensible effects, in our practices, plans and activities.** This principle I call the Principle of the Determination of Sense. [5, p. 271]

In the view here defended, spontaneous linguistic understanding *is not* modular (see [3] for the opposite position); it uses information of different sources and its domain is not well specified, so that the two main criteria for modularity (informational encapsulation and domain specificity) are not satisfied. Our actions are parts of plans, and this holds for speech acts too. We are obliged all the time to realize adjustments in an infinity of different situations. To be successful most of the time in these processes of adjustments is what makes us rational agents. The abilities to speak and understand discourse is just part of that larger process. Those who believe that intelligent machines will be able to simulate perfectly the common use of language have all my sympathy, but I am not a believer.

References

[1] J. L. Austin. "Truth". In: *Proceedings of the Aristotelian Society.* reprinted in J. L. Austin, *Philosophical Papers*, Oxford: Claredon Press, 1979. 1950.

[2] A. Bezuidenhout. "Truth-Conditional Pragmatics". In: *Philosophical Perspectives* 16 (2002), pp. 105–134.

[3] E. Borg. *Minimal semantics*. Oxford University Press, 2004.

[4] D. Kaplan. "Demonstratives". In: *Themes from Kaplan*. Ed. by J. Perry J. Almog and H. Wettstein. Oxford: Oxford University Press, 1989, pp. 481–563.

[5] A. Leclerc. "Meanings, actions and agreements". In: *Manuscrito* 32.1 (2009), pp. 249–282.

[6] Hilary P. *The Threefold Cord: Mind, Body and World*. New York: Columbia University Press, 1999.

[7] J. Perry. "Frege on Demonstratives". In: *The problem of the essential indexical and other essays*. New York: Oxford University Press, 1993.

[8] F. Recanati. *Literal meaning*. Cambridge University Press, 2004.

[9] F. Recanati. *Truth-conditional pragmatics*. Oxford University Press, 2010.

[10] F. Recanati. "Compositionality, Semantic Flexibility and Context-dependence". In: *The Oxford handbook of compositionality*. Ed. by Markus Werning, Wolfram Hinzen, and Edouard Machery. Oxford University Press, 2012.

[11] C. Travis. "Meaning's role in truth". In: *Mind* 105.419 (1996), pp. 451–466.

[12] C. Travis. *Thought's footing: a theme in Wittgenstein's Philosophical investigations*. Oxford University Press, 2006.

How to Understand the Normativity of Logic in the Context of Logical Pluralism: a Pragmatist Proposal

Marcos Silva

> "*The concept of logical consequence is one of those whose introduction into the field of exact formal investigations was not a matter of arbitrary decision on the side of this or that investigator: in making precise the content of this concept, efforts were made to conform to the everyday 'pre-existing' way it is used.*"
>
> Tarski, 1933

> "*The mode of existence of a rule is as a generalization written in flesh and blood, or nerve and sinew, rather than in pen and ink.*"
>
> Sellars, 1951

1 Introduction

In *Grundgesetze der Arithmetik* (1893) [12], Frege states that "laws of logic ... are the most general laws, which prescribe universally the way in which one ought to think if one is to think at all". And Wittgenstein adds in the *Tractatus* (1921) [23] that "thought can never be of anything illogical, since, if it were, we should have to think illogically" (3.03), indeed: "That logic is a priori consists in the fact that we cannot think illogically" (5.4731). These remarks strongly suggests that logic plays a normative role in our rational life since it guides and orientates us on how we ought or ought not to think. There is something rationally defective when one asserts contradictory propositions or holds inconsistent beliefs or when one fails to commit to some direct logical consequences of one's belief system. Accordingly, logical laws seem to prescribe how we should think, and there has to be a connection between logical laws and the evaluation of reasoning, on the one side, and the very possibility of rationality, on the other side.

There is a traditional view about normativity in logic according to which logic provides the laws of thought, i.e. prescriptions about how we ought to think, as opposed to psychological laws, which describe how we actually think. Another approach, a realist one, advocates that logic should be taken as representing an eternal and independent structured reality to which we should conform our reasoning. Both lines of thought seem to defend a monist point of view concerning logic, as they are committed to a view of one unique legitimate logic as the canon for all thinking and reasoning. This monist approach is severely challenged in the contemporary context of a great multiplicity of alternative logical systems.

As an alternative to the monist point of view concerning logic, I defend an anti-realist and expressivist approach to the nature of logic inspired by Brandom's work (1994, 2000) [3, 4], especially his emphasis on a pragmatist normativity to understand rationality in general and formal systems in particular. I apply his pragmatist view about normativity in the context of logical pluralism. As a result, I propose that we ground the plurality of logical systems on our great plurality of ordinary inferential practices. A crucial point in this view is to reject the idea of 'laws of logic' representing a realm of supernatural facts or being grounded in any a priori-structured reason.

In this work, I introduce a pragmatist approach to logical consequence in terms of the preservation of commitments, which, in turn, provides a seminal platform for treating logical pluralism. The paper is divided into three parts. Section 2 discusses some philosophical problems concerning logical pluralism. Section 3 advances a pragmatist point of view to tackle the nature of logic. Section 4 addresses a connection between the normativity of logic and the normativity of morals to present the idea that logical necessity is a form of moral obligation.

2 On some philosophical problems concerning logical pluralism

It is important that I make some preliminary remarks on logical pluralism as a philosophical problem before I present and defend my pragmatist point of view.

We have nowadays different and legitimate non-classical logical systems with many different and interesting (local) applications. The existence of alternative logics raises the question of which one is correct or legitimate. Further, if any is correct, should it be taken as universally correct? It seems that, in the context of a great plurality of logics and revision of classical reasoning, traditional important features of logic as universality, a prioricity, neutrality, necessity should be revised.

According to Priest (2006) [19], logical monists believe there to be, regardless questions of regional applicability, a core or "canonical" application of logic. The canonical use of logic consists in determining "what follows from what—what premises support what conclusion—and why" (p. 196). Once we accept the existence of this canon application, or uniquely correct logic, we are committed to regard it as a paradigm for reasoning and to correct any deviation. It is only when the question is posed in these terms that the full force of the opposition between monists and pluralists concerning logic can be appreciated. The logical monist maintains that there is just one logic to play this core role, regardless of local applications, whereas the logical pluralist holds that no logic plays an absolutely central role. Accordingly, a leading question in this context is in which sense and why (just) one logic should be taken as the cannon for reasoning.

Another philosophical problem concerning logical pluralism relates to the question whether the determination of a logical system is a matter of convention and arbitrary choice. In other words, does the debate between monists and pluralist concern the mere introduction of different formalisms and the choice of some determined goals and different applications? How can one be a pluralist concerning the relation of logical consequence without being a conventionalist? Although the conventionalist approach may do justice to the plasticity concerning logical systems, it obscures some important philosophical problems regarding the connection of logic to rationality. If logic is just a matter of properly and clearly introducing a formal system, we may overlook the comprehensiveness of the question of the nature and bounds of rationality.

We may advance our investigation by addressing the legitimacy of non-classical reasoning. If we also reason non-classically, what does it mean to be rational? In other words, is it rational to reason non-classically? I hold it may be even more rational to reason non-classically, for instance in connection with paraconsistent way-outs concerning contradictory information or rules, in reasoning based on the refusal of indirect proofs in constructive mathematics, and when we want to tackle the vagueness of ordinary language with many-valued logics. In fact, we have heterogeneous philosophical motivations for alternative logical systems to deal with, for instance, vagueness, constructive math, conflicting information, discursive dynamics, quantum phenomena, etc. Accordingly, one may ask if it would be possible to integrate very heterogeneous philosophical motivations for non-classical reasoning under one philosophical program. Is it possible to have an integrated philosophical platform to unify those challenging different motivations?

Further, I think that the normative problem of when reasoning is correct is "easy" to tackle for logical monists, since for them any deviation from the canonical logic is not thinking rather than thinking something wrongly. However, how is the normativity of logic to be dealt with when we take logical pluralism seriously? My intuition is that it is not accidental that some notion of *correct reasoning* is to be found in any logical system. Normativity is not a marginal feature of logic. Logical concepts, for instance, connectives, contradictions, tautologies and logical principles do not have to represent anything because they are *just* normative notions. On this view, the emphasis is on the connections between logic and *Handlungsformen* rather than between logic and independent structures and/or a priori categories of mental states.

In order to defend this point, I present, in the next sections, a pragmatist view to logical consequence, which constitutes a seminal platform to integrate different philosophical motivations for the emergence of non-classical logics. I defend the proposition that logical necessity should be understood as moral coercion. What is being preserving in a valid argument is not truth but the commitments we make when, for instance, we assert something or engage in any ruled practices. These commitments pervade our inferential practices and are made explicit by formal rules in logical systems. They control our reasoning and can be publically tested.

How could this pragmatist framework be helpful to understand the normativity in the case of logical pluralism? How could it give us philosophical and conceptual tools to understand our current discussions in logic?

A well-motivated approach to logical pluralism should be anti-realist and relativist. On a pluralist view, the notion of logical consequence does not need to represent any supernatural facts and does not ground its objectivity on an independent reality and it should be relative to the logical system in which we are reasoning. But how could we be anti-realist, pluralist and relativist concerning logic in a unified platform? I believe pragmatism may help us. I am aiming at applying Wittgensteinian and Brandomian lines of thought to the scenario of logical pluralism to emphasize the normative connections between judgement and action. As a consequence, the plurality of logics may very naturally be grounded on the plurality of our inferential practices.

For the plausibility of this pragmatist program, I favor proof-theoretical notions and not model-theoretical notions as models, interpretations, truth and satisfaction. This choice concerns my anti-realist commitment and motivates my pragmatist approach logically and philosophically. Note that norms which pervades our practices in the world constrain our freedom while binding us together in the phenomenon of rationality. It is almost trivial to hold that we may have different and conflicting principles and reason very differently in different contexts in our daily lives. Accordingly, logical rules do not relate to any particular state of affairs in the world but rather relate to *our* criteria or norms for evaluating descriptions and other actions in the world. Logic does not need to represent anything in reality, as logical rules show, express, make explicit, possible forms of representing the world. Logical systems express some of our public commitments and social norms in rational discussions and practices. During permanent exchanges with our rational fellows, we learn how to act and to make inferences. In our daily lives, we learn how to master a great multitude of heterogeneous logical inferences through special practical trainings.

The determination of a canon of deduction is not the determination of an absolute truth but neither is it a matter of plain choice. To attach any character of uniqueness or absolute truth to any particular system of logic is a philosophical misconception. The entities of formal logic are abstractions, which expresses our commitments in several inferential practices in daily material exchanges with our environment over a long period of (deontological) interactions with other rational beings. Formal logical systems are not just a matter of convenient strategy to a particular goal; they make explicit by, its formal rules, norms implicit in our inferential practices, rules we use to judge and control deviant thinking in our daily lives.

3 A pragmatist point of view to tackle the nature of logic

What I am proposing is an "anthropological approach" to logic as an essential normative phenomenon. In other words, it is to take logic as a human phenomenon. Once logical vocabulary is treated as normative, that is, as determining the ways in which we conform to a framework of prohibitions and authorizations, our focus should be on the nature of rules and norms itself, as we are favoring a proof-theoretical approach to logical consequence. It is because we want to focus the normative character of logic in the context of logical pluralism that we should embrace this form of anti-realism. It is crucial to observe that it makes no sense to expect that there are prohibitions/authorizations (rules and instructions) in the world without human beings (or other rational agents). There is no such a thing as a "real or true rule" in the nature radically independent of our practices.

Wittgenstein's Tractarian anti-realist intuition that "logical constants do not represent" (4.0312) can be fruitfully applied to logical principles too. We can combine it with another Tractarian line of thought: There must be a radical difference between facts (here: empirical matters) and values (here: normative matters). If the nature of logic is to be taken as normative, in this view, it cannot be grounded on empirical facts. The problem is not just that *if* something is normative it is not empirical and vice versa; but rather: *when* something plays a normative role, it cannot provide empirical information, because it defines the very criteria to determine truth and falsehood. Rules cannot be false or true as a legitimate proposition, which describes facts. Rules are not regular propositions because of their content or their logical form, but because of their special normative role in our ways of acting in the world.

Given that philosophical motivations for non-classical logics are very heterogeneous (eg. Brouwer's strong idealism, Priest's dialetheism), what is original in our approach to logical pluralism? The recent literature about logical pluralism (see Shapiro (2014) [20] and Beal and Restall (2006) [1]) very often neglects relevant philosophical issues concerning the nature of logic, in particular its connections with our actions and practices related to its normativity. A notable exception is Field (2009) [11]. Field also defends an anti-realist and relativist approach, where the core concern in classical logic (and many non-classical logics too) is to be characterized in terms of legitimacy of belief, not in terms of necessary truth preservation. Here I propose an expressivist approach in which we are preserving commitments, and not legitimate beliefs or truths.

Another important difference in a pragmatist approach to alternatives can be presented. While Dummett's anti-realism (1978, 1981) [9, 10], for instance, emphasizes epistemic notions, I want to emphasize normative and deontological notions for social agency, cooperation, collaborative and regulative joint activities, including abilities and capacities to normatively engage in games. Note that the programs (epistemic and normative) are independent, but compatible.

Also in opposition to Bensusan *et. all* (2015) [2], who address a highly metaphysical enterprise concerning logical pluralism with their ontology of galaxies, I would like to go pragmatically instead of ontologically to develop an integrated platform for logical pluralism.

To understand the nature of logical constants, Brandom (1994, 2000) [3, 4] and Peregrin (2014) [18] emphasize normative vocabulary, as prohibitions and authorizations, violations, permissions, obligations, respect, obedience. However, they do not advance a robust proposal concerning logical principles and the legitimacy of alternative logics. In fact, some lines are open to logical pluralism as suggested than developed in Brandom (2008) [5].

Against this background, our problem could re-stated as follows: How could some tenets in a pragmatist tradition (especially Wittgenstein's, Brandom's, Peregrin's) help us or contribute to the discussions on the nature of non-classical logics, logical pluralism and its challenge to classical rationality? How can they help us to face the normativity challenge within a great plurality of logical systems nowadays?

It is salient that normative vocabulary very often and surreptitiously appears in the conflict among logics (Silva 2015) [21], regarding obedience, commitment, entitlement, authorization, obligation and the rejection or approval of movements in a system. Other normative vocabulary may also be used, such as engagement, authority, obligation, duty and responsibility concerning logical laws, as they can be held as (consolidated) rules (norms of representation) which express some implicit norms in our practices. Further, in doing so, they can be used to evaluate and control our descriptions and practices.

On the present view the objectivity of logic comes from its public conventions, agreements and stipulations, its regular use over a complicated history of social interactions and the normativity connected to our daily inferential practices (*Verbot und Erlaubnis*). This view rejects any independent and eternal abstract structure and the need to postulate any transcendental mental structures to accommodate normativity of logic and logical pluralism. Logic's conditions of possibility, or transcendentality, as well as its objectivity and inter-subjectivity, are traceable to social interactions, since we do not postulate facts, only the criteria with which facts are measured and evaluated[1].

If the concept of consequence fluctuates in ordinary reasoning, then there are several ways to make it precise: several possible definitions, none of which is excluded by common usage of the language of everyday life. For instance, Tarski's well-known definition of logical consequence is not the only possible way to make the concept of consequence precise. The point of my pragmatist approach is to extend this view to ground logical pluralism, not by focusing on the plurality of deductions made in ordinary language, but by emphasizing the social and deontological features which ground the plurality of inferential practices in our daily lives and the material interactions with other rational agents and the material challenges of our environment. Our being in the world in dynamic interactions with other beings and with material offerings are presupposed in any understanding of our grasp of concepts and the inferential practices involved in it. Our material inferential practices are a path

---

[1] Compare different normative and factual uses of "this is red" and "this is one meter long" in PI §50.

to the late, more sophisticated and intellectually-laden development of logical systems which express those inferential commitments in our daily practices.

In this pragmatist approach, the normative force of logic does not merely constrain reasoning, it applies to all thinking, understood as a conceptual activity. As concepts demand mastery of conditions of application, in order to understand a concept one has to be able to master the *use* of this concept. To know how to use a concept one must master its incompatibilities and affinities with other concepts, the conditions of its applications and the inferences governing these applications. In other words, one has to master both material exclusions and inferences concerning other concepts, as our practices are already inferentially robust. Our logical systems express with explicit rules some of the norms governing our inferential practices in the world so we can publically test and control diverse agents' reasoning.

Accordingly, we have to learn to make inferences. We have to be educated in how to infer, i.e. trained by and among peers and authorities. Training is more important than explaining while learning to infer. Logical education is a social enterprise rather than an intellectual one. Theoretical or practical *oughts* in inferences are best analyzed not as operators acting on propositions but as rules expressing relations between agents and their performances in the world. Rules in logical systems make explicit implicit norms of our inferential practices by addressing in a systematic and formal way how we ought to proceed in thought and reasoning.

Roughly speaking, both sophisticated formal logic and our daily inferential practices are normative. I hold formal logic is regulative (as in the case of driving); once we get them straight we can publically control and correct other's reasoning. This means that a rational agent should accept that her thinking must be able to be evaluated in the light of the laws of logic. However, the important normative role here to ground logical pluralism is other than mere regulation. The regulative role of rules in a logical system reflect some *constitutive norms* of thinking (as chess rules are constitutive for playing chess). Constitutive norms expressed in regulative rules in logical systems govern our rational interactions with our peers. Within logical systems, deviations are simply prohibited. Analogously, when our concerns turn to the constitutive norms in our daily inferential practices, deviations are defective, not just a matter of prohibition: one stops thinking when one stops obeying constitutive rules.

In this expressivist and anti-realist approach to logical pluralism, the question is not how logical consequence in formal systems (both classical and non-classical) normatively constrains reasoning, but how our reasoning as a product of a long history of social interaction constrain the building of logical systems. Formal systems regulate the authorizations, prohibitions and obligations governing certain kinds of social interactions. Accordingly, formal logic's normative impact on mental activities is derivative.

Formal logic is as relevant to ordinary thinking as the driving laws are to knowing how to drive. We can both drive and reason without knowing formal rules. However, ordinary thinking is crucial to formal logic, because the latter expresses nuances of the former. To introduce a formalism is not just a matter of choice. It is a matter of expressing in a systematic and precise manner some lines of inferential practices that we already use and master in ordinary reasoning, and, as a consequence, of making explicit how we were trained and educated, so we can publically control our practices. In other words, those norms in our daily inferential practices are expressed in our formalisms. The validity of a deduction (according to one's preferred logic) is a question of fact, while its normative power is derivative from our daily practices. The normativity of our daily commitments in ordinary inferential practices comes before their logical consequences in a formal logical system since the latter expresses the former.

I agree with Wittgenstein, when he suggests that normativity in the case of logical monism seems easily justified. Deviation is not reasoning, since it is perceived as queer; incomprehensible; a joke; a sign of madness (OC 553; 347; 463; 467) [22].

As a consequence of our pragmatist proposal to logical pluralism, the logical "must be", "have to be", "necessarily" etc. should be taken as "ought to". Common wisdom would say that the former are logical notions and the latter a deontological notion. I maintain that they do not differ in nature. Both notions should be taken as normative and should be grounded on our ruled practices in the world, in our ways of acting, that is, on our moral behavior (concerning prohibitions and authorizations) as rational beings.

4 Logic and morality

Before investigating some relevant connections between logic and morality, I would like to sum up some of the main lines of my Brandomian approach to logic and focus on some Kantian advancement concerning a Cartesian tradition. This advances our pragmatist approach.

To make sense out of the idea that in logic we transform practical norms into principles for public control, justification, correction and development, I would restate three points in the pragmatist proposal for logical pluralism. First, we use a proof-theoretical approach favoring procedures, methods of constructions and rules. Our focus is on rules of inferences rather than truth, models and satisfaction, i.e. on practical know how mastered in our conceptual activities through education, training and the rule-governed use of language. Second, we adopt an expressivist approach to logic, i.e. we hold that, logic does not represent anything, but it expresses what we do when we reason in various different ways under different environmental pressures. Logic's main purpose is justification and exposition, which makes sense against a background of dialectical practices and public debates, or social interaction among rational agents, where interlocutors explain and debate what they themselves already know. (Brandom (1994, 2000) [3, 4]; Dutilh-Novaes (2015) [16]). Third, our approach is an inferentialist in that it treats our practices as inferentially articulated, so they can be tested and controlled through norms.

In line with those three points, in a Kantian tradition, the mind should not be viewed as a vehicle of ideas representing objects in the world, but taken as a capacity that can be trained. The focus in cognition shifts from the representation as a mental image to the rule-governed act of representing something and from the theoretical "know that" to the mastering of rule-governed practices and actions, such as the act of judging or of inferring. The traditional reference to some special objects gives way to the examination of the conceptual connections between our judgments and the material inferences which connect them. Intellect is no longer taken to be a passive activity of mirroring the world but as an active power of illuminating the conceptual organization of our perceived world. Against this background, what are we? What is, at last, rationality and the role that logic plays in it?

The answer we are favoring is anti-Cartesian. Mind and cognition are not entities outside of the world, consuming and manipulating representations, but a special capacity of situated and embodied animals of *deontologically* engaging in a permanent and dynamical exchange with their environment and other animals. I emphasize "deontologically", because rationality should be regarded as an aspect of morality. We are rational because we set and identify rules that prohibit and authorize moves in our practices and modulate, in self-reflection, our behavior using them.

As Brandom (1994) [3] states:

> The key concept is obligation by a rule. It is tempting, but misleading, to understand Kant's use of the notion of necessity anachronistically, in terms of contemporary discussions of alethic modality. It is misleading because Kant's concerns

are at base normative, in the sense that the fundamental categories are those of deontic modality, of commitment and entitlement, rather than of alethic modality, of necessity and possibility as those terms are used today. Kant's commitment to the primacy of the practical consists in seeing both theoretical and practical consciousness, cognitive and conative activity, in these ultimately normative terms. (p.10)

As Brandom puts it, what is special about us is not what we have inside our minds but what we can do in the world. Concerning logic, the correct question is not what is the true logic. Rather it is: what is the best (inferential) practice in a particular context and under particular circumstances, which shows how we perform activities in a cooperative manner with our rational peers?

We do not have at our disposal a plurality of rationalities, since we have a plurality of logics. The problem in a pragmatist point of view is that to be rational is indeed to dynamically coordinate and master various heterogeneous ways of reasoning, conforming them to innumerous environmental pressures and social contexts.

In this approach to rationality, it is important to note that we not simply emphasize our cognition as incorporated and always situated. Nor do we focus on what "we" do as animals subjected to the contingencies of our natural history. Rather we are understood to be animals engaging *deontologically* with other animals and the world. That means, that we are consumers of rules and norms and we are able to identify very finely prohibitions and authorizations identify and we conform our behavior to them to evaluate our own actions and other rational agents' actions. We give each other rules and criteria to evaluating conceptual activities. This connection between morals, logic and rationality deserves more attention.

The interpretation to be developed here to ground logical pluralism in our inferentially thick daily practices holds: 1) that rational obligation is moral obligation and, in particular, 2) that logical necessity is a kind of moral coercion, based on the normative notions of rules (prohibition, authorization, authority, commitment etc.)

This strategy connects logic to morals in a manner opposed to Carnap's tolerance principle, which can be taken as one of the first attempts to ground logical pluralism. Carnap (1937) [6] states that "in logic there are no morals. Everyone is at liberty to build his own logic, i.e. his own language, as he wishes. All that is required of him is that, if he wishes to discuss it, he must state his methods clearly, and give syntactical rules instead of philosophical arguments" (§17).

Prima facie, Carnap's principle has two readings. First, it may be taken negatively in a way that leads straightforwardly to tolerance. From a logical point of view, we cannot say that logical system is forbidden. (Goedel's intuitionist logic is a good example of tolerance in this sense). Second, it may be taken positively in a way that leads to a radical freedom. It holds that we are permitted to introduce any syntactical device when constructing a logical systems; the only constraint is a methodological one: rules should be presented as clearly as possible. Both readings allow plasticity and innovations is logic and are in line with the multiplicity of logical systems, but also open the way to conventionalism and arbitrariness and obscure crucial connections between logic and the nature and limits of reason. In order to avoid arbitrariness and radical conventionalism, we have to introduce some association between language and the "forms of life" understood as pervaded by deontological features.

Further, it is noteworthy that conventions alone do not coerce anybody. Mere configurations on a piece of paper do not compel us to draw any consequence. No disposition of signs has itself the normative power or pragmatic force to induce one to infer something.

Against this background, let's go back to our opening problems concerning the connections between logic, rationality and normativity. Why and how do we take reason as an authority and feel obliged to obey it? In virtue of what do we feel coerced by reason in our inferential practices, both practical and theoretical? The power of reason can be taken as guiding our decisions for practical life and as the power to compel one to accept the conclusion of a proof. So how can some forms of reasoning compel one to act and to infer?

The pragmatist working hypothesis here is that we need some sort of (moral) commitment to norms and rules to understand the nature of logical necessity. Rational beings are consumers of rules; they identify rules and feel compelled to obey them. The crucial point here is an anti-representationalist one: we give each other rules, criteria, authorizations, prohibitions, commitments etc. The important point is that logical systems do not encapsulate laws of truth but rather express clearly and explicitly norms which already inform our inferential practices, the object being that we enable us to correct ourselves and others.

We are logical beings in virtue of our way of *deontologically* acting in the world, or in a word, because of our morality and not vice-versa. Our practices are already inferentially articulated; we first master those inferences embodied in our social practices and make theories to express them with an eye to correcting deviations. It is crucial to note that when I am addressing morality in connection to logic, I am not dealing with ethical imperatives or maxims, but with "mores", that is, customs, conventions, ways of acting, ways of life, traditions, practices, habits etc.

In spite of Carnap, I am in good company defending connections between morals and logic. For example, Frege (1897) [13] relates the nature of logic to the philosophical discussion on moral and freedom when he states: "Logic has a closer affinity with ethics [than psychology] ... Here, too, we can talk of justification, and here, too, this is not simply a matter of relating what actually took place or of showing that things had to happen as they did and not in any other way" (Posthumous Writings, p. 4). Husserl, in the first volume of his *Logical Investigations*, also maintains that: "He [the logician] aims not at a physics, but an ethics of thinking." [14, p. 43]. Around the same period, but independently, Peirce also stated something very close to it:

> "The purpose and utility of logic [...] lies in its final achievement of a methodeutic for the guidance of thought; and from this point of view logic is the theory of the self-control of thought in order to realize its intention, which is truth. So regarded, logic may be called a special kind of ethics, if by ethics we mean the theory of the self-control of conduct in order to realize a deliberately adopted purpose. For inquiry is only a particular kind of conduct." ([17], MS 602:8)

Schlick, Waismann and Wittgenstein, in the beginning of the 1930's, discussed some topics in the line too [24, pp. 128,131,175] and Field (2009) [11] has systematically defended the normativity of logic in contemporary philosophy.

According to the proposal defended here, we should re-think the traditional question of the nature of logical necessity. In this pragmatist approach, what does it mean to say that B follows from A (in a given system)? In particular, what does "to follow from" mean? To implement our antirealist and expressivist proposal and understand the normativity of logic in the context of logical pluralism, we need to answer in what sense an inference compels us to judge the truth of a conclusion from the assumption of the truth of the premises.

Logical coercion has its roots in the rational obligation compelled by our determination to follow agreed rules. Logical consequence is a relation that makes explicit determined relations of authorization and prohibition inherent in practices of agents in communities in as much as our practices are always inferentially articulated.

A direct consequence is that different communities may have different norms. Further, we may have heterogeneous norms in the same communities. If truth is a moral command to assent, an individual who asserts something commit herself to the truth of the content of the assertion and all the consequences which can be drawn from her assertion. She may vary her reasoning according to pressures in her environment and to the nature of her needs, interests and tasks. Here we have an original interpretation of Frege's truth preservation. According to Brandom (1994, 2000) [3, 4] and Peregrin (2014) [18], there is an important line of inferentialism in the *Begriffschrift*. The main notion is neither truth nor reference. A notational system makes explicit inferential relations in terms of assertible content. Preservation of truth is preservation of commitments in asserting something, because to assert something is to commit to the truth of a propositional content and its consequences.

Carnielli and Rodrigues (2015, 2016) [7, 8] explore, among other things, alternative interpretations for validity. They investigate it as a matter of truth preservation for classical reasoning, as connected to the availability of a constructive proof for the intuitionist cases and, then, as related to the preservation of evidence for paraconsistent reasoning. In the present pragmatist approach, a proposal can be made to extend this framework with a fourth way of thinking of validity and logical consequence, namely, as preserving commitments in material reasoning which can be expressed in formal systems. On this way of thinking validity can be explored as a way to cover the other definitions of validity in a unique pragmatist approach to unify the philosophical motivations in grounding logical pluralism.

If one asserts that a shirt is (all-over) red, one is committed, for instance, to the fact of the shirt being something material and not something abstract, and not being blue or any other color. The talk of materially correct inferences is indeed intended to inforce a contrast with those that are formally correct (in the sense of logically valid). But the force of this contrast is just that the validity of inferences in virtue of their logical form is to be understood as a sophisticated, late-coming sort of propriety of inference founded and conceptually derivative in respect to a more primitive sort of propriety of inference. This is the refusal of the formalist approach to inference, for which the correctness of inference is intelligible only as formal logical validity, i.e., correctness in virtue of logical form. Calling the more primitive sort of propriety of inference materially correct simply registers the rejection of this order of explanation.

My view about logic is not foundationalist, even if I try to ground logical systems on our daily inferential practices, because the envisioned pragmatic ground is itself groundless. There is a core in our rationality, I accept this; some things are more fundamental than others. However, that core may change over time. It does not have to be absolute, only stable enough. There is a pre-theoretical field wherein our inferential practices and our theoretical investigations are embedded. The groundlessness of our practices is not an epistemic deficiency nor does it represent the victory of skepticism. Our inferential practices expressed by our great variety of logical systems for many different purposes deals with something so deep that it does not depend upon any rational justification.

According to Sharrock (2016) [15], "groundlessness here does not denote precariousness, but characterizes the rock solidity that makes our spades turn and allows us to 'stand fast'. The certainty on which all our knowledge is logically hinged. Moreover, though groundlessness means absence of justification or reasoning, it does not mean detachment from reality: our hinges are conditioned by how the world is, by 'very general facts of nature'; they are rooted, albeit not ratiocinatively, in our human form of life and in the various forms of human life." (p. 107).

5 Conclusion

What I discussed in this paper is part of a more general "anti-realist" and pragmatist enterprise about the nature of logical systems themselves and their relation to *our* practices, rationality and forms of lives. The focus on this pragmatist view to logical pluralism is that: the great plurality of our inferential practices in ordinary exchanges with our environment and with other rational beings renders a great plurality of logical systems, which can in turn be used to test, control and revise the same ordinary inferential practices that give rise to them.

As the main feature of logical systems is to express through explicit rules the norms which are implicit in our doings in the world, logic should be thought of as a crucial tool to human rationality. It permits public justification, correction and development of our conceptual practices. I hold the recent emergence of non-classical logics as an example of this kind of rational and reflexive development.

If the proposed pragmatist approach to logic is plausible, it has as a philosophically seminal result that our morality is not a consequence of our being logical; we are not theoretical creatures that act in the world but the other way around. Our theories and logical systems emerge from our deontological engagements in mundane activities with our environment and other rational agents. We are logical in virtue of our morality and not vice-versa. Practices and language are already inferentially articulated. We are not talking about ethical imperatives, but "*mores*", that is, customs, conventions, ways of life, traditions, practices, habits, which pervade our form of life. They are full of a great variety of tacit commitments, which set a dense network of prohibitions, incompatibilities, entitlements and authorizations.

If I am right to be a logical pluralist means to be a moral pluralist. We may have different logical systems because we have different inferential practices in reacting to the world and to the demands of a rational and cooperative life.

A fine taxonomy of rules is important and should be developed in future works. However, in this paper the distinction between rule and facts is more vital. That is, the emphasis on deontological vocabulary (commitments, authorizations and prohibitions) as opposed to representational vocabulary (truth, satisfaction, model) is to be taken as crucial. Once we start to think seriously about logic as founded on ruled practices, we realize that: Neither a metaphysical nor a transcendental subject is required, but rather individuals or human beings engaged in deontological practices (as "Regel haben mit *Verbot* und *Erlaubnis* zu tun", [24, p. 175]).

There is a tension in my approach between leaving things as they are and the possibility of criticizing standards and norms. I do not intend to solve it in this work. This problem within the context of logical principles is even more difficult to tackle, because criticism of logical principles presupposes logical principles [21]. It must, however, be possible to raise criticisms and to reconsider principles even in logic.

References

[1] J. Beal and G. Restall. *Logical Pluralism*. Oxford: Oxford University Press, 2006.

[2] H. Bensusan, A. Costa-Leite, and E. G. De Souza. "Logics and their galaxies". In: *The Road to Universal Logic, volume II*. Springer, 2015.

[3] R. Brandom. *Making It Explicit: Reasoning, Representing, and Discursive Commitment*. Cambridge, MA: Harvard University Press, 1994.

[4] R. Brandom. *Articulating Reasons*. Cambridge, MA: Harvard University Press, 2000.

[5] R. Brandom. *Between Saying and Doing: Towards an Analytic Pragmatism*. Oxford: Oxford University Press, 2008.

[6] R. Carnap. *The Logical Syntax of Language*. London: Kegan Paul, 1937.

[7] W. Carnielli and A. Rodrigues. "Towards a philosophical understanding of the logics of formal inconsistency". In: *Manuscrito* 38.2 (2015), pp. 155–184.

[8] W. Carnielli and A. Rodrigues. "Paraconsistency and duality: between ontological and epistemological views". In: *The Logica Yearbook 2015* (2016), pp. 1–30.

[9] M. Dummett. *Truth and Other Enigmas*. Harvard University Press, 1978.

[10] M. Dummett. *Frege: Philosophy of Language*. 2nd ed. Cambridge, MA: Harvard University Press, 1981.

[11] H. Field. "Pluralism in logic". In: *The Review of Symbolic Logic* 2.2 (2009), pp. 342–359.

[12] G. Frege. *Grundgesetze der Arithmetik*. Vol. Band I/II. Jena: Verlag Hermann Pohle, 1893.

[13] G. Frege. *Posthumous Writings*. Ed. by H. Hermes, F. Kambartel, and F. Kaulbach. Trans. by P. Long and R. White. Chicago: U. of Chicago Press, 1979.

[14] E. Husserl. *Logical Investigations*. Vol. 1. London: Routledge, 2001.

[15] D. Moyal-Sharrock. "The Animal in Epistemology: Wittgenstein's Enactivist Solution to the Problem of Regress Hinge Epistemology". In: *International Journal for the Study of Skepticism* 6 (2016), pp. 97–119.

[16] C. Dutilh Novaes. "A Dialogical, Multi-Agent Account of the Normativity of Logic". In: *Dialectica* 69.4 (2015), pp. 587–609.

[17] C. S. Peirce. "An Outline Classification of the Sciences'". In: *Collected Papers of Charles Sanders Peirce*. Ed. by Charles Hartshorne, Paul Weiss, and Arthur W. Burks. Cambridge, Massachusetts: Harvard University Press, 1902, pp. 1931–1958.

[18] J. Peregrin. *Inferentialism: why rules matter*. Hampshire: Palgrave Macmilliam, 2014.

[19] G. Priest. *Doubt Truth to be a Liar*. Oxford: Oxford University Press, 2006.

[20] S. Shapiro. *Varieties of Logic*. Oxford University Press, 2014.

[21] M. Silva. "Persuasion over convincement: On the role of conversion in logical conflicts between realists and anti-realists". In: *Wittgenstein e seus aspectos*. Ed. by Arley R. Moreno. Campinas: CLE, 2015, pp. 143–166.

[22] L. Wittgenstein. *On Certainty*. Ed. by G. H. von Wright & G. E. Anscombe. Edição bilíngue. Londres: Basil Blackwell, 1969.

[23] L. Wittgenstein. *Tractatus Logico-philosophicus. Tagebücher 1914-16. Philosophische Untersuchungen. Werkausgabe Band 1.* Frankfurt am Main: Suhrkamp, 1984.

[24] L. Wittgenstein. *Wittgenstein und der Wiener Kreis (1929-1932).* Werkausgabe Band 3. Frankfurt am Main: Suhrkamp, 1984.

Language, Tools and Machines
CARLOS EDUARDO FISCH DE BRITO

> *Human thinking is individual improvisation enmeshed in a sociocultural matrix.*
> — M. Tomasello, *A Natural History of Human Thinking*

Traditional accounts of language and meaning take the natural point of view that the basic function of language is to provide us with the necessary means to be able to talk and think about the world. As the story goes, propositions stand somehow in correspondence with states of affairs of the world. And this relation explains both how propositions acquire their meanings, and how they give us intellectual access to objective aspects of reality. Moreover, since propositions can be freely manipulated in thought, by playing with their meanings we gain intellectual access to non-actual, but possible states of the world. This representationalist point of view is indeed very attractive, as it fits quite well our everyday experience of talking to other people, thinking about what have come to pass, imagining and planning our future activities. So much so that it seems difficult to conceive an alternative approach to understand language.

But, the pragmatic accounts of language and meaning achieve just that by shifting the attention from the intellectual musings of the individual to the practical activities of social groups. The move should be understood in two dimensions. First, the pragmatist abandons the paradigm of the observer who describes or represents aspects of the world in language, in favor of the more engaged perspective of the person who is involved in some practical doing and cares about its results. She notes that most of our collective activities are mediated by the use of language. And, in these situations, the role of the language constructs is better understood as that of helping to direct the development of the activity, rather than describing anything. The meanings of the linguistic expressions would then be given by the effects they have on the ongoing

---

Copyright © 2019 by Carlos Eduardo Fisch de Brito. All rights reserved.

activity. Such observations naturally lead to the now familiar conceptions of social activities as language games, propositions as moves in the game, and meaning as use. The second dimension has to do with how the mind relates to language. Here, the pragmatist rejects the idea that our intellectual capabilities have any sort of privileged access to the meaning of concepts and propositions. To the extent that meaning is given by use, mastering concepts and grasping the meaning of propositions amount to using them properly. And this comes about as the result of learning how to apply them in the intended circumstances, and learning how they relate to each other inferentially. Again, with the pragmatic turn we obtain a very attractive point of view on language and cognition indeed, as it fits quite well our everyday experience of doing things together with other people, playing roles in social activities, and engaging in the practice of giving and asking for reasons. So much so that it seems difficult to conceive still another alternative way of understanding language. But this is the task with which we want to entertain ourselves in this essay.

The shift of perspective proposed by the pragmatic tradition helps to bring our views on cognition and the mind closer to our understanding of the natural phenomena. And we can identify at least two particular contributions of pragmatism to this end. First, it replaces the somewhat mysterious conception of a proposition as a picture by the more concrete idea that concepts and propositions are like tools that we must learn how to handle and to put to proper use in practice. Second, with the new emphasis on practical activities, knowledge tends to become assimilated to skills and abilities, in contrast with the traditional view in terms of information and (true) belief. Despite this approximation to the point of view of the sciences, pragmatism still falls short of a complete naturalization of cognition — indeed, this is not one of the goals of the movement, and some authors explicitly reject this possibility. And there is a sense in which pragmatism actually shares something of the representationalist point of view: a tool is still something that must be in the hands of someone who knows how to use it properly. That is, although the pragmatist sustains that the collective practical activities are the right setting to understand human cognition, and although she moves the emphasis from knowing that to knowing how, she still puts too much reliance on the cognitive competences of the individual by expecting that he knows what he is doing.

There certainly are instances of collaborative activity in which the components of the group do have a clear awareness of the overall activity and their particular individual contributions to it. But this is by no means the typical case, and neither the most interesting one. The distinctive mark of group activity in more natural, non-contrived situations is the fact that people don't know exactly what is going on and must be told what to do. Or, better, no one knows exactly what is going on but keep telling things to each other. Think of group hunting, where unexpected events happen all the time changing the current situation, bringing about scenarios of danger and emergency, but also opportunities for action and success. In those settings, no one tells you what to do. They just shout "go!", then you go, and from there on you will be on your own, responding to whatever happens the best way you can. At this point, the objection may be raised that the situation we have just described is only possible because you know how to proper respond to the shout "go!", namely, by actually going. Yes, indeed, but this is so much less than what else is going on. Clearly, in this kind of situation there are individual cognitive competences being deployed, and those competences are being articulated through communication between the partners. But no partner is telling what the other should do, and they actually don't know exactly how the other is going to respond. They just trust that the other will respond well. And, when you respond to the call of you partners, you don't know exactly why they are calling you into action, and what exactly they expect from you. You just trust them and move on. And then things can go wrong, of course. But then someone will shout "no!", and you will stop, return or find a refuge. And this is also a reason why you may trust your fellows and throw yourself into action when they call. The bottom line of this discussion is that collaborative activity can work fine without knowledge — however one conceives it — at the top. Individual competences articulated through communication, plus some awareness of the common goal and the capacity to judge whether things are going well or not, from the part of the individuals, is all that is required. Knowledge can actually arise after the fact, when the partners get acquainted with the ways the other usually responds and are thus able to standardize their interactions. In this way, successful group dynamics can be rehearsed in the future. At this point, we enter the story of the pragmatist. But we want to focus in the previous situation.

We will attempt to gain insight into this mode of group activity by looking at machines, examining how machines work, how we work with machines, and how machines are constructed. Looking at machines helps in this case because almost everyone would agree that a machine does not know what it is supposed to do, a machine does not know how it accomplishes its task, a machine does not know why it does things in the way it does. Still, that does not prevent the machine from doing its job. And it is also fair to say that in almost every case the person who operates the machine does not know how the machine works, he does not know why the machine does things the way it does, and he does not have to describe — or otherwise indicate — to the machine what it is supposed to do. He just puts the machine into operation and trusts that it will do its job right. And there is a sense in which the machine trusts the user too. Most often, when we put ourserlves to analyze machines, our attention gets attracted to the rigidity of the parts and components, and we circle around the idea that, by offering resistance to certain kinds of movements, the parts define the functionality of the machine. But then we lose sight of the fact that in actual operation the movement flows through the paths of no resistance. Components push each other and let themselves be moved about. And, when properly used, the machine offers no resistance to the user and gets thrown into operation.

Now consider the experience of learning how to use a tool. Typically, you sort of know what the tool is used for, but you have never had such tool in your hands before. You take the tool to see how it feels, and then you start playing with it. For a while all attempts lead nowhere. And then you realize that you did something right. The tool somehow yielded the right movement. You have found a path of no resistance and were able to make some progress. Here we may as well ask: "But, where was the resistance coming from?" Well, from the material, of course, and also from the tool, if not properly handled. But, there was resistance coming from you too. Sometimes, when you are busy with your attempts, it is precisely by relaxing some muscles and letting the movement happen that you find out the proper way of doing things. What happens at this moment is that the tool takes over the control for a while; the specific shape of the tool guides the movement. You don't know how it happens, you don't need to know. But now you are beginning to learn how to use the tool. The point of this example is to call attention to the fact that there are many happenings intermingled with our doings. That is, brief moments in which things go by themselves, outside of our control and awareness. And, as the example shows, those moments can be the crucial steps in which the real work is being done. Learning how to use a tool involves finding such moments among all the many happenings which yield no progress and, in this sense, offer resistance. Learning how to use a tool involves getting acquainted with such happenings and developing trust.

The experience we have just described, of starting off movements or actions and expecting them to work, is by no means something peculiar to the process of learning. Something that is later replaced by a more thorough awareness of the situation, as you become an expert. The difference between the novice and the competent worker or the skilled artisan lies somewhere else. Knowing how to use a tool amounts to being able to bring about those moments in which the tool can do work for you. But, once you trust your abilities and you trust the tool, you have no concern for them anymore, and your attention can move somewhere else. The task at hand actually demands your attention, there is always something to accomplish. In well-known domains of activity, work is basically routine. Each particular situation requires a proper response, which then produces another situation. Awareness must be present for the right decision to be made, but you are typically confronted with something familiar and have only a few standard ways of responding to it. The activity evolves like the operation of a machine, little resistance is met during the process. In this sense, there is maximum control, but it is not necessarily you who keeps things under control. You have the right tools at your disposal, you are working with the right materials, everything around you is in the right place, and you are following the right ways of doing things. So, things seldom go seriously wrong. Here we are back to the realm of the pragmatist – there is plenty of practical knowledge. There is knowledge at the top, and there is knowledge at the bottom. But, to the extent that the worker is only responsible for making local interventions, this knowledge is not his own. He knows what, why and how to do things just in a very limited sense. But that is fine, so long as everything else is in the proper place to ensure smooth operation. Now, since things seldom go wrong, everything is quite predictable. Given a familiar situation, one sort of knows what to expect if he decides to take a certain course of action. And if the new situation is again familiar, he can put himself to think about what he could eventually do next. Not much is required for thinking, in this case. Just the ability to respond to imagined situations, and then foresee the results of such responses. This is possible because you have somehow internalized the regularities of your activity. This is knowledge, no doubt. But not considerably more knowledge than you had in the previous situation. The elements that support those regularities are still out of reach. Nevertheless, this knowledge is enough to allow you to disengage yourself from the actual manipulation of objects and materials. By thinking and pondering about what may happen, you anticipate problems, find paths of no resistance and work is performed very efficiently. Here we enter the realm of the representationalist.

Next, we will contrast the proficient activity of the competent worker with the creative activity of the skilled artisan. In a sense, routine is also part of the experience of the artisan. We assume her to be just as skillful and capable as the worker with respect to abilities and use of tools. Thus sometimes the activity just flows. Nothing seems to go wrong, one movement follows another, and there is little sense of resistance. Quite similar to the activity of the worker. But the difference shows up when the artisan gets stuck. The worker usually has a clear, well-defined goal to achieve. And, although at times he may be unsure about which course of action to take, the situation is never completely new, there are not many ways to proceed and he can always try to elaborate a plan. The nature of creative activities, however, is such that the artisan has only a vague idea in her mind to guide her progress. And even this vague idea is subject to change, as the working situation keeps opening new perspectives and possibilities for manipulation. In a setting where goals and ways to achieve them are not settled, materials, tools and routines also cannot be right, and you cannot expect much help from them either. Without familiar situations and standard ways to respond to them, you cannot calculate your actions. Imagined situations do not present themselves to us in the same way as actual ones. You cannot play with them. Practical knowledge just reflects the regularities of our activities, the ways we are used to respond to situations and the ways situations usually evolve with such responses. So, you also cannot expect much help from thinking. The only alternative is to keep experimenting and monitoring the results. And sometimes starting things all over again from scratch. The quality of the worker's and the artisan's awareness is also quite distinct. The worker's mind moves ahead of the activity, actively inspecting possibilities, focused on what is relevant, anticipating decisions and looking for efficient ways to accomplish the task. The artisan's mind, on the other hand, must be broadly open to the present situation. Not knowing, it should be alert to recognize signs of progress which may come from anywhere. In this sense, the creative activity of the skilled artisan is very much like the dynamics of group hunting. There is an important difference: in one case there is one global awareness, while in the other case the awareness is distributed and local. And there is an important similarity: in both situations the activity evolves in basically the same way, putting things into movement, letting them go for a while, and monitoring the results.

But, before we return to the discussion of group hunting and other social activities, it is useful to consider one last topic: the creation of tools and the construction of machines. When you play with tools, the opportunity also arises for you to investigate and inquire about how the tool does work for you. In the typical scenario, you pay attention to the moments in which things seem to go well, and try to see what is going on. In one such case, you may realize, say, that the precise way in which you grasped the tool was the key for the right movement to happen. And then you may decide to change the form of the tool so that ti becomes natural to grasp the tool in that particular way. Now you have a different tool, which is more efficient to work with. Or else it may happen that, while concentrated on your work and not particularly concerned with tools and surrounding conditions, you find yourself confronted with a difficult situation. Looking around for help, you find an object close by, grab it and use it to hold things in place. With things thus fixed, the movement then follows with ease. Now you can see what was the source of the difficulty, and you decide to keep the object to use it again in the future. These examples are not significantly different from the experience of learning how to use a tool. With repeated encounters with the difficulties of your activity, you eventually find the paths of no resistance which allow you to make progress. And then you may try to fix the conditions that lead to them, be it the objects and materials you are working with, the specific shape and features of your tools, or the abilities and skills you end up developing. Once the conditions are fixed, you can actually rely and count on those paths. If you realize, in a given instance, that you can move things in the direction of a certain condition, then you know that you will be able to push the situation a little further still, through a path of no resistance. In this way, you concatenate paths to form longer ones, and you combine conditions to articulate your steps and organize your activity. With time, the activity may get reorganized, as you realize that a difficult task which used to require great ability or a sophisticated tool, can also be accomplished through the combination of simpler movements and simpler tools. And finally you discover that you can take yourself out of the way, once you see that the role you play in the activity, with the exhertion of your force, skills and abilities, can also be performed by a clever arrangement of motors and parts. Those are but a few basic principles underlying the construction of machines.

Now it is time to summarize our findings and move on to the presentation of our main ideas. The notion of practical knowledge that has emerged from this discussion corresponds to basically anything that (systematically) guides or directs our activities through paths of no resistance. That is, anything that (systematically) contributes to move our activities toward the goals that we care about. We have seen that this kind of practical knowledge can be located anywhere, from the objects and materials, to our skills and abilities, and the tools and machines we use. We have also seen that, by focusing our attention on each of those things, in the context of the activity, we may discover the ways in which they can contribute to get us closer to the goal. And then we may try to fix appropriate conditions so that this contribution becomes systematic. This is how practical knowledge is acquired. Finally, we have seen that the several kinds of practical knowledge can replace each other. It is possible to work with inadequate materials, up to a certain extent, so long as you have sufficient skill. You may be able to accomplish a difficult task without having great abilities, if you have suitable tools at your disposal. And it may be possible to replace a sophisticated, specialized tool by a number of simpler ones, if you learn how to articulate their uses to achieve the same functionality. Now, there is the case where the activity develops with plenty of knowledge. Everything is in the right place, the steps smoothly follow one another, and the worker may even move ahead of the activity, in thought, choosing the paths to be taken before the opportunity arises to actually take them. In this situation, it may happen that the practical knowledge possessed by the worker himself is quite limited — say, if he is unable to bring things back to track in case something goes wrong. Here, the doubt could be raised whether he knows what he is doing, despite the fact that he is actually doing it. Lastly, there is the case we are particularly interested in, where the activity develops without knowledge. We have the case in which knowledge is lacking at the bottom, where you may have a plan or good idea about how to achieve the goal, but are still struggling with the objects and materials, learning how they respond to your manipulations. And we have the case in which knowledge is lacking at the top, where you are basically lost, with not much practical sense of direction, apart from some awareness of the goal you want to achieve. In both cases, the lack of knowledge does not hinder the progress of the activity, but you have to keep experimenting with the situation, paying attention to the movements that yield progress, backtracking whenever necessary, trying to find the paths of no resistance.

The similarity between group hunting and the creative work of the skilled artisan lies in the fact that both activities are carried out without knowledge at the top. And this means, we claim, that both activities are governed by the same sort of principles or mechanism: putting things into movement, letting them go for a while, monitoring the results, and moving on from there or trying to find another route to the goal. But, there is also and important difference between the two activities. The skilled artisan has a global view of the situation, which allows her to judge whether things are going well or not, and to identify opportunities to make movements which may work. In the case of group hunting, the activity must unfold through the interaction of a number of centers of awareness. No one has a global view of the situation, but they are able to perform local actions and to make local judgments. And communication allows to connect the judgment of one to the actions of the other. In this way, the group is able to emulate the mode of operation of a single person. Now, in settings where there is a single focal point of action, we already have a good idea of how practical knowledge helps to guide the activity, and how practical knowledge is acquired and modified. But, what about distributed settings like group hunting? What seems clear is that the individuals of the group are faced with a much harder cognitive task. The fact that one must operate with very limited knowledge and very limited control is quite vivid here. The activity evolves by itself, as it were, with the independent local judgments and actions of the individuals, and the changes of the environment. Hence the urge to acquire some sort of control through interaction and communication. And the first step toward control is the introduction a new layer of resistance. Imagine a group of primitive men hunting together, shouting and making all sorts of noise. With the shouts and the noise, the group effectively becomes a second source of sensorial stimuli in the environment. The attention of each individual becomes divided between the local impressions of the neighborhood and the remote stimuli coming from the partners. Their immediate impulses are thus constrained and the group is held together – no one is let loose. The second step is to learn how to break that resistance. To identify opportunities for action, and find paths of no resistance which bring the group closer to the common goal. Clearly, the situation requires the development of a whole new set of skills and abilities. But the actions are not restricted to things you do by yourself. You must learn to identify opportunities for others and call them into action too. In a sense, this is not much different from learning how to use a tool, where you have to find out the ways in which the tool can do work for you. And, of course, you must also learn to let yourself be used as a tool. The key to the success is trust. In this way, individuals can guide each other and let themselves be guided. And the entire group operates like a machine. But then, as we well know, things can go wrong. The danger may not be apparent from your point of view, and you are not aware of the fact, but others are monitoring your activity. And by shouting at you they introduce resistance to hold you back. Hence, control is acquired by playing with the new layer of resistance. Releasing and restoring it skillfully, to steer the group activity towards the goal.

Things can actually become more complex than that. Someone may shout to you "Go!", but you reply back "No!". Now you are offering resistance to the movement. But that is not blind resistance, like that of inanimate objects and materials. Something to be worked around or overcome in some way or another. You know something that the other doesn't know. You can see it from where you are, but the other cannot see. And by saying "No!" you let your judgment connect with the judgment of the other. The other, of course, can make further attempts to get you moving. And the interaction would go on till some sort of common judgment gets stabilized and the activity can proceed. At this point the group is no just acting with the mechanical coherence of a machine. Actions are taken on the basis of judgments that rely on more than the local, individual impressions of each person. It looks as if a global awareness is emerging from the interaction among the individuals and controlling the activity. But, let us not be deceived by the appearances! No one knows what is going on. In the setting of the representationalist, people can exchange descriptions from their respective points of view, and everybody thus get an approximate picture of the global situation. And deliberation may then ensue to allow the group to decide on a common course of action. In the setting of the pragmatist, people can take pre-defined roles and use codes to coordinate their actions. They may not know what is happening at a global scale, but they have a strategy which can carry them through to the common goal. But in our setting there is no knowledge of either kind. People don't know what is going on globally, and they don't know how to proceed. Indeed, they don't need to know. The activity can successfully unfold wihtout knowledge at the top. By playing with their mutual resistances, they connect their judgments and take the group behavior to another level of performance.

The key insight in the discussion above is that the activity of a group can be effectively organized without being apprehended in its totality by a single mind. No one knows everything, and no one devises a plan to be executed by the group. Now this insight can be brought back to the activity of a single person. When dealing with novel or difficult problems, it can be very hard to get your mind around the situation. And things get worse when you do not have a unified conceptual scheme to accommodate the several aspects of the issue, or a strategy to integrate information coming from unrelated sources. The only way to handle the situation is to keep your attention moving from one place to another, trying to find opportunities to make progress. Indeed, here and there you take some steps that seem to work — you find paths of no resistance. But no isolated part will hold the key to unlock the entire problem. You may see a pattern in one place which gives you the idea to try a move in another place. Things may work out right allowing you to put some pieces together. And if they don't, you may take this information back to analyze the original situation again. Thus involved with the situation, you feel the tension of the conflicting facets of the problem and attempt to connect your judgments. After a while, with some luck and a good amount of persistence, you eventually find your way out of the puzzle. In this way, a person can use her divided mind to solve a difficult problem without really knowing how to do it in the beginning. And by doing so a single person emulates the mode of operation of a group.

Next we will briefly sketch a model of how cognition can work without knowledge at the top. The first observation is that, although there is no knowledge, there must be something at the top to stabilize and give coherence to the activity of the group, as the common goal alone may not be sufficient in the most difficult tasks. Moreover, we make no assumption about what is going on at the bottom, where there may be plenty of knowledge — informational, representational, pragmatic, etc. To present the model, we will give a schematic description of our agents, and then provide an account of what might be happening at the top. We assume that the agent has a goal in mind, and regulate his activity based on judgments regarding opportunities for action and the quality of their results. The agent has a limited awareness of the environment around him, but his attention can be directed to different aspects of the situation and the local impressions he obtains can affect his judgments.

In the context of the group, the activity of the agent is also regulated by the interactions with the others. Here we assume that the agents can make judgments regarding opportunities of action for others. The communication, however, is restricted to messages that can only induce or refrain the other from acting (e.g., "Go!", "No!"). That is, the agents don't tell each other what to do, or give indications of why they are making the call. Each agent has to realize that by himself, and respond accordingly or not. By monitoring the activity of the other, the agents obtain the necessary feedback to make the regulation effective. But we also consider another form of feedback, whereby an agent can explicitly refuse to accept the call of the other (e.g., by saying "No!"). Here, presumably, the agent has strong confidence on his own judgment. And, by making it plainly clear, he is effectively attempting to regulate the judgments of the other. Due to the extreme simplicity of the messages, we call this the 1-bit model of communication.

Now, the effect of the interactions among the agents on the overall activity of the group can be interpreted in two ways. First, if we return to the example of the primitive hunting men, where everybody hears everybody else, the shouts and noise produced by each individual contribute to maintain a global state which is shared by all components of the group. Here, the conditions are pretty much the same as in the 1-bit model of communication described above, so we assume that the global state has no propositional content. Similarly to the goal, the function of the shared global state is to provide a common reference point which can help to give cohesion to the actions of the individuals. On the other hand, a difference can also be pointed out with respect to the goal, as the global state can be updated and will typically change during the development of the action, although at a different time scale when compared to the immediate events of the scene. Second, if we now focus on

the peer-to-peer interactions, where one agent regulates the action or judgment of another agent, we note that paths of causal influence will tend to form, organizing the dynamics of the group activity. The connections that establish the paths are basically the responses of the agents to the local impressions from the environment and the remote stimuli from the partners. As the paths can span several individuals, they can handle aspects or regularities of the situation which are out of the reach of a single individual — be it in time, space or complexity. This is the key for the solution of difficult problems. The agents, of course, are not aware of the coherences that arise in their activity, and they do not control them. They are also unaware of the distributed process that constructs and maintains the shared global state, and they do not control it either. There is no knowledge at the top. The group is controlled by dynamical structures which emerge at the time of the activity.

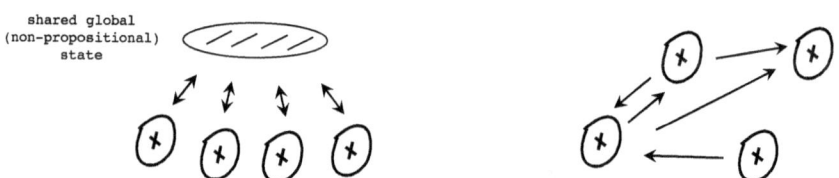

Next, making a slight change of perspective, we can interpret the 1-bit model of communication in mechanistic terms, rather than informationally. Since each message basically represents a stimulus or an inhibition to action (or a refusal to accept the call), the communication can be interpreted as a mechanical interaction, with the agents pushing and pulling each other. And, since at the topmost level the interaction is mechanical, it is natural to interpret the activity of the group as the operation of a machine. Taking the analogy a little further, the shared global state could be seen as the frame of the machine (or a higher level structure), where the agents can get a support for their actions, and pushes and pulls. On the other hand, the paths of causal influence that arise during the activity are clearly the paths of no resistance through which the machine operates. And it is at this point that the analogy breaks down. In a typical machine, the paths of no resistance are already fixed by the rigidity of the parts and structural connections. In the case of the group activity, however, the behavior of the agents is not fixed, and the connections that give rise to the paths depend on the judgment of the agents, which reflect the development of the action. So, in the dynamics of the group, the paths of no resistance are assembled and disassembled in real time, in the effort to move the activity closer to the goal. This is not quite how a typical machine works. In any case, the machine-like behavior that emerges and organizes the group explains how the activity can successfully develop without knowledge at the top.

The conclusion we want to draw from this discussion is that the function of the language is not just that of an instrument which allows us to codify and deploy knowledge – be it the theoretical 'knowing that' of the representationalist, or the practical 'knowing how' of the pragmatist. Language, we claim, also has the important function of supporting intelligent behavior in situations where knowledge is not yet available. The two paradigmatic cases being novel situations and situations with arbitrarily high complexity. Both cases are characterized by a fragmentary and initially incoherent apprehension of aspects of the situation from the part of the agent or agents. Indeed, this is inevitable if we take into account the several limitations of human cognition: the essentially local character of perception and action, the limited capacity of holding a number of objects simultaneously in awareness, the relatively short span of attention, etc. The picture becomes most clear when we analyze the unorganized activity of a group of people trying to solve a task which is beyond the capacity of any single individual. As we have seen, the key to success then is the possibility of connecting the judgments and actions of the distributed agents in a dynamics of mutual regulation. And in principle the same strategy can be applied by a single person to connect her own judgments and actions in the attempt to solve a difficult problem. The image we used to illustrate this idea was that of a machine that is put together at the time of operation. In this sense, the role of language is to support the causal relations that allow the emergence of a higher level dynamical structure which organizes the activity of the group. Now, the natural objection that could be raised at this point is that we are working with a very crude notion of language, which some would in fact not consider to be language at all. But then we may recall that, as the paths of no resistance that structure the activity are found, and the participants of the group begin to notice them, they will try to fix the conditions that lead to those paths. In this case, the conditions may take the form of primitive language constructs to be used in specific circumstances. Once fixed, the constructs would correspond to schematic causal relations that allow the machine to be assembled much more efficiently, as the agents would know some moves that usually work. And the schematic relations also enable the construction of more complex machines. If a new member joins the group at this point, he will have to learn how to use the language constructs properly in order to effectively colaborate with the others. Hence, despite being crude, the notion of language we are working with can provide a basis on top of which knowledge can be acquired and fixed. And it makes sense to have the same instrument supporting the discovery, codification and deployment of knowledge. Of course, much work has still to be done to ellucidate the processes underlying the building up of knowledge, and the appearance of representation and inference, on this model. But we will leave that for another opportunity. Here we just hope to have given the general ideas for an account of language that fits our everyday experiences of dealing with novel situations and solving difficult problems without previous knowledge, and being creative.

On the Limits of Language, Thought and Experience

CANDIDA DE SOUSA MELO AND DANIEL VANDERVEKEN

1 Introduction

An important task of philosophy of mind is to determine the nature and limits of human thoughts and experiences. Many philosophers throughout history, including Plato, Aristotle, René Descartes, Gottfried Leibniz, Immanuel Kant, Ludwig Wittgenstein, Gilbert Ryle, Paul Grice and P.F. Strawson were concerned with this task whose achievement continues to be a challenge. A linguistic turn, however, occurred with the first Wittgenstein who attempted in his *Tractatus logico-philosophicus* [28] to determine the boundaries of thought and of experience from those of language. There are new results in speech act theory which allow us to develop Wittgenstein's perspective and better explain the nature of limits imposed by language on thought and experience.

According to the founders of speech act theory semantic properties of linguistic symbols come from the intentionality of speakers who use them. As J.R. Searle pointed out in *Intentionality* [10], philosophy of mind underlies for that reason philosophy of language. Because the units of meaning and communication are *intrinsically intentional* illocutionary acts, in order to mean and communicate human agents must make attempts to perform illocutions. These attempts are conscious and voluntary speech acts[1] which are constitutive of meaningful utterances. According to our philosophy of mind, human agents have by nature *consciousness* which gives them an identity, *intentionality* which directs them towards objects and facts of the world and enables them to think and act, and *rationality* which gives them theoretical and practical reasons for their attitudes and actions. These three mental properties are constitutive of our mental and linguistic competence.

By adopting the principles that intentionality underlies language use

---

Copyright © 2019 by Candida de Sousa Melo and Daniel Vanderveken. All rights reserved.

[1] Our ideas are based on D. Vanderveken's logic of action [20, 24].

and that the basic units of meaning in the use of language are entire illocutionary acts with felicity conditions rather than isolated propositions with truth conditions, one can show that all our conceptual thoughts are provided with a proper direction of fit between the mind and the world[2]. Austin and Searle distinguished in their discourse typology different types of speech acts like acts of utterance, reference and predication, acts of expression of senses and attitudes, acts of presupposition and illocutionary and perlocutionary acts. As we pointed out in [14, 15, 20, 24] pointed out, *attempts to perform illocutionary acts* are speech acts of a new type whose importance is capital in the theory of meaning and understanding. For speakers mean and intend to communicate, whenever they make these attempts, no matter whether they succeed or fail. We have analyzed the felicity conditions of these constitutive acts of meaning in the expression and communication of conceptual thoughts. We will now use our conceptual apparatus in order to explicate the forms of expression of conceptual thoughts and draw consequences of the principle of expressibility linked to felicity conditions of elementary illocutionary acts.

2 On the nature of conceptual thoughts

In [10] Searle emphasized that both elementary *states* and *acts* of *conceptual thoughts* have a similar formal structure and are logically related. On one hand, *propositional attitudes* like beliefs, desires and intentions are states of conceptual thought of the form $M(P)$, where M is a psychological *mode* and P a propositional content. On the other hand, elementary *illocutionary acts* like assertions, commitments and directives are acts of conceptual thoughts of the form $F(P)$, where F is an illocutionary *force* and P a propositional content. As Searle pointed out, our conscious beliefs and desires are in a network of many unconscious attitudes. Unlike attitudes, our *acts of conceptual thought* are always *conscious*. We link consciously and voluntarily their propositional content to the world with an *illocutionary force* that is always *mental* and linguistic in the case of verbal utterances.

Searle and Vanderveken follow Frege's theory of *indirect reference*: human agents can only *indirectly refer* to objects of the world by subsuming them under senses. Like Frege in[5] they claim that there is no reference without predication. Whoever refers to objects predicates of them attributes and determines truth conditions of propositions. In performing elementary illocutionary acts, speakers express propositional contents, in

---

[2]For a more complete account of our principles see Searle [10] and de Sousa Melo [12].

having propositional attitudes they mentally apprehend them. Furthermore like Frege in [6] Searle and Vanderveken advocate that references, senses and forces are all components of sentence meaning. Agents of meaningful utterances of elementary sentences always relate to the world the propositional contents that they express with forces[3].

In *Foundations of Illocutionary Logic* [11], Searle and Vanderveken analyzed the felicity conditions of first level illocutionary acts. Unlike Frege and Austin who just gave a list of important forces, they decomposed illocutionary *forces* into their components (*illocutionary point*, *mode of achievement* of illocutionary point, *degree of strength* and *propositional content*, *preparatory* and *sincerity conditions*). They also formulated a *recursive definition of the set of all possible forces*. The *primitive forces* are the five simplest forces with an illocutionary point: the force of *assertion* expressed by the declarative sentential type; the force of a *commitment to an action* named by the performative verb "commit"; the force of a *linguistic attempt to get the hearer to act* expressed by the imperative sentential type; the force of *declaration* expressed in performative utterances, and the force of *expression* of an attitude realized in the exclamatory sentential type. Other complex forces are obtained by adding to primitive forces special modes of achievement of illocutionary point, propositional content, preparatory or sincerity conditions, or by increasing or decreasing the degree of strength.

According to them the felicity conditions of elementary illocutions are entirely determined by the components of their force and propositional content. Thus in order to *succeed in performing an illocutionary act* of the form $F(P)$ in a context of utterance, the speaker must *express* a proposition P which satisfies the *propositional content conditions* of the force F in that context and *achieve* the *illocutionary point* of force F on that proposition P with the proper *mode of achievement* of that force; he or she must *presuppose* all propositions determined by the *preparatory conditions* of F with respect to P in the context; and *express* with the *degree of strength* of F all attitudes of the form $M(P)$ whose modes M belong to the *sincerity conditions* of F. For example, in order to make a promise, a speaker must express a proposition which represents his or her action in the present or the future (propositional content conditions) and put himself or herself under the obligation to do that action (mode of achievement of the commissive point of a promise), that agent must presuppose that he or she can do that action and that it is good for the hearer (preparatory conditions) and express a strong intention to do it

---

[3]The term of force is due to Frege [6]. J.L. Austin [1] was much inspired by Frege's philosophy.

(sincerity conditions).

As Searle and Vanderveken pointed out, the *satisfaction* of elementary illocutionary acts requires the *truth* of their propositional content. In traditional philosophy *truth* is based on *correspondence* since Aristotle: a proposition is *true* at a moment when it *corresponds* to a fact which exists then in the world. The words used and the ideas expressed by the agent of a satisfied elementary illocution must then correspond to the objects of reference as they are in the world. However we live in an indeterminist world. Each moment of time has a unique past but many different possible historic continuations. A future proposition can be true according to a possible historic continuation of a moment of utterance and false according to another possible historic continuation of that moment. Like W. of Ockham [9] we consider that future propositions are true at a moment when they represents how things will be in *the real historic continuation* of that moment.

Now the satisfaction of many elementary illocutionary acts requires more than correspondence. For their satisfaction requires that the *correspondence* between words and ideas of the agent, on one hand, and objects of the world, on the other hand, *be established according to the* appropriate *direction of fit* of their force. The four different possible directions of fit between mind and reality underlie the four directions of fit between language and reality[4]. Like Descartes we distinguish cognitive and volitive propositional attitudes. *Cognitive attitudes* are attitudes like belief, conviction, faith, confidence, knowledge, certainty, pride, vanity, surprise, amazement, prevision and expectation which contain a belief in the truth of their propositional content. *Volitive attitudes* are attitudes like desire, wish, will, intention, project, hope, satisfaction, pleasure, enjoyment, delight, gladness, joy and amusement which contain a desire of existence of a represented fact.

Firstly, *cognitive attitudes* and *assertive illocutionary acts* have the *mind-to-world* (or *ideas-to-things*) *direction of fit*. Our mind must then think objects as they really are or will turn to be in the world in order to form a satisfied conceptual thought. Cognitive attitudes and assertive illocutions are satisfied when they are *true*. Their satisfaction only requires that our ideas correspond to things in the world. Assertive illocutionary acts whose agent uses words have the *language-to-world* (or *words-to-things*) *direction of fit*. Their speaker must use words and express ideas corresponding to things.

Secondly, *volitive attitudes* and *commissive and directive illocutions* have the opposite *world-to-mind* (or *things-to-ideas*) *direction of fit*. In

---
[4]See de Sousa Melo [12].

order that these thoughts be satisfied, the objects of the world towards which our mind is directed must be or turn to be as we represent them. Objects must then correspond to the agent's ideas. Commissive and directive illocutionary acts whose agent uses words have the *world-to-language* (or *things-to-words*) *direction of fit*. Their speaker or hearer must change things of the world so that they come to correspond to ideas expressed by used words[5].

Other predicates than the truth predicate apply to satisfied conceptual thoughts with the *world-to-mind* direction of fit. Our desires are *satisfied* when they are *realized* and our intentions when they are *executed*. Our commitments and promises are *satisfied* when we *keep* them and our directives when they are *obeyed*.

Each direction of fit between mind and the world determines which side is at fault in case of insatisfaction. When a belief or a judgment turns out to be false, the agent is at fault, not the world. He or she should have thought differently about objects. In that case, the agent easily corrects the situation in changing his or her ideas. However, when a desire, a commitment or a directive turns out to be unsatisfied, it is not the agent but the world which is at fault. Other facts should have occurred. The agent sometimes corrects the situation in abandoning or changing his or her desire, commitment or directive. Otherwise that agent is unsatisfied and can keep a regret. In illocutionary logic all acts of conceptual thought with the *world-to-mind* direction of fit have *self-referential* conditions of satisfaction. One *keeps* a commitment if and only if one carries out the action to which one is committed in order to keep that commitment. Someone who has forgotten a past promise does not keep that promise when he or she does the promised action. Similarly, in order to obey a directive one must realize the directed action in order to obey that directive. When one does a commanded action without having the least intention to follow the will of a commanding agent, one does not then obey his command. The satisfaction of commissive and directive acts requires their successful performance. There are a lot of true assertions that have not yet been made. However all satisfied commissive and directive illocutions have been made. For they are always a *practical reason* why the action represented by their propositional content has been carried out.

Thirdly, all declaratory illocutions have the *double direction of fit between mind and world* (or *ideas and things*). Any agent of a successful declaration changes the world by way of doing the action that he or she

---

[5]Commissive and directive illocutions assign respectively to the speaker and to the hearer the role of transforming the world in order to fit the propositional content.

represents by virtue of his or her declaration. At that moment, the agent changes represented things of the world so that they correspond to his or her ideas. Successful declarations are always satisfied. Verbal declarations have the *double direction of fit between language and world* (or *words and things*). When we make a verbal declaration like a gift, we do things with our words (e.g. we give an object that we name just by way of saying that we give it). All performative utterances are declarations by the speaker that he or she performs the illocution named by their main verb by virtue of their utterance. Not all declarations are verbal. As de Sousa Melo [12] pointed out, there are purely mental declarations. We can discover and define mentally a new concept without saying a word just by way of thinking. There are no states of conceptual thoughts with the double direction of fit. Only acts of thought can change objects of reality.

All *expressive illocutionary acts* (like thanks, apologies, complains) have the *empty direction of fit between mind and world* and *between language and world*. Their agents just want to express their propositional attitudes about the ways in which things are in the world. They do not want to establish a correspondence even when they express a belief or a desire. For they *presuppose* then that things are in the world as they represent them. These illocutions have the special preparatory condition that their propositional content is true. So are attitudes like *satisfaction, contentment, joy, gladness, gratitude, surprise* and *pride*. Their agents do not want to represent how things are nor to have them changed. Expressive illocutions and attitudes with the empty direction of fit have no proper conditions of satisfaction. They are just *appropriate* or not. They are *inappropriate* when the represented fact does not exist or when their psychological mode does not fit that fact. No one should thank and be grateful to someone who has not done anything or done a bad thing.

In recent years Vanderveken has developed in [21, 22, 26] a general logic of all cognitive and volitive attitudes that accounts for the fact that we, human agents, are neither logically omniscient nor perfectly rational. We ignore many necessarily true propositions and are sometimes inconsistent. Vanderveken decomposed psychological modes into their various components (their basic psychological *category of cognition* or *volition*, their special *cognitive or volitive way*, their special *propositional content* and *preparatory conditions*). He also formulated a *recursive definition of the set of all possible psychological modes*. The two primitive psychological modes are the modes of *belief* and *desire* which are the simplest modes. All other more complex psychological modes are obtained by adding special cognitive or volitive ways or new propositional content and preparatory conditions to the primitive modes. The possession and satisfaction conditions of propositional attitudes are entirely determined by the components of their mode and their propositional content. Thus an agent *possesses a cognitive (or volitive) attitude of the form $M(P)$ at a moment* when that agent then *believes* (or *desires*) the propositional content P, he or she feels that belief (or desire) that P *in the cognitive (or volitive way)* proper to psychological mode M, the *proposition P then satisfies propositional content conditions* of mode M, and finally he or she *then presupposes* and *believes all* propositions determined by the *preparatory conditions* of M with respect to the content P. For example, an agent has an *intention that P* at a moment when proposition P represents a present or future action of that agent at that moment (propositional content conditions of an intention), that agent desires so much that action that he or she is disposed to do it (special volitive way of intention) and moreover he or she then presupposes and believes to be able to do the action (special preparatory condition). Whoever has an unconscious desire that P has unconsciously in mind some attributes or concepts of P. But he or she could express them thanks to his language and determine how he or she would prefer things to be in the world.

Like all commissive and directive acts, there are a lot of *volitive attitudes* such as *will, intentions, projects, pretensions* and *ambitions* whose satisfaction requires more than correspondence. These attitudes have a psychological mode with a special volitive way that requires that things fit the agent's ideas because the agent wants them in that way. They also have the special preparatory condition that the agent has *means* in order to realize his or her desire. Such volitive attitudes have *self-referential satisfaction conditions* like the illocutions (orders, commands, pledges and promises) which express them. Their satisfaction requires that the represented fact turns to be existent in the world in order to satisfy their agent. So in order to execute a prior intention, one must do more than realize the intended action in the real future; one must realize that action because of that previous intention. If one does not act for that reason, (if one has forgotten the intention or does not act freely), one does not then execute the prior intention. Searle and Vanderveken explain in their philosophy of mind self-referential satisfaction conditions of volitive attitudes and commissive and directive illocutions by relying on *intentional causation*. The agent's attitude or action is then a *practical reason* why the represented fact turns to be existent. Incidentally, there are volitive attitudes whose satisfaction only requires correspondence. Agents having a wish, a hope or an expectance desire the existence of the fact represented by the propositional content no matter how that fact turns to be existent in the world. Such volitive attitudes of agents at a moment are or will be satisfied if and only if their content is or turns to be true in the historic continuation of that moment, no matter for which reason. Things correspond or turn to correspond then to ideas of the desiring agent, no matter what caused their existence. Wishes are volitive attitudes whose satisfaction generally depends on the course of nature or on the good will of someone else. Speakers who make a request just express a wish that the hearer will grant their request.

According to illocutionary logic, elementary acts and states of conceptual thoughts are logically linked. First of all, illocutions are *intrinsically intentional* acts. Agents of illocutions always *desire* and *intend* to perform their illocutions. Moreover they *express* propositional attitudes of certain modes in performing elementary illocutionary acts. They express beliefs in making judgments and assertions, intentions in making attempts and commitments, desires in giving directives, and regrets by making apologies. This is why illocutions have *sincerity conditions*. Sincere speakers have the attitudes that they express.

Furthermore, Searle and Vanderveken advocate a general *principle of expressibility of conceptual thoughts*. Any agent of a conceptual thought directed towards a fact can in principle express linguistically that thought in the attempted performance of an illocutionary act. The principle of expressibility holds for both states and acts of conceptual thoughts. Thus every propositional attitude (whether conscious or unconscious) and every act of conceptual thought (even a purely mental act) can in principle be expressed by means of language in attempting to perform a public illocutionary act. They follow Wittgenstein's linguistic turn in the *Tractatus* [28] : "Everything that can be thought can be expressed. Everything that can be expressed can be expressed clearly" (§ 4.116). In their view any agent of a conceptual thought can determine under which conditions his or her thought is satisfied. Otherwise, that conceptual thought would not be *well determined*[6]. So it would not be a real thought with a content[7]. Thus any agent of a belief or of a purely mental judgment can determine how things are in the world in case his or her thought is true. Whoever has, forms or expresses a desire knows what must happen in the world in order that his or her desire be realized. Whoever has, forms or expresses an intention knows how to execute that intention. Our natural languages have limited expressive powers. We discover and express new concepts, when we define new theoretical notions in science. In that case we must use new words or give a new meaning to existing words in order to express our new conceptual thoughts. As one can expect, satisfied elementary illocutions have satisfied propositional attitudes corresponding

---

[6]Wittgenstein already pointed out in 1914-16 that meaning is well determined. « It seems clear that what we mean must always be sharp ». See page 68 of [27].

[7]In our perspective, the principle of expressibility is analytical. See [13],

to their point. Whoever makes a true assertion expresses a true belief, whoever keeps a promise executes his or her intention. Whoever grants a request satisfies the wish expressed by the agent of that request.

2.1 On the conditions of possibility of conceptual thoughts

According to standard philosophy of language we cannot speak and make a meaningful use of a natural or formal language without thinking. We can of course think without speaking publicly. However, when we think we are in principle able to express our conceptual thoughts by means of *our language*. In Wittgenstein's terminology, each agent has *his or her language* which determines the set of well-formed sentences that he or she is able to use. In the *Tractatus*, "my language" is the "language that I understand" (§ 5.62). My language whose lexicon is restricted has of course contingent limits that I can transcend in learning the meaning of new words that I still do not understand.

However as the first Wittgenstein pointed out, our languages have necessary limits that logic imposes because the deep structure of all possible human languages is logical. Logic is *a priori* and *transcendental* (§ 6.13)[8]. Language is not only the vehicle of our conceptual thoughts directed towards objects and facts of the world. It is also the structure without which there would be no thought. As Wittgenstein said "... What we cannot think, we cannot think of; we cannot therefore say more what we cannot think" (§ 5.61). Moreover language is the indispensable means of *representing* facts of our experience. For we can represent any fact that we experience by having or forming an expressible thought directed towards that fact. So logic dominates both our thoughts (§ 5.4731) and our experiences (§ 6.13). The laws of illocutionary logic are therefore *a priori* and transcendental in the sense of Kant's *Critique* [8][9]. The theory of success of illocutionary logic fixes limits to the meaningful use of language, which restricts our possible thoughts, just as its theory of satisfaction fixes limits to the world, which restrict our possible experiences. By stating valid laws of felicity illocutionary logic shows *a priori* forms of human conceptual thoughts and experiences. Searle and Vanderveken revise Wittgenstein's linguistic turn.

Given our principle of expressibility, all linguistically inexpressible conceptual thoughts are by hypothesis *impossible* thoughts. As Wittgenstein pointed out, possible thoughts can be expressed clearly by using a well formed sentence. However, according to speech act theory, not all expressible thoughts are possible, because there are a lot of paradoxical *illocutionarily inconsistent sentences* that no one can never use literally

---

[8]See E. Stenius [16] on that matter.
[9]On the idea that illocutionary logic is transcendental, see the conclusions of [17].

with success. So are well-formed sentences like "The present statement is false", "I promise to do this and I do not intend to do it", "I request you to go but you cannot go" and "I order you to have done this last weekend". They express *non-performable illocutionary acts* that we cannot succeed in performing. Some illocutionarily inconsistent sentences correspond to well-known philosophical paradoxes. Sentences "Disobey the present directive!" and "I will violate the present promise" express the liar's paradox. Sentences "Every assertion is true" and "I promise to do everything" correspond to the dogmatist's paradox and "All assertions are false" and "Disobey all directives!" to the sceptic's paradox. Sentences "I promise to come and I am unable to come" and "Please come and I do not wish you to come" express Moore's paradox. A first level illocutionary act is *performable* when it can be performed by an agent in at least one possible context of utterance. Some illocutions are performable only by certain agents. Thus commitments must be personal. Only I can personally promise that I will do something. No one else can. Some illocutions are performable only at certain moments. I can promise today to help you tomorrow. The day after tomorrow I will be unable to make again that promise. *Non-performable illocutions* have impossible conditions of success. No agent can promise to act in the past. One can only commit oneself to a present or future action at a moment of time. This is a basic law governing the commissive point.

In order to well determine the limits of thought from those of language, one must take into account not only the logical form of sentences but also the success conditions of illocutionary acts expressed by using them. In addition to impossible conceptual thoughts that one could not express by using a well-formed sentence of a natural language, there are non-performable illocutions expressed by well-formed illocutionarily inconsistent sentences of English. The novelty of speech act theory with respect to Wittgenstein lies in the advocated intermediary role of illocutionary acts. We must in principle be able to express *our possible thoughts* by means of a meaningful use of *our language* in *attempting* to perform illocutions expressing them. This is how the limits of thought show themselves within the limits of language according to us. Now what happens when speakers use illocutionarily inconsistent sentences? Either they do not really understand what they say (what is the expressed literal illocutionary act) or they mean something else (they do not attempt then to perform the literal illocutionary act but another non literal illocution[10]).

---

[10] An utterance is *literal* when one only means to perform primarily the illocution expressed by the sentence that he or she uses in the context of utterance. As Austin

We are *imperfectly rational* but we always remain minimally rational[11]. Our desires and intentions are detachable. We sometimes intend to carry out certain actions and believe that these actions will have certain bad effects without having the intention to produce these effects. However we coordinate minimally our cognitive and volitive attitudes. Whenever we intend to do an action, whether verbal or not, we always at least believe that we can do it. Otherwise, we would not really intend and attempt to do it[12]. There are of course impossible actions that we sometimes wrongly believe to be able to do. However we never attempt to perform non-performable illocutionary acts. For in the particular case of illocutions we know *a priori* by virtue of our linguistic competence whether they are performable or not just by understanding their felicity conditions. So we never intend nor attempt to perform in the first person non-performable illocutions. Unlike many other intentional actions, illocutionary acts are performable if and only if one can attempt to perform them. The limits that logic imposes on the successful performance of illocutions are those that it imposes on attempts to perform them and on acts of meaning.

2.2 On the conditions of possibility of human experience

One general principle of philosophy of mind is that the human mind is directed towards objects and facts of the world. As F. Brentano [2] and E. Husserl [7] pointed out, *intentionality* is a constitutive property of our mind by virtue of which we are directed towards reality. All our states and acts of conceptual thoughts have intentionality as an essential feature. Moreover any fact of *lived experience* of an agent can be represented by a conceptual thought of that agent whose propositional content is *true*. We often believe in the existence of facts that we experience and can describe them. We also feel volitive attitudes directed at them. We are satisfied or insatisfied with their existence. Every fact that an agent could experience can therefore be represented by a possible *satisfiable conceptual thought* of that agent. A *conceptual thought* is *satisfiable* when it is satisfied at least at one moment of time of a possible course of history of the world. For example, my prevision and prediction that Quebec will be independent are satisfiable; they are true according to at least one possible historic continuation of this moment. In our

---

[1] pointed out, many utterances are non literal and some non serious. On non literal meaning and meaning in discourse see [19, 23].

[11] The notion of minimal rationality comes from C. Cherniak [3]. Vanderveken's logic of language, action and attitudes formulates the laws of minimal rationality. See [17, 24, 26].

[12] Unlike intentions attempts are not states but actions that we make in order to achieve an objective.

terminology, *truth-conditionally consistent sentences* express satisfiable illocutionary acts. Their propositional content is logically *possible*: it is true in at least one possible circumstance.

3 On the limits of conceptual thoughts and human experience

Searle and Vanderveken [11] stated basic laws governing the success and satisfaction conditions of first level illocutionary acts. Vanderveken [18] made a generally complete and sound axiomatization of the logic of elementary illocutionary acts in Volume 2 of *Meaning and Speech Acts*. Using the laws of illocutionary logic we will now present the respective limits imposed by language on thought and experience.

3.1 On the impossibility of certain conceptual thoughts

Given our definitions of success of illocutions and of possession of attitudes one can establish the following classification of impossible conceptual thoughts.

First, certain elementary conceptual thoughts are impossible just because their complex force or mode is itself impossible. Impossible forces and impossible psychological modes are obtained from primitive forces and modes by adding incompatible components. Thus illocutions whose force has two incompatible modes of achievement of their illocutionary point or two incompatible psychological modes in their sincerity conditions are non-performable. For no one can achieve an illocutionary point with an impossible mode of achievement and no one can have nor express incompatible attitudes. Similarly illocutions whose force has incompatible sincerity and preparatory conditions are non-performable. For no one can simultaneously presuppose propositions and express attitudes that enter respectively in their preparatory and sincerity conditions. This is why the imperative sentence "Please, whether you like it or not, come!" the exclamatory sentence "How glad and sad I am that my friend is dead!" and the declaratory sentence "Alas and fortunately Trump has won the last American elections!" are illocutionarily inconsistent. Along the same lines, there are a lot of complex impossible psychological modes which contain incompatible cognitive or volitive ways and preparatory or propositional content conditions. No agent can possess propositional attitudes with such modes.

The second kind of impossible conceptual thoughts are those whose force or mode is incompatible with their propositional content: *Conceptual thoughts whose force or mode cannot be applied on their propositional content are impossible*. So are thoughts like predictions and previsions whose propositional content is past. Their propositional content cannot satisfy the propositional content conditions of their force or mode because one can only predict and foresee propositions which are future at the moment of thought.

The third kind of impossible conceptual thoughts are thoughts with a non-empty direction of fit whose propositional content is an obvious contradiction. In predicative propositional logic an *obvious contradiction* is a necessarily false proposition that agents know *a priori* to be false by virtue of their linguistic competence. Unlike traditional modal and intensional logics, predicative logic does not identify all propositions with the same truth conditions. It analyzes the structure of constituents of propositions and takes into account the acts of predication that we make in expressing them and in understanding their truth conditions. Each proposition has a finite *structure of propositional constituents*. It predicates *properties* or *relations* of *objects subsumed under concepts*. In our view to understand a proposition is to understand which attributes objects must possess in a possible circumstance in order that this proposition be true in that circumstance. Our illocutionary acts and attitudes are directed towards *objects under a concept* rather than towards pure objects. By recognizing the indispensable role of concepts in reference, predicative logic explains referential opacity and accounts for thoughts directed towards inexistent objects. Predicative logic also proposes a *better explication of truth-conditions*. We understand most propositions without knowing in which possible circumstances they are true, because we ignore *real denotations* of their attributes and concepts in many circumstances. One can refer to a friend's mother without knowing who she is. However we can at least think of persons who could be that mother. So in any possible use and interpretation of language, there are according to illocutionary logic many *possible denotation assignments to attributes and concepts* in addition to the standard *real denotation assignment* that associates with each propositional constituent its *actual denotation* in every possible circumstance[13].

In predicative logic[14] truth is relative to both possible circumstances and possible denotation assignments. An elementary proposition predicating an extensional property of an object under a concept is true in a circumstance according to a possible denotation assignment when ac-

---

[13]*Possible* denotation assignments are functions associating with each concept a unique object or no object at all in every possible circumstance and with each property and circumstance a set of objects under concepts. According to the real denotation assignment, my friend's mother is his real mother. According to other possible denotation assignments, his mother can be another person. All possible denotation assignments respect *meaning postulates* that we internalize in learning language.

[14]For more information on predicative logic see [25].

cording to that assignment the object which falls under that concept has that property in that circumstance. In understanding most propositions we do not know whether they are true or false. We just know that their truth in a circumstance is compatible with certain possible denotation assignments to their constituents. Of course, in order to be *true in a circumstance* a proposition has to be *true in that circumstance according to the real denotation assignment*. Now, in predicative logic *identical* propositions must have the same attributes and concepts and be true in the same circumstances according to the same possible denotation assignments. Our finer criterion of propositional identity explains why many propositions with the same truth conditions have a different cognitive or volitive value. One can have in mind a proposition without having in mind another logically equivalent proposition whose attributes or concepts are different. At the time of his wedding with Jocasta, Oedipus did not want to marry his mother. He wanted to marry the queen of Thebes. Predicative logic distinguishes logically equivalent propositions that we do not understand to be equivalent: they are not true according to the same possible denotation assignments to their senses. Few necessarily true propositions are *obvious tautologies* that we know *a priori*. A proposition is *necessarily true* (or *necessarily false*) when it is true (or false) in every possible circumstance according to *the real denotation assignment*. In order to be *obviously tautological* (or *obviously contradictory*), a proposition must be true (or false) in every possible circumstance according to *every possible* denotation assignment to its constituent senses. Unlike the proposition that Oedipus' mother is a woman, the necessarily true proposition that she is Jocasta is not an obvious tautology.

We often believe and assert necessarily false propositions representing objectively impossible facts. Unlike traditional logic, predicative logic explicates subjective and objective impossibilities. A proposition is *objectively impossible* when it is false in all possible circumstances according to *the real denotation assignment*. In order to be *subjectively impossible* it must be false in all circumstances according to all possible denotation assignments. Unlike many necessarily false propositions, for example, that whales are fishes, only obvious contradictions, like the proposition that whales are not animals, represent facts whose existence is impossible both subjectively and objectively.

The impossibility of the third kind of conceptual thoughts is due to the fact that their non-empty direction of fit is incompatible with their analytic unsatisfiability. Because we, human agents, are minimally rational, we can only have beliefs or desires directed towards facts whose existence is possible according to us. As we pointed out, because of our minimal rationality, we never intend to perform non-performable illocutionary acts for we know that they are non-performable. This is also why we never intend to perform illocutionary acts with a non-empty direction of fit that are analytically unsatisfiable. For we know in that case that there is no possible correspondence between our ideas or words, on one hand, and objects of the world, on the other hand. And we could not then intend to mislead the hearer who is also aware of their unsatisfiability. Illocutionary logic states a valid *law of minimal consistency of human speakers* who cannot even intend to achieve the assertive, commissive, directive and declaratory illocutionary points on an obviously contradictory propositional content. Sentences of the form $f(p)$, whose marker f expresses a force with a non-empty direction of fit and whose clause p expresses an obvious contradiction are both illocutionarily and truth-conditionally inconsistent. So are sentences like "You will walk without moving", "Walk without moving!" and "I order you to walk without moving!"

3.2 On the unsatisfiability of certain conceptual thoughts

When illocutionary logic states valid laws of satisfaction for illocutions, it fixes limits to reality and to our experience. By hypothesis we could never experience impossible facts represented by unsatisfiable conceptual thoughts with a necessarily false propositional content. Moreover, as we pointed out, we can in principle represent facts that we experience by forming a *satisfiable conceptual thought* expressible in our language. Certain kinds of unsatisfiable and impossible conceptual thoughts are therefore logically related. Many impossible conceptual thoughts are unsatisfiable. So are conceptual thoughts with a non-empty direction of fit whose content is an obvious contradiction. Impossible thoughts with the world-to-mind or the world-to-language directions of fit are unsatisfiable because of their self-referential conditions of satisfaction. Only possible intentions can be executed and possible directives can be obeyed. Furthermore all unsatisfiable illocutions with the double direction of fit are non-performable, because their success requires their truth.

4 Conclusion

As we pointed out, the two sets of possible and of satisfiable conceptual thoughts do not coincide. We, human agents, are imperfectly rational. We are often inconsistent and we do not make all logically valid inferences. We have necessarily false beliefs and unrealizable desires. We also successfully use truth-conditionally inconsistent sentences not only in ordinary conversation but even in scientific discourse. Before Gödel we thought that all valid laws of arithmetic are provable. We moreover believe and assert propositions without believing and asserting many logical consequences of these propositions. The limits that the logic of language imposes on our possible experiences are different from those that it imposes on our possible thoughts. Some limits imposed to thought do not limit experience. Many impossible thoughts of the first and the second kinds are satisfiable. For their impossibility is only due to their internal structure and has nothing to do with their correspondence with the world or the inconsistency of their propositional content. Many past propositions are true. However, we cannot foresee nor predict them. Similarly, some limits imposed to the world do not limit our thoughts. Many unsatisfiable illocutions are performable. We can tell stories where agents go back in time.

We must therefore revise Wittgenstein's linguistic turn and reformulate his considerations on the limits of language, thought and reality. Illocutionary logic distinguishes four limit cases of conceptual thoughts. Firstly, there are *impossible conceptual thoughts* that no agent can have or form in the first person. As Wittgenstein pointed out, logic can only fix limits to thoughts *indirectly* by fixing limits to linguistic expressions of thought. Otherwise one would have to think what cannot be thought in fixing these limits. In our view the impossibility of certain conceptual thoughts is shown in language by so-called illocutionarily inconsistent sentences which express non-performable illocutionary acts. On our account illocutionarily inconsistent sentences are both useless and meaningless. We can never make a successful literal utterance of these sentences. Their utterances are *analytically unsuccessful*. We cannot even mean what we say when we utter them. Illocutionarily inconsistent sentences show that *there are limits to thought*. Incidentally we are able to think about certain impossible thoughts. Illocutionary logic even describes the logical form of non-performable illocutionary acts.

Secondly, there are *unsatisfiable conceptual thoughts* that represent facts that no agent could ever experience. So-called truth-conditionally inconsistent sentences which express unsatisfiable illocutionary acts show that *there are limits to our experience*. As we pointed out, some impossible thoughts are satisfiable and some possible thoughts are unsatisfiable. This is why many illocutionarily inconsistent sentences are truth conditionally consistent and many truth conditionally inconsistent sentences are illocutionarily consistent.

Thirdly, there are conceptual thoughts which are both impossible and unsatisfiable. So are thoughts with a non-empty direction of fit whose propositional content is an obvious contradiction. They are expressed by sentences which are both illocutionarily and truth conditionally inconsistent. Such sentences show that *there are common limits to thought and to experience.*

Fourthly, there are *analytically satisfied* conceptual thoughts whose are satisfied in all possible circumstances according to all possible denotation assignments. So are beliefs and assertions whose propositional content is an obvious tautology. They are expressed by sentences whose literal utterances are analytically satisfied like "Mothers are women" which show that there is an *a priori order in the world*. Some facts exist in all possible worlds, no matter whether we experience or think about them or not. Moreover it is not possible to experience certain facts without experiencing other facts. This is shown in language by the fact that certain sentences *strongly truth-conditionally entail* other sentences. The satisfaction conditions of illocutionary acts expressed by the first sentences are stronger than the satisfaction conditions of illocutionary acts expressed by the other sentences. For example the imperative sentence "Please come!" strongly truth-conditionally entail the declarative sentence "You can come".

Incidentally, according to the logic of ramified time, we, human agents, are free and our states and acts of conceptual thought are not determined. At certain moments we do not think and we do not make any attempt. In order to think and act, we must exist. Moreover we can fail to achieve our objectives. So there are no illocutionary acts which are necessarily successful. However, attempts to perform illocutionary acts of certain forms are necessarily successful and satisfied. Whoever makes a literal utterance of declarative sentences like "I think" or "I speak" succeeds in making his or her assertion and makes that assertion true. So is Descartes famous sentence "Cogito, ergo sum" ("I think, therefore I am") in [4]. All its literal utterances are analytically successful and satisfied. Moreover it is not possible to have or form certain conceptual thoughts without having or forming other thoughts. There is also *a priori order of thoughts*. Certain attitudes strongly commit their agent to other attitudes and certain illocutions strongly commit their agent to other illocutions. This is shown in language by the fact that certain sentences *strongly illocutionarily entail* other sentences: one cannot perform the illocutionary act expressed by the first sentences in a context without *eo ipso* performing the illocutions expressed by the other sentences in the same context. For example, the imperative sentence "Please come back here tomorrow morning or tomorrow afternoon!" strongly illocutionarily entail the sentence "Come back here tomorrow!" The *a priori* order of thought is different from the *a priori* order of the world. This is shown in language by the fact that the two notions of strong truth conditional entailment and strong illocutionary entailment do not coincide in extension.

References

[1] J. L. Austin. *How to Do Things with Words*. Oxford: Clarendon Press, 1986.

[2] F. Brentano. *Psychology from an Empirical Standpoint*. London: Routlede and Kegan Paul, 1973.

[3] C. Cherniak. *Minimal Rationality*. MIT Press, Bradford Books, 1986.

[4] R. Descartes. *Principles of Philosophy*. Dordrecht: Reidel, 1983.

[5] G. Frege. "On Sense and Reference". In: *Translations from the Philosophical Writings of Gottlob Frege*. Ed. by P. Geach & M. Black. Oxford: Blackwell, 1970, pp. 56–78.

[6] G. Frege. "Thoughts". In: *G. Frege Logical Investigations*. New Haven: Yale University Press, 1977, pp. 1–30.

[7] E. Husserl. *Logical Investigations*. London: Routledge & Kegan Paul, 1970.

[8] I. Kant. *Critique of Pure Reason*. Cambridge: Cambridge University Press, 1999.

[9] W. Ockham. *Tractatus de Praedestinatione*. 1321–23. Reedited by the Franciscan in Institute Edition in 1945.

[10] J. R. Searle. *Intentionality*. Cambridge: Cambridge University Press, 1983.

[11] J. R. Searle and D. Vanderveken. *Foundations of Illocutionary Logic*. Cambridge: Cambridge University Press., 1985.

[12] C. de Sousa Melo. "Possible Directions of Fit between Mind, Language and the World". In: *Essays in Speech Act Theory*. Ed. by D. Vanderveken and S. Kubo. Amsterdam and Philadelphia: John Benjamins, 2001, pp. 109–117.

[13] C. de Sousa Melo. "De la relation entre la pensée et le langage selon trois approches philosophiques majeures". In: *Philosophical Review Manuscrito* 29.2 (2006): *Language and Thought*, pp. 597–636.

[14] C. de Sousa Melo. "Signification et Action". In: 48.4 (2009), pp. 801–811.

[15] C. de Sousa Melo. "Intentionality and Meaning in Natural Languages". In: *The Journal of Intercultural Studies* 39 (2014), pp. 75–91.

[16] E. Stenius. *Wittgenstein's Tractatus*. Ithaca: Cornell University Press, 1960.

[17] D. Vanderveken. *Meaning and Speech Acts, Volume 1: Principles of Language Use.* Cambridge: Cambridge University Press, 1990.

[18] D. Vanderveken. *Meaning and Speech Acts, Volume 2: Formal Semantics of Success and Satisfaction.* Cambridge: Cambridge University Press, 1991.

[19] D. Vanderveken. "Formal Pragmatics of Non Literal Meaning". In: *Linguistische Berichte* 8 (1997): *Pragmatik*, pp. 324–341.

[20] D. Vanderveken. "Attempt, Success and Action Generation: A Logical Study of Intentional Action". In: *Logic, Thought and Action.* Ed. by D. Vanderveken. Dordrecht: Springer, 2005, pp. 316–342.

[21] D. Vanderveken. "A General Logic of Propositional Attitudes". In: *Dialogues, Logics and Other Strange Things.* Ed. by C. Dégrémont, L. Keiff, and H. Rückert. London: College Publications, 2008, pp. 449–483.

[22] D. Vanderveken. "Beliefs, Desires and Minimal Rationality". In: *Uppsala Philosophical Studies* 57 (2009): *Logic, Ethics and All That Jazz*, pp. 357–372.

[23] D. Vanderveken. "Towards a Formal Pragmatics of Discourse". In: *International Review of Pragmatics* 5.1 (2013), pp. 34–69.

[24] D. Vanderveken. "Intentionality and Minimal Rationality in the Logic of Action". In: *Nuel Belnap on Indeterminism and Free Action.* Ed. by T. Müller. Dordrecht: Springer, 2014, pp. 315–341.

[25] D. Vanderveken. "Quantification and Predication in Modal Predicative Propositional Logic". In: *Logique et Analyse* 229 (2015): *In the Memory of Paul Gochet*, pp. 35–55.

[26] D. Vanderveken. "On the Intentionality and Imperfect but Minimal Rationality of Human Speakers". In: *The Journal of Intercultural Studies* 40 (2016), pp. 1–32.

[27] L. Wittgenstein. *Notebooks 1914-16.* Ed. by G. H. von Wright & G. E. M. Anscombe. Oxford: Basil Blackwell, 1961.

[28] L. Wittgenstein. *Tractatus logico-philosophicus.* London: Routledge & Kegan Paul, 1961.

Principia Ethica Illocutionary Acts in Deontic Logic

Daniel Vanderveken

Ethics deals with norms and values and especially with what is good and bad. G.E. Moore stated basic principles of ethical reasoning in his Principia Ethica [17]. Many illocutionary acts like evaluations, pledges, directives, recommendations, prohibitions, authorizations and permissions are capital in ethics. Thanks to illocutionary logic founded by J.R. Searle and D. Vanderveken in [22] and Vanderveken's general semantics [23, 24], one can now interpret all kinds of sentences that are used in ethics including imperative, performative, exclamatory and ought sentences and analyze the felicity conditions of capital illocutions of ethics. One can also formalize valid practical and theoretical inferences made in the conduct of ethical discourses thanks to Vanderveken's typology of discourse [25, 30]. From an illocutionary point of view, any evaluation according to which it is good (or bad) to carry out an action commits the speaker to giving the directive "Carry out (or do not carry out) that action!" My main purpose here will be to use the resources of illocutionary logic in order to analyze capital first level illocutions of ethics and give a better account of moral obligation, permission and forbiddance than the traditional deontic logic of G.H. von Wright [33, 34]. First level actions and attitudes are actions and attitudes of individual agents at a single moment of time. I have integrated in illocutionary logic a new logic of first level attitudes and actions that takes into account the intentionality and imperfect but minimal rationality of human agents [26, 27, 28, 29, 31]. I will criticize standard deontic logic and formulate the foundations of a more general and adequate deontic logic *Principia Ethica* which can state the different kinds of valid principles of ethics and eliminate well known deontic paradoxes. The ideography of my *Principia Ethica* can express first level attitudes, actions and illocutionary acts of all logical forms. So my deontic logic will not only state logically valid assertions governing moral obligations and good and bad actions. It also gives us universally good directives that we ought to follow. Imperative

Copyright © 2019 by Daniel Vanderveken. All rights reserved.

formulas that express conditional directives of the form "If an action is bad, do not do it!" are provable. Furthermore my deontic logic states valid declarations like interdictions according to which bad actions are prohibited.

1 Issues and theoretical objectives

The primary *units of meaning and communication* in a context of utterance are not isolated true or false propositions but rather speech acts of the type called by J.L. Austin [2] *illocutionary acts* which have *felicity* rather than *truth conditions*. Most basic first level illocutionary acts are *elementary illocutionary acts* of the form $F(P)$ like assertions, promises, directives, gifts and thanks. As G. Frege [10] pointed out, they have a *force F* and a *propositional content P*. Requests, recommendations, commands, interdictions, promises, pledges, submissions, vows, decisions and pardons are capital elementary illocutions of ethics. By nature illocutionary acts are *intrinsically intentional actions* that agents perform voluntarily by making a mental and a verbal *attempt*. Their main felicity conditions are their *success and* their *satisfaction conditions*. According to speech act theory, in order to be *happy* speakers who attempt to perform an illocution should *succeed* and perform a *non defective* illocution which should moreover be or turn to be *satisfied* in the world. In order to *succeed*, speakers must express attempted illocutionary acts by using appropriate words in an adequate context. A successful illocution is non defective when the speaker is sincere and only makes true presuppositions. Speakers generally *relate* propositional contents of elementary illocutions to the world with the aim of establishing a *correspondence* between words and things from a certain direction of fit. This is why most illocutions have satisfaction conditions. The notion of *satisfaction* is a generalization of that of truth needed to cover all forces. An *assertion* is *satisfied* when it is *true*; a *promise* when it is *kept*; a *command* when it is *obeyed* and a *blessing* when it places the hearer in a state of God's grace.

As John Searle and I [22] pointed out in Foundations of Illocutionary Logic [22], the principal component of each force is its *primary illocutionary point* that determines the *direction of fit* of illocutions with that force, that is to say from which direction the correspondence must be achieved between words and things. The *five illocutionary points* are: the *assertive*, the *commissive*, the *directive*, the *declaratory* and the *expressive points* which correspond to the four possible directions of fit that exist between words and things in language use. *Illocutionary acts with the assertive point* (e.g. assertions, acknowledgements, testimonies, predictions and blames) *have the words-to-things direction of fit*. Their point is *to represent how things are in the world*. In the case of assertive utterances, the used words must correspond to the objects of reference as they stand or will stand in the world. Assertive illocutions are *satisfied* when their propositional content represents a fact which exists or will exist in the world of the utterance. *Illocutionary acts with the commissive or directive point have the things-to-words direction of fit*. The *point of commissive illocutions* like pledges, promises, threats, vows and renunciations *is to commit the speaker to an action*, while the *point of directive illocutions* like requests, solicitations, prayers, commands and recommendations *is to make an attempt to get the hearer to do an action* in the world. In the case of such illocutions, the objects of reference in the world have to be changed by one protagonist of the utterance in order to correspond to the words used. The responsibility for changing the world lies with the speaker in the case of commissives and with the hearer in the case of directives. Thus commissive and directive illocutions are *satisfied* when their speaker or hearer transforms the world in order to fit their propositional content.

Illocutions with the declaratory illocutionary point (e.g. decisions, gifts, pardons, condemnations and blessings) *have the double direction of fit*. Their point is to get the world to match the propositional content by saying that the propositional content matches the world. In successful declarations, objects of reference are changed by the speaker at the moment of utterance in order to correspond to used words in the very utterance of these words. As Austin said, in making successful performative utterances, agents *do things with words*. Every successful declaration is satisfied because the speaker performs at the moment of utterance the represented action by way of representing himself as performing then that very action. Successful declarations are assertions that always make their propositional content true by virtue of the utterance. They have both the *things-to-words and* the *words-to-things direction of fit*. For some elementary illocutions, there is no question of success or failure of fit. *Illocutions with the expressive point* like thanks, apologies, congratulations and boasts *have the empty direction of fit*. Their point is just *to express propositional attitudes of the speaker about the fact* represented by the propositional content. *Propositional attitudes* are individual attitudes of the form *M(P)* like beliefs, previsions, convictions, desires, wishes and intentions which have a *psychological mode M* and a *propositional content P*. In purely expressive utterances, speakers do not attempt to establish a correspondence between words and things. They just want to manifest verbally their propositional attitudes about the ways in which objects are in the world. They presuppose then the existence of the fact that inspires their attitude. So purely expressive illocutions do not have proper conditions of satisfaction. Expressive illocutionary acts are just *appropriate* or *inappropriate*. An expressive illocution is *appropriate* when the speaker expresses an attitude whose mode is right for the fact that inspires that attitude and when that fact is really existent. It is inappropriate to thank someone for an action that was very bad or not carried out.

A lot of complex first level illocutionary acts are not reducible to elementary illocutions. *Acts of illocutionary denegation* like refusals and disapprovals are of the form $\neg F(P)$; their aim is to make explicit the non-performance by the speaker of an illocution $F(P)$. A *refusal* is the illocutionary denegation of an acceptance. *Conditional illocutionary acts* like offers are of the form $(P \Rightarrow F(Q))$; their aim is to perform an illocutionary act $F(Q)$ not categorically, but on the condition that a proposition P is true. An *offer* is a promise that is conditional on the hearer's acceptance. *Conjunctions of illocutionary acts* like warnings are of the form $(F_1(P_1) \& F_2(P_2))$; their aim is to perform simultaneously two illocutionary acts $F_1(P_1)$ and $F_2(P_2)$. A *warning* is the conjunction of an assertion that something is the case and of a suggestion to the hearer to do something about it. Complex first level illocutions that are capital in ethics are conditional directives, acts of granting permission, authorizations, offers and warnings. An act of illocutionary denegation $\neg F(P)$ is *satisfied* when the agent does not perform the denegated illocution $F(P)$. So any successful illocutionary denegation is satisfied. Whoever performs a conditional illocutionary act of the form $(P \Rightarrow F(Q))$ performs categorically the illocution $F(Q)$ when the antecedent proposition P is true. So a conditional illocution $(P \Rightarrow F(Q))$ *is satisfied* if and only if the categorical illocution $F(Q)$ is also satisfied when the antecedent proposition P is true. In order to satisfy an accepted offer of help the speaker must keep his promise. Finally a conjunction of two illocutions is satisfied when the two illocutions are satisfied.

Assertions can be false, promises violated and commands disobeyed. However one needs a unified theory of *success, satisfaction and truth* in illocutionary logic. Whoever attempts to perform an illocution knows under which conditions that illocution is satisfied. Moreover, the satisfaction of elementary illocutions requires the *truth* of their propositional content. Whoever follows a recommendation does the recommended action. Some illocutions have stronger felicity conditions than others. Certain illocutions *strongly commit the speaker to* other illocutions: One cannot perform these illocutions without *eo ipso* performing the others. Recommendations contain advice. Some illocutions have stronger satisfaction conditions than others. When a speaker keeps his promise the assertion that he does the promised action turns to be true. Certain illocutions cannot be performed unless others are satisfied. Whoever gives his pardon to someone makes the true assertion that he or she is forgiven. Literal *performative utterances* are declarations according to which the speaker performs the illocution named by their main performative verb. In understanding felicity conditions, interlocutors make practical and theoretical valid inferences. They understand that certain illocutions cannot be felicitous unless others are. We need a *recursive unified theory of felicity* that accounts for all this.

2 Progress and methodology

Until now speech act theory has mainly tended to study first level illocutions. Searle and I [22] have stated the principles of a theory of felicity for first level illocutions. We have decomposed each force into several components, namely: its primary *illocutionary point*, its *proper mode of achievement of* that *point*, its proper *propositional content conditions*, its *preparatory* and sincerity *conditions*, and its *degree of strength*. We also recursively defined the set of all possible illocutionary forces and success conditions of elementary illocutions. The five *primitive forces* are the simplest forces with an illocutionary point. These are: the force of *assertion* expressed by the declarative sentential type; the primitive force of a *commitment to an action* named by the performative verb "commit"; the primitive force of a *linguistic attempt to get the hearer to act* expressed by the imperative sentential type; the force of *declaration* expressed in performative utterances, and the primitive force of *expression* of a speaker's attitude expressed by the exclamatory sentential type. All primitive forces are universal.

All other more complex forces are obtained by adding to primitive forces new linguistically significant modes of achievement of illocutionary point, special propositional content conditions, preparatory or sincerity conditions, or by increasing or decreasing the degree of strength. Thus a *request* is a directive with a special courteous mode of achievement of the directive point: the speaker then gives option of refusal to the hearer. A *question* is a request that the hearer gives an *answer* to that question (special propositional content condition). An act of *praying* is a strong request with a humble and earnest mode of achievement and the special sincerity condition that the speaker expresses respect to the hearer. A directive *suggestion* is a weak attempt to get the hearer to act (weak degree of strength). A *recommendation* is a directive suggestion with the preparatory condition that the recommended action is both good for the hearer and good in general. When added components are transcendent, complex forces are universal. The force of recommendation is universal for one can express that actions are good in general and good for someone in all human languages. All capital elementary illocutions of ethics have a complex force with special components.

As Searle and I pointed out, the felicity conditions of elementary illocutions are entirely determined by the components of their force and their propositional content. Agents obey *constitutive rules* determined by force components in their attempted performance of elementary illocutions. Thus whoever *attempts to perform* an elementary illocution of the form $F(P)$ in a context of utterance must first of all *attempt to achieve the illocutionary point of the force F on the proposition P* with the required *mode of achievement* of that force. In order to achieve a force F on a proposition P the agent has of course *to express* that *force* and that *proposition* in the context of utterance. So the agent must attempt then *to express* a *proposition P* that satisfies the *propositional content conditions* of the force F in the context. Thirdly he or she must also *presuppose all propositions determined by* the *preparatory conditions* of force F for the propositional content P and finally he or she must *attempt to express with the degree of strength of F all attitudes* of the form $M(P)$ *whose psychological mode M is determined by the sincerity conditions* of force F. For example, whoever attempts to make a *promise* must address an utterance to a hearer with the intention of committing himself or herself to carrying out an action (commissive illocutionary point and propositional content condition of a promise) while putting himself or herself under the obligation to carry out that action (special mode of realization of the commissive point peculiar to promise) and presupposing moreover that the promised action is then good for that hearer (special preparatory condition of promise).

By definition, an attempted illocution of the form $F(P)$ is *successfully performed* in a context of utterance when the speaker succeeds in achieving the illocutionary point of the force F on the content P with the required mode of achievement and the proposition P satisfies the propositional content conditions of force F in the context, when he or she presupposes well all propositions determined by preparatory conditions of illocution $F(P)$ and when he or she succeeds in expressing with the required degree of strength all propositional attitudes $M(P)$ determined by the sincerity conditions of force F. Whenever the speaker uses wrong words and fails to express the force or propositional content of an attempted illocution, he or she fails to communicate to hearers his or her intention to perform that illocution. It happens that speakers presuppose false propositions or express attitudes that they do not have. An attempted first level illocution is *non-defective* in a context when it is successful and all its preparatory and sincerity conditions are then fulfilled. Austin with his notion of *felicity-condition* did not distinguish clearly between successful utterances that are defective and utterances that are not even successful. On one hand there are limits to success. No one can successfully advise a hearer to have done something. For one can only achieve the directive point on propositions that represent a present or future action of the hearer. Any attempt to give a directive with a past propositional content would be a failure. This is why we never try to give such directives. Such limits to success show themselves in language. There are no successful literal imperative utterances whose main verb names a past action of the hearer. One can succeed in giving advice which is bad for the hearer (failure of a preparatory condition) or that one does not wish to be followed (failure of sincerity conditions). In illocutionary logic, *felicity conditions* of illocutionary acts are the sum of their success, preparatory, sincerity and satisfaction conditions.

Searle and I analyzed in [22] the meaning of English force-markers and performative verbs. Most force markers contain modifiers of sentential type expressing particular force components. So sentences with the same syntactic type can express illocutions with different forces. Imperative sentences like "Do it please!", "Do it whether you like it or not!" and "Do it, it is good for you!" respectively express a *request*, a *peremptory directive* and an *advice*. Unfortunately we did not analyze in detail the nature of propositional contents of illocutions and their satisfaction-conditions. Speech act theory requires a *finer criterion of propositional identity* than logical equivalence. For many propositions with the same truth conditions are not the contents of illocutions with the same forces and of propositional attitudes with the same modes. We can assert and believe that Natal is a city without asserting and believing that it is a city and not a sphere. However the two propositions are logically equivalent: they are true in exactly the same possible circumstances. In order to explicate this, I have formulated in the past decades a non-classical *natural predicative logic of propositions* that takes into account the acts of predication that we make in expressing and understanding propositions. My predicative approach respects the double nature of propositions which are both *senses of sentences* and *contents of elementary illocutions* and *propositional attitudes*.

I have first formulated in Meaning and Speech Acts [23, 24] a simple *predicative propositional logic* that analyses elementary propositions representing atomic facts and their truth functions. However speakers are oriented towards past, present and future facts, they distinguish between actual and possible facts and make generalisations. They attempt to perform their successful illocutions and express then all kinds of cognitive and volitive attitudes. We need to represent modal, general, past, present and future facts as well as first level actions and attitudes of agents in the ideography of illocutionary logic in order to formulate the theory of felicity. In the past fifteen years I have added to predicative propositional logic historic and logical *modalities* and ramified *time, generalization*, first level *actions* and *attitudes*. See [26, 27, 28, 29, 31, 32].

In order that an elementary illocutionary act $F(P)$ be satisfied in a context of utterance, its propositional content P must be true in the circumstance of utterance. In philosophical logic, expressed propositions are true or false in *possible circumstances* that contain a moment of time. Now we, human agents, live in an *indeterminist* world where the future is open. According to *indeterminism*, the ways in which things are at a moment are not entirely determined by the ways in which they were before. Our actions and attitudes are not determined. Whenever we do or think something, we could have done or thought something else, or nothing at all. We need a *ramified conception of time* in order to account for indeterminism and freedom. In branching time each *moment* represents a *complete possible state of the actual world at a given instant*. There is a single causal route to the past but multiple future routes. For several incompatible moments of time might directly follow a moment of utterance in the future of the world.

Consequently, the set of moments of time is a *tree-like frame* of the following form:

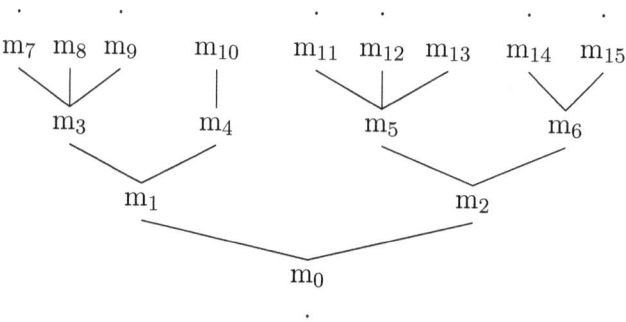

Figure 1.

A maximal chain h of moments of time is called a *history*. It represents a *possible course of history of our world*. As Belnap [4, 5] pointed out, *possible circumstances*, where propositions are true or false, are pairs of a moment of time and of a history to which that moment belongs. Thanks to histories temporal logic can analyze modal notions like *settled truth* and the different kinds of *causal, historic* and *universal necessity* and *possibility*. Certain propositions are true at a moment according to all histories. Their truth is then *settled at that moment* no matter how the world continues. So are past propositions and propositions according to which agents perform first level illocutions and attitudes. Whoever attempts to perform an elementary illocution either succeeds or fails to perform that illocution at the moment of utterance. He or she *succeeds* when all success conditions are fulfilled at that moment and *fails* otherwise. Unlike success and failure, the *satisfaction* or the *insatisfaction* of attempted elementary illocutions oriented towards the future is *not* settled at a moment of utterance. Contrary to the past, the future is open. The world can continue in various ways after most moments of utterance. The truth or the falsehood of future propositional contents of elementary illocutions is not settled at these moments. It depends on what will be their actual historical continuation. It can be true at that moment according to one possible historic continuation of that moment and false at that moment according to another historic continuation.

As W. of Ockham [18] pointed out, if the world continues after a moment, it will continue in a unique way. Like Ockham, I believe that each non-final moment will have a unique real historic continuation even if that continuation is still undetermined at that very moment. Indeterminism cannot prevent that uniqueness. According to standard philosophy of mind human agents, who are directed by virtue of their intentionality towards things and facts of the world, are intrinsically oriented at each moment of their life towards the *real continuation* of the world. We ignore now how the world will continue but we are intrinsically oriented at each moment towards the real continuation of that moment and we always distinguish conceptually that real from other possible continuations whenever our actions or thoughts are directed towards the future. Whoever requests or wishes future help requests or wishes to be helped in the real future. So in my approach our elementary illocutions and propositional attitudes at each moment have or will have a certain satisfaction value even if that satisfaction value is still undetermined when they are oriented towards the future. In order to keep a present promise and execute a present intention to help someone later, one must help that person in the real continuation of the world. Other possible historic continuations do not matter. All possible circumstances whether actual or not have a proper *world of utterance* whose course of history contains the real historic continuation of their moment of utterance.

In order that an elementary illocution be *satisfied*, it is not enough that its propositional content is or turns to be true and corresponds to an existing fact in the real historic continuation of the moment of utterance. The correspondence between words and things must be established according to the proper *direction of fit* of its force. When an illocution has the things-to-words direction of fit, it is or it will be satisfied if and only if its propositional content is or turns to be true *because of* its performance. Whoever follows a recommendation must do what is recommended because of that recommendation. There is often an over determination of reasons. An agent can do a recommended action for several reasons. But one of his reasons is that he intends then to follow that recommendation. Whoever has forgotten a previous recommendation cannot follow it.

In ramified temporal logic, two *moments of time* are *coinstantaneous* when they belong to the same instant. Coinstantaneous moments are on the same horizontal line in each tree-like frame. Each moment of time is of course coinstantaneous with itself. So in the last figure m_3, m_4, m_5 and m_6 are four coinstantaneous moments. Two *possible circumstances* are *coinstantaneous* when their moments are coinstantaneous. One analyzes *historic necessity* by quantifying over coinstantaneous moments. The proposition that *P is then necessary* (in symbols $\Box P$) is true at a circumstance when the truth of P is settled at all coinstantaneous circumstances. The notion of historic necessity is stronger than that of settled truth. The historically necessary fact is not only established but inevitable. According to philosophy there are no inevitable actions and intentions. So when an agent carries out an action at a moment he or she could then not have carried it out. The notions of *logical* or *universal necessity* are stronger than historic necessity. The proposition that *P is universally necessary* (in symbols: $\blacksquare P$) is true in a circumstance when P is true in all possible circumstances. In that case the fact represented is objectively necessary and always inevitable. The inexistence of that fact is then *objectively impossible*.

As Belnap pointed out, the possible causes and effects so to speak of actions of any agent at a moment are limited to those which are possible outcomes of the way the world has been up to that moment. In order to explicate *historical relevance* we must then consider coinstantaneous moments having the same past. Such moments are called *alternative moments* in my logic. Thus m_7, m_8 and m_9 are alternative moments in figure 1. Moments m_{11}, m_{12}, m_{13} are also alternative moments in that figure. A proposition is *causally necessary* in a circumstance when that proposition is true in all possible circumstances which are compatible with the laws of nature that prevail in the world of that circumstance. So when a circumstance is causally compatible with another circumstance all the facts that exist in that circumstance respect the laws of nature prevailing in the other circumstance.

3 The need to integrate illocutionary logic in a new logic of action and attitudes

Like R. Montague [16], I believe that pragmatics should use the resources of formalisms in order to establish a rigorous *theory of meaning and use*. Natural languages can be learned by human agents whose cognitive abilities are creative and limited. Formalisms enable us to construct better models of linguistic competence and understanding. However one must revise basic hypotheses of standard logics of *propositions, attitudes* and *actions* in order to analyze adequately felicity-conditions of illocutions. Speech act theory requires an explication of the *intentionality* and *imperfect* but *minimal rationality* of interlocutors and of the *generation of their speech acts*.

Standard logics of attitudes based on J. Hintikka's approach [13, 12] approach and the standard logic of action of of N. Belnap and M. Perloff [3, 4, 5] are defective. In Hintikka's epistemic logic, human agents are *logically omniscient*. They know and believe all logically necessary propositions. For *possible circumstances are compatible with the truth of agents' beliefs* and the *realization of their desires* at each moment of time. To each agent a and moment m there corresponds in each model of standard epistemic logic a unique set of possible circumstances $Belief_m^a$ where all beliefs of that agent at that moment are true. An agent a believes a proposition P at a *moment m* in a model when that proposition P is true in all circumstances of the set $Belief_m^a$. On this account, we, human agents, know all logically necessary propositions and our beliefs are closed under logical implication. Whoever believes a proposition P also believes all propositions that P logically implies. However we ignore many essential properties of objects which persist in the world. *Essential properties* are properties that objects really have in all possible circumstances. Human agents, have the essential property of having certain parents. We learn *a posteriori* empiric essential properties and we ignore some of them. Certain adopted children are wrong about the identity of their parents. Moreover according to the standard approach agents are either perfectly rational or they are totally irrational. But as the Greek philosophers pointed out, it is paradoxical to believe that every proposition is true.

As Nuel Belnap pointed out, we need a *ramified* conception of time and modalities in order to account for physical indeterminism and the apparent freedom of agents. However Belnap's indeterminist logic of action ignores the very intentionality of agents and it also does not analyze adequately their agentive commitments. My formal logic of attitudes and actions is compatible with standard philosophy of mind and of action. On one hand, my logic of attitudes deals with all cognitive and volitive attitudes and with complex attitudes like conditional attitudes and psychological denegations and conjunctions of propositional attitudes. In performing acts of illocutionary denegations like refusals speakers express denegations of attitudes such as discords. A discord is the psychological denegation of the attitude of agreement. In performing conditional illocutions like offers speakers express conditional attitudes (conditional intentions). My logic also explicates formally why we are neither logically omniscient nor perfectly rational. It explains why human agents *a priori* know certain logically necessary truths and always remain *minimally rational* [7]. They make certain logical practical and theoretical inferences. My logic of first level actions revises and enriches Belnap's standard logic of action in order to take into account the *intentionality and minimal rationality of human agents*. It explicates the nature of our *intentional* and *basic* actions and it classifies the different *kinds of action generation* in order to state adequate logical laws of *agentive commitment*.

In my *predicative propositional logic*, each proposition has a finite *structure of constituents*. In expressing propositions, we predicate properties or relations of *objects subsumed under concepts*. We understand a proposition when we understand which attributes objects of reference must possess in a possible circumstance in order that this proposition be true in that circumstance. As G. Frege [9] pointed out, we always *refer to* objects by subsuming them under senses. We cannot directly have in mind *individual objects* like material bodies and persons. We rather have in mind *concepts* of individuals (called *individual concepts*) and we *indirectly* refer to individuals and predicate attributes of them through these concepts. So our attitudes and illocutions are directed towards *individuals under a concept* rather than towards pure individuals. By recognizing the indispensable role of concepts in reference and predication, predicative logic accounts for attitudes and actions directed towards inexistent objects. It also explains why our attitudes and intentional actions can be directed towards an individual object under a concept without being directed towards the same individual under another concept.

We understand most propositions without knowing in which possible circumstances they are true, because we ignore *real denotations* of most attributes and concepts in many circumstances. One can refer to a colleague's girlfriend without knowing who she is. However we can always in principle think of persons who could be his girlfriend. So in any possible use and interpretation of natural language, there are a lot of *possible denotation assignments to attributes and concepts* in addition to the standard *real denotation assignment* of traditional logic which associates with each propositional constituent its *real denotation* in every possible circumstance. Possible and real denotation assignments to senses in models are functions of the same type which associate with each individual concept a unique individual or no individual at all and with each property of individuals a unique set of individuals under concepts in every considered possible circumstance. According to the real denotation assignment, my colleague's girlfriend is the woman who is his real girlfriend in every circumstance when there is such a person. According to other possible denotation assignments, his girlfriend can be another woman or nobody in that circumstance. However all possible denotation assignments respect *meaning postulates* in all models. According to any interpretation, a girlfriend is a woman.

In my approach, the truth definition is then relative to both possible circumstances and denotation assignments. In understanding propositions we in general do not know whether they are true or false. We just know that their truth in a circumstance is compatible with certain possible denotation assignments to their concepts and attributes. Thus an elementary proposition predicating an extensional property of an object under a concept is true in a circumstance according to a denotation assignment when according to that assignment the individual which falls under that concept has that property in that circumstance. Otherwise, it is false in that circumstance according to that assignment. Of course, *truth* is based on *real correspondence*. In order to be *true in a possible circumstance* a proposition has to be *true in that circumstance according to the real denotation assignment*. So among all possible truth conditions of a proposition, there are its *real Carnapian truth conditions* which correspond to the set of possible circumstances where it is true according to the real denotation assignment.

Traditional propositional logic whose models only assign real denotations to senses deal with *objective possibilities* and *necessities*. Thanks to possible denotation assignments predicative logic can now distinguish subjective and objective possibilities. A proposition is *subjectively possible* when it is true in a possible circumstance according to at least one possible denotation assignment. In order to be *objectively possible* a proposition has to be true in a circumstance according to the real denotation assignment. Many subjective possibilities are not objective. We ignore *the real denotation* of most concepts and attributes in many circumstances. But we can in principle think of denotations that they could have. There are a lot of *possible subjective denotations*. Now when we have in mind concepts and attributes, only certain possible denotation assignments to these senses *are then compatible with* the truth of our beliefs. Suppose that according to you my mother was born in France.

In that case, possible denotation assignments according to which she was born in Belgium are then incompatible with your beliefs. In my approach, *possible denotation assignments rather than possible circumstances are compatible with the truth of our beliefs*. This is how I account for subjective possibilities. By definition the real denotation assignment always respects the essential properties of objects. It associates with each individual concept an object which possesses all its essential properties. When all possible denotations assignments compatible with the truth of the beliefs of an agent violate an essential property of an individual, that agent has then inconsistent beliefs.

In predicative logic, two propositions are *identical* when they contain the same elementary propositions and they are true in the same circumstances according to the same possible denotation assignments. Such a finer criterion of propositional identity explains why many logically equivalent propositions have a different cognitive and volitive values. Propositions whose expression requires different predications have a different structure of constituents. We do not express and understand them at the same moments. So are the two different logically necessary propositions that Natal is a city and that Natal is a city and not a sphere. My identity criterion also distinguishes logically equivalent propositions that we do not understand to be true in the same possible circumstances: they are indeed not true according to the same possible denotation assignments to their constituents. So the two logically necessary propositions that all whales are animals and that all whales are mammals are different. We discovered at a certain time that whales are mammals. Before we used to believe that they were fishes. Few necessarily true propositions are *obvious tautologies* that we know *a priori*. By hypothesis, a proposition is *necessarily true* when it is true in every possible circumstance according to the real denotation assignment. In order to be *obviously tautological*, a proposition P has to be true in every circumstance according to every possible denotation assignment to its constituents (in symbols: *Tautologically P*). Unlike the proposition that whales are animals, the necessarily true proposition that whales are mammals is not an obvious tautology. It is false according to possible denotation assignments. The notion of *tautological obviousness* is the strongest modal notion. A tautologically obvious fact is indeed analytically inevitable subjectively as well as objectively.

Here is my new approach in the logic of attitudes and actions. By nature attitudes of human agents are directed towards objects and facts of the world. No agent can have a propositional attitude without having in mind all attributes and concepts of its content. Otherwise, the agent would be unable to determine under which conditions his or her attitude is satisfied. As Wittgenstein and Searle pointed out, an attitude with entirely undetermined satisfaction conditions would be an attitude without content. It would not be an attitude at all. Each agent has in mind a possibly empty set of attributes and concepts at each moment. Whoever has an *unconscious attitude* does not have consciously in mind all its concepts and attributes but he or she is able to express them.

In my view, *possible denotation assignments to senses* rather than possible circumstances *are compatible with the satisfaction of our attitudes.* So there corresponds to each agent a and moment m in each model $\mathcal{M}$ a unique set $Belief_m^a\ \mathcal{M}$ of possible denotation assignments to attributes and concepts that are compatible with the truth of beliefs of that agent at that moment. When the agent a has no attribute in mind at the moment m according to the model, $Belief_m^a\ \mathcal{M}$ is the set Val of all possible denotation assignments to senses. In that case, that agent has then no attitudes. Otherwise, $Belief_m^a\ \mathcal{M}$ contains *some but not all* possible denotation assignments. Whoever has in mind senses respects meaning postulates. In each model, an agent a *believes a proposition* at a moment m when he or she has then in mind all its concepts and attributes and that proposition is true at that moment according to all possible denotation assignments of $Belief_m^a\ \mathcal{M}$.

I have explicated desire in the same way. To each agent a and moment m there corresponds in each model $\mathcal{M}$ a unique non empty set $Desire_m^a\ \mathcal{M}$ of possible denotation assignments to attributes and concepts that are compatible with the realization of all desires of that agent at that moment. There is an important difference between desire and believe. We can believe, but we cannot desire, that objects are such and such without believing that they could be otherwise. For *any desire contains a preference.* Whoever desires the existence of a fact distinguishes then two different ways in which objects could be in the real world. In a first preferred way, objects are in the world as the agent desires, in a second way, they are not. The agent's desire is realized in the first case, it is not in the second case. Thus in order that an agent a *desires the fact represented by a proposition* P at a moment m, it is not enough that he or she has then in mind all its senses and that the proposition P is true at that moment according to all possible denotation assignments of $Desire_m^a\ \mathcal{M}$ compatible with what he or she desires then. The proposition P must also be false in a circumstance according to that agent at m. Otherwise he would not then prefer that fact.

My explication of belief and desire accounts for the fact that we *are neither logically omniscient nor perfectly rational.* No human agent can express all possible concepts and attributes in *his* or *her* own language. We cannot have in mind all possible propositions. So we ignore a lot of obvious tautologies and logically necessary truths. We ignore which objects possess many properties in most circumstances, especially in future circumstances. So many possible denotation assignments associating different denotations to properties are compatible with the truth of our beliefs. Furthermore the real denotation assignment is often incompatible with the truth of our beliefs and the realization of our desires. We have many false beliefs and unsatisfied desires. Finally, a lot of possible denotation assignments compatible with the truth of our beliefs and the realization of our desires violate essential properties of objects. This is why we have *necessarily false beliefs* and *unsatisfiable desires*. However *we never are totally irrational* but we always remain *minimally rational*. First of all, we cannot believe or desire everything or nothing at all, because some possible denotation assignments are always compatible with the satisfaction of our beliefs and desires. So the sophist and sceptic paradoxes are eliminated. Moreover, we are *minimally logically omniscient*, in the sense that we cannot understand an obvious tautology without knowing that it is universally true. There are ways in which objects could never be according to us. We can neither believe nor desire facts whose existence is subjectively impossible. Some contemporary logicians still wish that arithmetic were complete (a necessarily false proposition). But they could not believe nor desire both the completeness and the incompleteness of arithmetic (an obvious contradiction). Agents sometimes desire the existence of inevitable facts. However they cannot desire the existence of facts that they know to be impossible nor desire subjectively inevitable facts. One can desire to drink or desire not to drink. But no one can desire to drink or not drink.

Finally, because we respect meaning postulates in understanding propositions, our attitudes are logically related. Certain attitudes *commit their agent to* have other attitudes. We cannot desire to drink a cup of white or green tea without desiring to drink a cup of tea. There is a new relation of *strong implication* much finer than that of logical implication that enables illocutionary logic to analyze *psychological* and *illocutionary commitment* and *agentive commitment*. A proposition P *logically implies* another Q when it is universally necessary that P implies Q. A proposition *strongly implies* another when it contains all its concepts and attributes and it cannot be true in a circumstance according to a denotation assignment unless the other proposition is also true in that circumstance according to that assignment. *Strong implication* is *finite, tautological, paraconsistent, decidable* and *a priori known*. Whoever believes a proposition P does not believe many propositions containing new senses that P logically implies. But he or she believes all the propositions that P strongly implies because he or she knows that P could not be true otherwise. All rules of elimination of natural deduction generate strong implication because all the senses of their conclusion are expressed by their premises. However rules like the rule of introduction of disjunction do not generate strong implication when their conclusion expresses a new sense. So a desire to drink does not contain a desire to drink or eat.

Unlike contemporary logic and philosophy which only consider a few paradigmatic attitudes like belief, knowledge, desire and intention, R. Descartes analyzed a lot of attitudes in *Les passions de l'âme* [8]. Could we use Cartesian analysis to develop a larger theory of all attitudes? Descartes uses the two basic categories of *cognition* and *volition* to analyze attitudes. Every propositional attitude is either cognitive or volitive. Examples of cognitive psychological modes are belief, conviction, persuasion, faith, confidence, knowledge, certainty, presumption, pride, haughtiness, pretension, vanity, arrogance, surprise, amazement, stupefaction, prevision, anticipation and expectation. Such cognitive attitudes contain a belief in the truth of their propositional content. Examples of volitive psychological modes are desire, wish, will, intention, ambition, project, hope, aspiration, satisfaction, pleasure, enjoyment, delight, gladness, joy, elation and amusement. Such volitive attitudes contain a desire of existence of the fact represented by their propositional content.

Descartes tends to reduce propositional attitudes to sums of beliefs and desires. However our intentions are much more than a desire to do an action with a belief that we can do it. In my analysis [27, 28], psychological modes have other components than the *basic categories of cognition* and *volition*. Complex modes like knowledge, wish and intention have a special *way of believing or desiring*, proper *conditions on their propositional content* or proper *preparatory conditions*. Thanks to these other components I have *recursively defined the set of all psychological modes* of attitude. As Descartes anticipated, the two *primitive psychological modes* are those of *belief* and *desire*. They are the simplest cognitive and volitive modes. Other more complex modes are obtained from the two primitives by adding to them special cognitive and volitive ways, special conditions on the propositional content or special preparatory conditions. We feel our beliefs and desires in a lot of ways. Many psychological modes require a special *cognitive or volitive way* of believing or desiring. Thus, *knowledge* is a belief based on strong evidence that guarantees truth. Whoever has an *intention* feels such a strong desire that he or she is disposed to *act* in order to realize that desire. Like forces, modes can have proper *propositional content* and *preparatory conditions*. Intentions are *desires to carry out a present or future action*. Any agent of a propositional attitude or of an elementary illocution *presupposes* certain propositions which depend on the psychological mode or illocutionary force. These presuppositions are determined by preparatory conditions. Thus *engagements* and *intentions* have the preparatory condition that the agent is then able to do the action represented by their content. In the illocutionary case the speaker can lie in order to mislead the hearer. However the agent who has an attitude cannot lie to him or herself. Whoever has an intention presupposes and believes that he or she can execute that intention.

By way of performing individual actions at a moment, agents bring about facts in the world. The logical constant of Belnap's logic of action is the connective *stit* ("sees to it that") which serves to express propositions according to which an agent *a does that P* (in symbols [*a stit P*]. Because attempts are constitutive of intentional actions, my logic of action contains a second logical constant *Tries* in order to express propositions of the form [*a Tries P*] according to which the agent *a attempts* to do that *P*. See [26, 31]. Unlike intentions which are attitudes that we have, *attempts* are actions that we make. They are *personal, intentional, conscious, free* and *successful*. Every agent can only make his or her individual attempts. When two agents succeed in doing the same action (the same elementary illocution), they do it thanks to different personal attempts. Attempts are *intrinsically intentional*. There are no involuntary attempts. Each attempt is free. Whoever attempts to make an attempt makes it. Direct attempts by an agent to move parts of one's body are real *basic actions* in the sense of A. Goldman [11]. When an agent *forms* the present intention to make a direct movement, an attempt is caused by the very formation of that intention. Attempts are *means* to achieve *ends*. Whoever makes an attempt makes that attempt in order to achieve an objective. The agent can succeed or fail to reach his objective. When the agent succeeds, his attempt is *satisfied*. Otherwise it is *unsatisfied*. In order to make a *satisfied* attempt, one must make a good attempt in a right circumstance.

Agents can repeat individual actions of the same type at different successive moments in a possible course of the world. They can also perform actions of the same type at alternative moments. When a player is in a checkmate position at a moment in a chess game, that player will lose the game at all directly posterior alternative moments where he or she makes a move. Moments of time are logically related by virtue of actions of agents. From a logical point of view, to each agent *a* and moment *m* there always corresponds in every model $\mathcal{M}$ the set $Action^a_m$ $\mathcal{M}$ of coinstantaneous moments *m'* which are *compatible with all the actions* that the agent *a* performs at the moment *m*. They are all, as B. Chellas says, "under the control of - or responsive to the actions of" that agent at that moment [6, p. 490]. Our actions and attitudes are not physically determined and evitable. An agent *acts at a moment m* in a model if and only if at least one moment coinstantaneous with m is not compatible with his or her actions at that moment. In my approach, the relation of compatibility with actions is a relation of equivalence. When a moment is compatible with all actions of an agent at another moment, that agent performs exactly the same actions at these two moments. So when an agent a carries out the action represented by a proposition P at a moment m, that proposition P is true at all moments $m' \in Action_m^a \; \mathcal{M}$ coinstantaneous with m according to any history.[1] What an agent does at a moment depends on how the world has been up to that moment. This is why the relation of compatibility with actions satisfies the so called *historical relevance condition*. Only alternative moments having the same past as m can belong to $Action_m^a \; \mathcal{M}$. Moreover, as Belnap said, *the world goes on*. Agents act in the same world. So the sets $Action_m^a \; \mathcal{M}$ and $Action_m^b \; \mathcal{M}$ corresponding to different agents a and b have a non empty intersection. Thanks to the compatibility function, logic can start to analyze individual actions. In order that the proposition that *agent a sees to it that P be* true in a circumstance m/h according to a model $\mathcal{M}$ the truth of proposition P must be established at all moments $m' \in Action_m^a \; \mathcal{M}$ compatible with the actions of agent a at m, and the proposition P must be historically contingent. No agent could bring about a fact whose existence is historically impossible or necessary. For historically inevitable facts exist at a moment no matter what we do.

My logic of action accounts for the minimal rationality of agents who are neither perfectly rational nor entirely irrational. We can intend and attempt to do impossible actions that we believe possible. However we can neither intend nor attempt to do subjectively impossible actions, just as we cannot desire subjectively impossible facts. Satisfaction conditions of desires, intentions and attempts are of course logically related in all models of my logic of action. To each agent a and moment m there correspond in every model $\mathcal{M}$ two non empty sets $Intention_m^a \; \mathcal{M}$ and $Attempt_m^a \; \mathcal{M}$ of possible denotation assignments to senses that represent satisfaction conditions of intentions and attempts. $Intention_m^a \; \mathcal{M}$ is the set of all possible denotation assignments which are compatible with the execution of all intentions of that agent at that moment in $\mathcal{M}$. $Attempt_m^a \; \mathcal{M}$ is the set of all denotation assignments which are compatible with the satisfaction of his or her attempts at that moment in $\mathcal{M}$. Attempts like intentions have the *world-to-mind direction of fit*. Only realized attempts can be satisfied. Consequently all denotation assignments to

---

[1] Of course the same actions of that agent can have different physical effects (events that are not actions) in the world at different moments which are compatible with what he or she does at that moment.

senses compatible with the satisfaction of attempts of an agent at a moment are compatible with their realisation at that very moment. Thus $Attempt_m^a \, \mathcal{M} \subseteq Action_m^a \, \mathcal{M}$. In my view, any agent of an attempt *forms the present intention* to make then his or her attempt. Because that attempt has a present or future objective, he or she also forms the intention to achieve that objective at that moment or later in the real historic continuation. Formally, $Attempt_m^a \, \mathcal{M} \subseteq Intention_m^a \, \mathcal{M} \subseteq Desire_m^a \, \mathcal{M}$. Moreover, because attempts are actions, each agent makes the same attempts at all moments compatible with his or her actions. Thus $Attempt_m^a \, \mathcal{M} = Attempt_{m'}^a \, \mathcal{M}$ when $m' \in Action_m^a \, \mathcal{M}$. There is no action without attempt. Consequently the agent a does not act at the moment m according to when $Attempt_m^a \, \mathcal{M} = Val$. Different agents can attempt to reach the same end. However no agent can make the attempt of another agent. One does something irreducibly personal when one makes an attempt. No one else can form one's intention of making that attempt. So $Attempt_m^a \, \mathcal{M} \neq Attempt_m^b \, \mathcal{M}$ when $a \neq b$ and agent a or b acts at moment m.

In my approach, agents do not act when they do not make any attempt. Thus $Action_m^a \, \mathcal{M}$ is the set of all moments which are coinstantaneous with m when all possible denotation assignments to senses belong to the set $Attempt_m^a \, \mathcal{M}$. Moreover unintentional actions are always *generated* by intentional actions of their agent and they could in principle have been attempted. So mistakes and failures are not real actions but rather events that happen to us. For our mistakes and failures could not be intentional. Moreover our actions are evitable. We cannot make utterances without agitating particles. Such inevitable agitations are not real actions but pure events in the world. All individual actions of an agent at a moment are *generated* by his or her *basic individual action* at that moment which is the primary direct attempt that he or she makes then.

4 How to revise deontic logic within illocutionary logic

Let me now use the principles that I have advocated in order to propose a substantial revision of traditional deontic logic within my logic of attitudes, actions and illocutions. The various kinds of causal, historic and universal necessity and possibility are important for philosophical logic. I agree with G. H. von Wright [33, 34] that we need modalities in deontic logic. Now deontic logic is primarily concerned with the deontic modality of *moral obligation*. Gottfried Leibniz [14] was the first to propose to apply modal logic to ethics. According to Leibniz, what is *morally obligatory* is *what a good man necessarily does*; what is *morally permitted* is *what a good man can possibly do* and what is *morally forbidden* is *what a good man cannot possibly do*.

Unfortunately G. H. von Wright's modal approach does not account for Leibniz principles. For his *deontic logic* has not enough expressive capacities. The ideal modal object language of von Wright's *standard propositional deontic logic* SDL [33] of 1951 is much too simple: it only adds the single logical constant of moral *obligation* **O** to the two truth connectives of *negation* $\sim$ and *material implication* $\rightarrow$ of elementary logic. Formulas of the form **O**p of SDL with a propositional formula p mean that it is *morally obligatory* that p. There are no other modal connectives for settled truth, necessity or possibility in SDL. There are also no connectives for action and for goodness in SDL. Moral obligation is the only primitive deontic notion of his standard deontic logic. According to von Wright one can derive the notions of moral permission, moral prohibition and moral option from moral obligation and truth functions just as one derives possibility, impossibility and contingency from necessity in modal logic. In SDL *Permitted* $p =_{\text{def}} \sim\!\!\boldsymbol{O}\!\!\sim\!\!p$; *Forbidden* $p =_{\text{def}} \boldsymbol{O}\!\!\sim\!\!p$; and *Optional* $p =_{\text{def}} (\sim\!\!\boldsymbol{O}p \ \& \ \sim\!\!\boldsymbol{O}\!\!\sim\!\!p)$. On my view von Wright is wrong on that matter. For acts of forbidding, prohibitions, interdictions and permissions are *illocutions* that speakers perform in making utterances addressed to hearers. To *forbid* someone to do something is just to *order* him or her *not to do* it. Thus *forbid* $p =_{\text{def}} order \sim\!\!p$. A *forbiddance* is an order with a negative propositional content which has the special mode of achievement of the directive point that the speaker invokes a position of force or authority over the hearer. *Prohibitions* are acts of forbidding an action not only at the moment of utterance but over a long period of time (special propositional content condition). To *interdict* an action is to *declare that it is prohibited*. To *permit* someone to do something is just to make clear that one does not forbid him or her to do it. A *permission* is the *illocutionary denegation of an act of forbidding*. Thus *permit* $p =_{\text{def}} \neg forbid p$. Illocutionary negation $\neg$ is very different from propositional negation $\sim$. It does not obey the law of double negation nor the law of excluded middle. $\neg\neg F(P) \neq F(P)$. The denegation of a permission to do something is not an act of forbidding to do it. The performance of the illocutionary denegation of an illocution requires more than the non-performance of that illocution. It is false that every speaker either forbids or permits any action.

One needs the conceptual apparatus of illocutionary logic in order to analyze the felicity-conditions of acts of forbiddance, prohibition, interdiction and permission. They cannot be defined in terms of agents' obligations. On the contrary, obligations are imposed on agents by the very performance of all kinds of illocutionary acts. In speech act theory any *evaluation* according to which it is good (or bad) to carry out an action is a statement or a declaration that commits the speaker to giving to hearers the directive *Carry out* (or *Refrain from carrying out*) that action! In the assertive case the speaker makes a *statement* that predicates a moral value to a certain action. In the declaratory case, he or she *gives* that moral value to that action. No matter whether it is assertive or declaratory, any *evaluation* according to which an action is good (or bad) contains a *recommendation* to the hearer to do (or to refrain from doing) that action. Moreover *certain commissive illocutions* like promises and vows *impose obligations on speakers*, just as *directive illocutions* like commands and prohibitions and declaratory illocutions like interdictions *impose obligations on hearers*. Promises and vows have a special mode of achievement of the commissive point: their agents put themselves under the obligation to do what they commit themselves to doing. A lot of assertive, commissive, directive and declaratory illocutions impose therefore categorical obligations on agents. Furthermore we often perform conditionally such illocutions. When we make conditional promises, directives and interdictions, we then impose conditional obligations on agents. A capital illocution of ethics is the fundamental conditional directive: If it is good to do something, do it! It imposes on interlocutors the conditional obligation to realize an action if it is good.[2]

The deontic modality of moral obligation is intrinsically related to what is a good action (or the action of a good person). A *moral obligation* is what *a good man should necessarily do*. Only a bad person can do what is *morally* forbidden. In order to express Leibniz's principles, we must add to the lexicon of deontic logic the two logical constants *Good* and *Bad* expressing the notions of goodness and badness as well as the standard *stit* connective of action. We need in illocutionary and deontic logics formulas of the form *Good p* and *Bad p* meaning respectively that p is good and that p is bad and formulas of the form $[a\ stit\ p]$ meaning that agent a sees to it that p. Any illocution is indeed an action. Moreover commissive, directive and declaratory illocutions always have propositional contents that represent actions of their speaker or hearer. Furthermore competent speakers are essentially *persons* who can distinguish good and bad actions.

In moral philosophy, *what is good or bad* is intrinsically related to *action* or behavior (which is a long standing action). Like other analytic philosophers I distinguish three types of facts in the world: *states of affairs, events* and *actions*. All actions are special events which are generated by an agent provided with intentionality. Pure states of affairs or events entirely independent of agents' intentionality are neither good nor bad in ethics. Only certain *actions* and *behaviors* of agents or their

---

[2] In 1968 von Wright [34] introduces conditional obligation in deontic logic.

consequences can be good or bad. Moreover many actions do not have a moral value. So it is wrong to adopt the definition: *Bad p* $=_{df}$ ¬*Good p*. We need both connectives in deontic logic. von Wright (1968) himself recognizes the relevance of the *stit* connective of action in deontic logic where he admits that the *content of moral norms* are *actions* in [34]. In his view the convenient way to interpret formulas of the form *Op* is "It is obligatory (or one ought) to see to it that *p*." In order to enrich deontic logic von Wright elaborated a logic of action which is unfortunately mainly restricted to a logic of temporal change. Actions are events that change the state of the world. Many present actions cause posterior effects in the world. However von Wright ignores the link between actions and the intentionality of agents and he does not consider that present actions can cause effects at the very moment of their performance. Agents of satisfied successful attempts reach their objectives at the moment of their attempts. It must be logically valid that (*a stit p*) → *p* in the logic of action.

Let us now consider the *axiomatic system* of von Wright's standard deontic logic SDL of 1951. SDL has *three axiom schemas* and *two rules of inference*:

Ax1. The axiom schema of *truth functional tautologies*: All formulas of the object-language which are standard tautologies are axioms of SDL.

Ax2. The axiom schema of distribution of obligation with respect to material implication: All formulas of the form $O(p \to q) \to (Op \to Oq)$ are axioms.

Ax3. The axiom schema according to which any obligation contains a permission: All formulas of the form $Op \to \sim O\sim p$ are axioms.

R1. The standard rule of *Modus Ponens*: From two formulas of SDL of the forms *p* and (*p* → *q*) one can infer the new formula *q* of SLD.

R2. The rule of *Deontic Necessitation*: From any theorem *p* of SDL one can infer the new theorem *Op* in SLD.

I will now criticize von Wright's axiomatization. His first two axiom schemas of SDL are well selected: they are schemas of logically valid formulas.[3] Thanks to axiom schema 1 SDL can prove all truth functional tautologies which are well-known basic logically necessary truths. All truth functional tautologies are obvious tautologies in my predicative propositional logic. Thanks to axiom schema 2 SDL can also prove all valid formulas of the form (*Op* & *Oq*) → *O*(*p* & *q*). Clearly when it is both obligatory to see to it that *p* and to see to it that *q*, it is obligatory

---

[3] According to standard terminology, a formula of a logic is *logically valid* when it is true in all possible circumstances according to all models of that logic. A formula of a logic is a *theorem* of that logic when it can be derived from its axioms using the rules of inference of its axiomatic system.

to see to it that both p and q. The same axiom schema $\Box(p \to q) \to (\Box p \to \Box q)$ holds in modal logic for all kinds of necessity. However, as E. J. Lemmon [15] pointed out, axiom schema 3 states a dubious law which prevents standard deontic logic from representing *moral dilemmas*. Obligations can be in conflict. Agents can be obliged for different reasons to realize incompatible actions or actions with more or less moral value. Agents could also be obliged to realize too many actions at a given moment. In such cases one can doubt that every moral obligation is a moral permission. Agents should then not be permitted to perform all obligatory actions but only those which have moral priority. Laws of deontic logic of the form $(\boldsymbol{O}p \to \sim\boldsymbol{O}\sim p)$ are not well established like laws of modal logic like $\Box p \to \Diamond p$. Furthermore the definition of permission of SDL is wrong. One must define permission from order and illocutionary and propositional negations: *permit [a stit p]* $=_{\text{def}}$ $\neg order \sim [a\ stit\ p]$. From this right definition one cannot infer the theorem $(\boldsymbol{O}p \to Permitted\ p)$ in illocutionary logic. One can just infer that a given order permits the hearer to realize the ordered action. In illocutionary logic an agent can be obliged to do a lot of actions (sometimes even incompatible actions) for many different reasons and because of illocutions of several agents. All these obligatory actions can moreover have a different moral value. So an agent is not permitted to do all obligatory actions and moreover he or she should not do all of them.

As one would expect, the deontic modality of obligation of SDL is weaker than any kind of necessity. Unlike necessity, moral obligation does not imply truth. Clearly we do not realize all morally obligatory actions. It also happens that we realize morally forbidden actions. Thus it should not be logically valid that $Op \to p$ and that $Forbidden\ p \to \sim p$.

A lot of well-known defects and paradoxes of standard deontic logic are due to its second rule of inference of deontic necessitation which entails that all tautological propositions are contents of moral obligations. Any truth functional tautology is indeed an axiom and consequently a theorem of SDL. So any formula of the form $\boldsymbol{O}(p \to p)$ is a theorem of SDL that one can derive by deontic necessitation from the theorem $(p \to p)$. And similarly for all other truth functional tautologies like $(p \lor \sim p)$ where $\lor$ is the truth functional connective of disjunction. According to SDL it is then morally obligatory to see to it that $(p \to p)$ and that $(p \lor \sim p)$. From a philosophical point of view, such theorems are absurd and logically false. For tautologies are by nature necessarily true no matter what we do. How could we then be morally obliged to make them true? von Wright (1968) is well aware of this defect of his axiomatic system. He admits that only *logically contingent actions* (actions that are neither logically necessary nor logically impossible) can have a deontic normative status. "One could build deontic logic in such a way that it accords with this view." But this would lead to "considerable formal complications and inconveniences" [34, p. 61]. According to Belnap and me, von Wright is wrong on this matter. Following traditional philosophy of action we think that no agent could bring about a fact which is necessarily existent. Any adequate logic of action must state the valid law that every action is evitable. One axiom schema of the standard logic of action is that $[a \; stit \; p] \rightarrow \sim \Box p$.

Furthermore because of its rule of deontic necessitation SDL can prove a lot of well-known *deontic paradoxes* that are inacceptable in ethics. As A. Ross [21] pointed out, theorems of the form $\boldsymbol{O}p \rightarrow \boldsymbol{O}(p \vee q)$ are paradoxes when q represents a bad action. According to A.N. Prior [20], one can also derive in SDL the so-called *paradox of the Good Samaritan*. One derived rule of inference of SDL: from a theorem $(p \rightarrow q)$ infer the new theorem $(\boldsymbol{O}p \rightarrow \boldsymbol{O}q)$ leads to paradoxes when p represents a good action (to help someone who has been attacked) whose performance entails a bad action (a previous attack). In that case if it is obligatory to do the good action (to help the attacked person), the bad action is then also obligatory (the person to help should have been attacked). One can moreover derive the *knowledge paradox* of L. Aqvist [1] who applies the paradox of the Good Samaritan to knowledge. It can be obligatory for someone to know that a bad action is or has been performed. In that case the obligatory knowledge entails the very performance of that bad action. One can then derive that the bad action is also obligatory. Finally according to A.N. Prior [19] one can also derive in SDL *paradoxes of derived obligations* of the form $(\boldsymbol{O}{\sim}p \rightarrow \boldsymbol{O}(p \rightarrow q)$ when p represents a forbidden action. Whoever does a forbidden action becomes obliged in SDL to do every action. Given its wrong definitions of illocutions deontic logic proves a lot of wrong theorems about capital illocutions of ethics and moral obligations. Speakers attempt to perform few illocutions (sometimes none) in contexts of utterance. So the law of excluded middle does not hold for the performance or non-performance of illocutions. However one can prove in SDL formulas of the form (*Forbidden p* $\vee \sim$*Forbidden p*), (*Permitted p* $\vee \sim$*Permitted p*) and even (*Forbidden p* $\vee$ *Permitted p*) which are wrong instances of the law of excluded middle.

In order to state logically valid laws of deontic logic, we need to integrate deontic logic within illocutionary logic and enrich its expressive abilities. We need force markers and illocutionary connectives. Many different kinds of sentences (declarative, imperative, performative, exclamatory, ought and conditional sentences) are used in natural language in order to perform capital illocutions of ethics. One should be able to translate in the ideography of deontic logic such sentences. In my view, ought sentences of the form "a *ought* to do p" are imperative sentences which give to an agent a the directive to do an action that one presupposes to be obligatory.

So my new deontic logic contains in its lexicon constants indispensable to form the markers ⊢, ⊥, !, ⊤, and ⊣ that express respectively the primitive assertive, commissive, directive, declaratory and expressive forces and components of more complex forces of ethics like the preparatory condition that an action is obligatory or good and the logical connectives ¬ and ⇒ of illocutionary negation and illocutionary conditional. In my ideography, ought sentences of the form "a *ought* to do p" have the complex directive marker $[\Sigma_{\text{obligatory}}]![a\ stit\ p]$. Clauses of the form $[a\ stit\ F(p)]$ with the force marker F express the proposition that the agent a *succeeds in performing* the illocutionary act of the form $F(p)$. In order to axiomatize the valid laws of deontic logic we need to add to the axioms of illocutionary logic stating valid laws governing first level attitudes, actions and illocutions new axioms stating specific universal laws of ethics. Our enriched deontic logic can now take into account the intentionality and minimal rationality of agents and state logically valid laws such as ⊢($Tautologically\ p \to \sim[a\ Tries\ [a\ stit\ p]])$ and ⊢($Tautologically \sim p \to \sim[a\ Tries\ [a\ stit\ p]])$. Such new laws serve to revise and eliminate wrong laws and previous paradoxes of standard deontic logic. One can prove in my deontic logic that ($Tautologically\ p \to \sim[\boldsymbol{O}[a\ stit\ p]])$ and that ($Bad\ q \to \sim[\boldsymbol{O}[a\ stit\ (p \lor q)]])$.

My new deontic logic does not only state axioms that are *logically valid assertions* governing moral obligations and good and bad actions like ⊢($\boldsymbol{O}p\ \&\ \boldsymbol{O}q) \to \boldsymbol{O}(p\ \&\ q)$, ⊢$Good\ (p \to q) \to ((Good\ p \to Good\ q))$ and ⊢($Good\ p \to \sim(\Box p \lor \sim \Diamond p)$). It also gives universally good directives to hearers that they ought to follow and also declare that bad actions are prohibited. Imperative and ought sentences of the form ![You *stit p*] and $[\Sigma_{\text{obligatory}}]![$You *stit p*] serve to express categorical directives in my symbolism. Some philosophers doubt that ethics can give categorical directives. My deontic logic gives at least conditional directives of the form: If an action is good, do it! And moreover: If an action is bad, do not do it! Conditional imperative formulas of the form ($Good\ [$You *stit p*]) ⇒ ![You *stit p*]) and ($Bad\ [$You *stit p*]) ⇒ ![You *stit* $\sim$[You *stit p*]]) are axioms of my deontic logic that express conditional directives that one ought to follow. See my next book **Speech Acts in Dialogue** for a complete formulation of my deontic logic.

References

[1] L. Aqvist. "Good Samaritans, Contrary-to-Duty Imperatives and Epistemic Obligations". In: *in issue 1 of Nous* (1967), pp. 361–369.

[2] J. L. Austin. *How to Do Things with Words*. Oxford: Clarendon Press, 1962.

[3] N. Belnap and M. Perloff. "Seeing to it that: A Canonical Form for Agentives". In: *Knowledge, Representation and defeasible Reasoning*. Ed. by H. B. Kyburg et al. Dordrecht: Kluwer, 1990, pp. 175–199.

[4] N. Belnap and M. Perloff. "The Way of the Agent". In: *Studia Logica. Logic of Action* 51.3/4 (1992), pp. 463–484.

[5] N. Belnap, M. Perloff, and Ming Xu. *Facing the Future Agents and Choices in Our Indeterminist World*. Oxford: Oxford University Press, 2001.

[6] B. F. Chellas. "Time and Modality in the Logic of Agency". In: *Studia Logica. Logic of Action* 51 (1992), pp. 485–517.

[7] C. Cherniak. *Minimal Rationality*. MIT Press, Bradford Books, 1986.

[8] R. Descartes. *Les passions de l'âme*. Paris. 1649. Reedited in R. Descartes (1953). *Œuvres et lettres*. La Pléiade, Gallimard. 1953.

[9] G. Frege. "Sinn und Bedeutung". In: ed. by English translation in P. Geach and M. Black (eds.) (1970). Oxford: Blackwell, 1892, pp. 56–78.

[10] G. Frege. *"Thoughts","Negation" and "Compound Thoughts"*. In: Logical Investigations. New Haven: Yale University Press, 1997.

[11] A. I. Goldman. *A Theory of Human Action*. Princeton University Press, 1970.

[12] J. Hintikka. *Knowledge and Belief*. Cornell University Press, 1962.

[13] J. Hintikka. "Semantics for Propositional Attitudes". In: *Reference and Modality*. Ed. by L. Linsky. Oxford: Oxford University Press, 1971, pp. 145–167.

[14] G. W. Leibniz. *Elementi Juris Natularis*. 1669–71. Translated in English by C. Jones *Elements of Natural Right* in C. Jones. *The Science of Right in Leibniz's Moral and Political Philosophy*. Bloomsbury Academic, 2013.

[15] E. J. Lemmon. "Moral Dilemmas". In: *Philosophical Papers* 71 (1962), pp. 139–158.

[16] R. Montague. *Formal Philosophy*. New Haven: Yale University Press, 1974.

[17] G. E. Moore. *Principia Ethica*. Cambridge University Press, 1903.

[18] W. of Ockham. *Tractatus de Praedestinatione*. 1321–23. Reedited by the Franciscan Institute Edition in 1945.

[19] A. N. Prior. "The Paradoxes of Derived Obligations". In: *in issue 63 of Mind* (1954), pp. 64–65.

[20] A. N. Prior. "Escapism". In: *Essays in Moral Philosophy*. Ed. by A.I. Melden. University of Washington Press, 1967, pp. 135–146.

[21] A. Ross. "Imperatives and Logic". In: *Theoria* (63 1941), pp. 53–71.

[22] J. R. Searle and D. Vanderveken. *Foundations of Illocutionary Logic*. Cambridge University Press, 1985.

[23] D. Vanderveken. *Meaning and Speech Acts. Principles of Language Use*. Vol. 1. Cambridge University Press, 1990.

[24] D. Vanderveken. *Meaning and Speech Acts. Formal Semantics of Success and Satisfaction*. Vol. 2. Cambridge: Cambridge University Press, 1991.

[25] D. Vanderveken. "Illocutionary logic and discourse typology". In: *Revue internationale de philosophie* 216 (2001), pp. 243–255.

[26] D. Vanderveken. "Attempt, Success and Action Generation: A Logical Study of Intentional Action". In: *Logic, Thought and Action*. Ed. by D. Vanderveken. Springer, 2005, pp. 316–342.

[27] D. Vanderveken. "Fondements de la logique des attitudes". In: *Philosophical Review Manuscrito* 29.2 (2006), pp. 597–636.

[28] D. Vanderveken. "A general logic of propositional attitudes". In: *Dialogues, Logic and Other Strange Things* 7 (2008). Ed. by C. Dégrémont, L. Keiff, and H. Rückert, pp. 449–483.

[29] D. Vanderveken. "Beliefs, desires and minimal rationality". In: 57 (2009).

[30] D. Vanderveken. "Towards a formal pragmatics of discourse". In: 5.1 (2013), pp. 34–69.

[31] D. Vanderveken. "Intentionality and Minimal Rationality in the Logic of Action". In: *Nuel Belnap on Indeterminism and Free Action*. Ed. by T. Müller. Springer, Cham, 2014, pp. 315–341.

[32] D. Vanderveken. "Quantification and Predication in Modal Predicative Propositional Logic". In: 229 (2015), pp. 35–55.

[33] G. H. Von Wright. "Deontic logic". In: *Mind* 60.237 (1951), pp. 1–15.

[34] G. H. Von Wright. "An essay in deontic logic and the general theory of action". In: *Acta Philosophica Fennica* 21 (1968).

How and Why to be a Minimalist
GUIDO IMAGUIRE

Tarcísio Pequeno is one of the most impressive thinkers I ever met. He is the new Socrates: for him, philosophy is not a collection of doctrines rigidly frozen in books and articles, but an activity. As far as I can see, he is not interested in defending a theory like realism, nominalism, pragmatism, or empiricism. He is only interested in the argumentative force of thoughts, arguments, and insights. However, we may say that there is a methodological attitude that is a constant in his approach to any problem of philosophy and logic. In this paper I try to explain and justify this methodological attitude: the minimalism, the principle according to which the less the better.

Putnam famously said that one philosopher's *modus ponens* is another philosopher's *modus tollens*. Indeed, many philosophers support Platonism appealing to arguments of the following kind

> A entails Platonism.
> A is obviously true.
> Therefore, Platonism is true.

Tarcísio Pequeno would probably reason in the opposite direction:

> A entails Platonism.
> Platonism is obviously false.
> Therefore, A is false.

I hope not to misrepresent Pequeno's thought when I assume that for him any claim that entails Platonism must be wrong. However, it is worth noticing that Pequeno was not particularly interested in defending nominalism or any other traditional metaphysical theory. His critical attitude towards Platonism has less to do with the medieval rejection of universals or the contemporary rejection of abstracts than properly with the rejection of the typical Platonistic attitude. By 'Platonistic attitude' I mean the methodological strategy of introducing entities for solving philosophical enigmas or, as we may say, the attitude of 'making

Copyright © 2019 by Guido Imaguire. All rights reserved.

makers'. Take for example Truthmakers. Some philosophers like Armstrong, Mulligan and Rodriguez-Pereyra feel the necessity of explaining the difference between true and false sentences. Their argument is based on a realistic intuition: true sentences must 'have something' that false sentence do not have, thus 'there must be' things in reality responsible for the truth of true sentences. In this way, these philosophers 'make' things that make sentences true, and call them 'Truthmakers'.

However, the opposition between *Platonism versus Nominalism* is misleading in one sense. It suggests that Platonism is a non-minimalistic theory. Indeed, it has become usual in metaphysics to put things this way: the Platonist is a philosopher who rejects the principle of parsimony, defended by the nominalist. In fact, I think that the genuine Platonist is like the nominalist a lover of simplicity as well, but a simplicity of a different nature (see [10]). Therefore, for avoiding this confusion I will use in this paper the more neutral terms 'minimalism' and 'non-minimalism' instead of 'nominalism' and 'Platonism'.

Pequeno is certainly not an adept of this attitude of making makers. And I think he is quite right in doing so. Why should we avoid this attitude? Why should we keep minimalist in our ontology? The usual aesthetical appeal to desert landscapes is too elusive to be taken seriously. A more substantial reason seems to be this: we may be minimalists because the introduction of new entities for accomplishing a work or for 'playing a role' very often deviates us from the hard but much more productive path of searching deeper in the nature of reality. By introducing new entities we are just creating new myths, proposing *ad hoc* solutions and stopping the development of genuine knowledge.

Let me explain this with an example of the history of philosophy—a caricatured example, admittedly. Some medieval philosophers were puzzled about the interaction of the two substances: the *res extensa* and the *res cogitans*. On the one hand, the human body, like any other matter, is extended in space and submitted to causal forces. One moving ball may collide with another ball making this start to move. Mind, on the other hand, is not extended in space and is not submitted to causal relations. But then, how can we explain that mind can cause my body to move, given the fact that the mind has no extension? When I decide in my mind to move my hand, I just move it—that simple is it. How is this possible? The non-minimalist route to explain this puzzle is to introduce a 'connection-maker', an entity which plays the role of connecting both substances. This way occasionalists appealed to God as the solution: whenever He sees that there is a wish to move a hand in somebody's mind, the omnipotent Being causes the corresponding movement. Of course, God was not introduced with the solely purpose of explaining the mysterious connection between mind and body. In that time He was used to explain almost everything—He was the big 'everything-maker'. In any case, this theory sounds incredible *ad hoc* to our 'enlightened' ears. Nowadays, we are still waiting for an answer for the mind-body puzzle. But we are convinced that, whatever the right answer is, it must be found in a more accurate analysis of the nature of both entities, mind and body, and not in the introduction of a third extraneous entity. In fact, the most promising solutions are based on the dissolution of two entities mind and body into one single unity. In a word: the less the better.

In the rest of this paper I will discuss five instances of these two ways of dealing with philosophical problems, the minimalist and the non-minimalist, and will show that in all of them, the minimalist approach is clearly superior.

1 Universals: the Similarity-Makers

Universals are the prototypical makers. They were introduced for playing a lot of different roles, many of them explained in Lewis classical *A New Work for a Theory of Universals* ([12]). But their original role was certainly that of grounding the similarity of different things. At least, according to a usual interpretation of Plato, he introduced universals, or 'pure forms', as he preferred to call them, in order to explain the *One Over Many Problem*, i.e. the problem about the possibility of different particulars sharing the same nature. Both Peter and John are humans, so 'they have something in common', viz., humanity. More generally, if a is F and b is F, we must conclude that a and b 'have something in common', viz. F-ness. In fact, this argument, usually called *One Over Many Argument*, yields a general schema for introducing (or 'reifying') different kinds of 'makers'. Thus, the Platonist explains similarity of nature in the same way we explain brotherhood: if Peter and John are brothers, then they have something in common, viz. their parents.

Is this a reasonable explanation? I think it isn't for many reasons I argued elsewhere. But let me mention just two reasons here. First, it has no real explanatory force. No ordinary man in the street, but only some philosophers (like e.g. [1, p. xiii]), will be really puzzled when confronted with the fact that this rose and that house are both red. And if we are successful in convincing someone that there is a reason for being puzzled, I do not think that he will be relieved when we explain: look, this rose and that house are both red because there is a realm of abstract entities beyond space and time, and in that realm there is an abstract entity which is the pure form of Redness, and when concrete things 'participate' in that pure form, they become red. Second, the explanation runs into a regress: if a is red because a participates in Redness, may we also say, by parity, that a participate in Redness because a and Redness participate in the relation of Participation, and this, on its turn, is the case because a, Redness and Participation participate in a higher order Participation, and so on?

But how can we explain the possibility of sameness of nature without appealing to Platonic forms? The notion of immanent properties, as suggested by Aristotle, is certainly a first step in the right direction. We do not have to assume a realm of abstract entities whose existence is independent of concrete particular things in order to explain that concrete things have certain properties. Properties that characterize things are 'inside' the concrete particulars. In this way, we reduce two realms to one single all-including realm. Everything that is, is in space and time. A metaphysical system with one realm instead of two realms is certainly an important step in the right direction.

However, this advance is limited. In the Aristotelian framework we still have two categories in this single realm: the necessarily one-located entities, viz. particulars, and the potentially multi-located entities, viz. properties. Only with the development of nominalism we achieve a more minimalistic stance. According to (any kind of) nominalism, there is only one kind of entities: particulars. Properties are not real, they are just *flatus vocis*. Thus, after reducing two realms to one, the traditional nominalist also succeed in reducing two categories to only one. But how can we explain sameness of nature? Do the red rose and the red house not have something in common? No, answers the nominalist. To say that they 'have something in common' is just a misleading manner of speaking. For Redness is not something things can 'have'. As [13] suggested, the way a is makes 'Fa' true, but ways that objects are are not additional entities alongside the objects that are those ways (for a similar view about properties as ways see [9, p. 126]). Everything that is, is a thing and is in some way, but the ways things are, do not exist. Many strategies may be developed in order to explain why things are the way they are, and how they can be similar and different at the same time. Predicate, concept, class, resemblance and ostrich or priority nominalism are some of the main options. Also trope theory, a new and innovative ontological theory, emerged in middle of the last century as an alternative solution.

It is not my purpose here to defend one of these possible solutions. The only point I want to stress is that any one of the nominalistic solutions is preferable over Platonism, for they do not appeal to extraneous abstract entities for explaining something that is fully located in space and time. In order to explain why a concrete particular is the way it is, it must be sufficient to scrutinize the nature of this very particular thing. This insight puts us in the right path of recognizing that the puzzle only emerged because we artificially split reality into two strange categories: bare undetermined particulars on the one side and free floating properties on the other side. Both entities considered in isolation are merely man-made creations with not ground in reality. Everything that is, is in many ways, and all ways are only ways things are. The reality is a composition of what Armstrong called 'thick particulars', i.e. fully determined or propertied particulars. Any minimalist will prefer an ontological system with one category instead of two.

But minimalism is addictive: once you feel the intellectual pleasure of reducing or eliminating entities, you will look for more to get rid of. So, for the new 'Harvard Nominalism' of [8], the reduction of two categories to the solely category of particulars does not go far enough. For one may accept that there are only particulars, but of two different kinds: concrete and abstract. For numbers, sets, classes, and other entities of mathematics are plausible abstract particulars. In fact, it is very appealing to ground the apparent objectivity of mathematical truths on the existence of such abstract mathematical entities. Indeed, the once abandoned Platonic realm of pure forms is restored in contemporary philosophy of mathematics camouflaged as Frege's 'Third Realm'. This realm was introduced for justifying the objectivity of mathematics and logic, our next topic.

2 Objectivity-Makers

There is a direct rout from the notion of existence to the notion of objectivity. Indeed, it is easy to assume the objectivity of the concrete world of space and time for it is independently existent. And it is very tempting to assume that everything that is not concrete, like thoughts and feelings, belong to the subjective realm of the mind. Since logic and mathematics are supposed to be objective and, at the same time, their entities are not concrete, Frege ([6]) assumed the existence of a realm of abstract entities, which are the references of logical and mathematical terms.

In fact, in philosophy of mathematics we are often confronted with arguments that intend to derive Platonism from the objective nature of mathematical truths.

> The objectivity of mathematical truths entails Platonism,
> Mathematical truths are obviously objective
> Therefore, Platonism is true.

Now, it is very hard to appeal to *modus tollens* for concluding the non-objectivity of mathematical truths. I suppose that even Pequeno wouldn't dare defend this strategy. Fortunately, there is still an option for the minimalist, viz. to reject the first premise, i.e. to refuse the claim that the objectivity of mathematical truths entails Platonism. There are different and quite exciting routes to defend objectivity of some truths without appealing to existence of corresponding entities. 'Objectivity without Objects' is a very suggestive title of Van Cleve's article that indicates one possible strategy. A much more difficult road is the one Pequeno would probably prefer: pragmatism or collectivism. Roughly, the clue of this approach is to ground objectivity on a kind of *ratio collective*. While subjective is everything that is grounded in the individual subject, objective is whatever transcend the individual subject and emerge as the result from the interaction of a plurality of subjects. Logic and mathematics are basically systems of rules. Thus, in order to understand the nature of these disciplines, we may understand the nature of their rules. Rules are basically instructions for acting and thinking according to some patterns. We may merely act according to some rules, but this does not entail what a pragmatism like [20] would consider a genuine case of rule-following (see [14]). Genuine rule-following occurs whenever an agent acts according to the rule *in virtue of the rule itself*, i.e. when the rule is the ground for the action. Logical, mathematical, linguistic and many other kinds of rules are rules we follow this way and so they constitute objective rules. As expected, objectivity is grounded beyond the subject and, fortunately, without the existential presupposition of a third realm of abstract entities.

The project of reducing all entities of the realm of abstract logical and mathematical entities is a huge one, and cannot be developed in a short paper. But the old nominalism about properties may indicate one further possible strategy. It would be certainly wrong to describe nominalism as a theory according to which properties do not exist *simpliciter*. A nominalist is not someone who disagrees with the realist about how to classify entities: Socrates is a particular and redness is a property. So, in a non-rigorous sense, we may truly say that 'there are' properties. The claim of the nominalist is that properties are not fundamental building blocks of reality. They are just derivative, just like the mind is derivative for a physicalist. A physicalist does not say that thoughts and feelings are not 'real'—in the sense that they are merely fictions as Sherlock Holmes—but only in the sense that they are not fundamental. Thus, if we define numbers in terms of properties or, more exactly, as second-order properties and at the same time claim that properties are just derivative entities, as a traditional nominalist, a plausible agenda for justifying nominalism in mathematics is found. Numbers are just special second-order properties, which are grounded in first-order properties that, on their turn, are grounded in particulars. So, if we accept [15] criterion of ontological commitment, according to which, we are committed to the entities we quantify over in the fundamental truths of our best theories, the mere existence of first and second-order predicates do not by itself force us to accept the existence of properties of any order, and so, of numbers.

But mathematics is grounded in logic, which is *per excellence* an abstract realm. So, logic is without doubt a huge challenge for the minimalist. How can someone be a minimalist in logic? Fortunately, even in logic the minimalist can find his own way to defend his stance. Let us see how.

3 Truth- and Validity-Makers

Truthmakers are usually defined as the entities in virtue of which sentences are true. So, given the true sentence S, the entity E is the truthmaker of S if and only if S is true in virtue of E existing (see [2, pp. 5, 7, 16-17], [18, p. 34] and [17]. Alternatively, we may say that the existence of E entails the truth of S: E is the truthmaker of S iff 'E exists' entails 'S is true'

The introduction of truthmakers is harmless insofar we do not necessarily have to introduce new entities in our ontology. One who believes in objects and properties, for instance, may simply assume that objects and properties are the truthmakers of sentences. Another one, who prefers the ontology of facts, will take facts as the truthmakers of true sentences. A nominalist of the kind presented in the end of section 1 e.g. may accept that thick particulars are the solely truthmakers of true sentences. Of course, one may reconsider his own ontology because of some constrains of the notion of truthmakers. But one can also certainly maintain his favourite ontological category as being the solely truthmakers.

Inspired in the idea of truthmaker Gregory Gaboardy suggested (in private communication) the introduction of a new kind of makers, which are called 'validity-makers' or 'validators' which could help us understand the notion of logical consequence. A validator is an entity that makes a given sequence of sentences a valid argument. Take the sequence (A) of sentences:

(S1) All men are mortal. —made true by (T1)
(S2) Socrates is a man. —made true by (T2)
(S3) Socrates is mortal. —made true by (T3)
} made valid by V(A)

We obviously recognize (A) as a sequence of three sentences of a very special kind: it is an instance of what logicians call 'a valid argument'.

Sentences (S1) and (S2) may be considered premises of the conclusion (S3). Since (S1)-(S3) are true, we may arguably suppose that we have three truthmakers (T1), (T2) and (T3) which make (S1), (S2) and (S3) respectively true.

Take now the sequence (B)

(S4) All men are mortal. —made true by (T4)
(S5) Tarcísio plays soccer at the beach. —made true by (T5)
(S6) Lancelot is a dog. —made true by (T6)

Also (B) is a sequence of true sentences. But, contrary to (A), (B) does not constitute an instance of a valid argument, although all sentences (S4) – (S6) are true and have therefore truthmakers. According to Gaboardi, we may say that the sequence (A) has something that sequence (B) does not have: a validator. The validator of (A), say 'V(A)' is that very thing in virtue of which A is valid.

I can easily imagine that some non-minimalist oriented philosophers may be sympathetic to validators and to the idea that such new species in our logical zoo may, in fact, be responsible for the validity of valid arguments. Validators are just cousins of one of the favourite entities of some realists in logic, viz., logical forms. But there are alternatives for the minimalist, and I think this is a very clear case in which not the addition, but a reduction of entities takes us further in the task of understanding the nature of the notion of logical consequence. Instead of assuming V(A) as a seventh entity in addition to the three sentences (S1)-(S3) and their three truthmakers (T1)-(T3), we may simply define validity by *the absence of an additional particular truthmaker for the conclusion*, i.e. (T3). With other words, what grounds the truth of the conclusion (S3) is neither a third truthmaker (T3) nor a validator V(A) of the sequence of sentences (S1) – (S3), but the truthmakers (T1) and (T2) together:

(S1) All men are mortal. —made true by (T1)
(S2) Socrates is a man. —made true by (T2)
(S3) Socrates is mortal. —made true by (T1) and (T2)

In the form of a slogan: the truthmakers of the premises are truthmakers enough. In our example (and for the sake of simplicity assuming an ontology of facts): the fact that all men are mortal and the fact that Socrates is a man are jointly truthmakers of 'Socrates is mortal'. Whenever the truthmakers of a set of sentences are sufficient to ground the truth of an additional sentence S, we may say that S is a logical consequence of the set of sentences. More formally:

$$S \text{ is a logical consequence of } P_1, \ldots, P_n$$
$$\text{iff}$$
$$\text{the joint truthmakers of } P_1, \ldots, P_n \text{ make } S \text{ true.}$$

It is not the aim of this paper to scrutinize the consequences of this definition, but it seems at first glance clear that this notion will be more suited to more restricted logical systems as relevant logic and paraconsistent logic, for we may hardly argue that the truthmakers of 'Lancelot is a dog' and 'Lancelot is not a dog' could ground the truth of 'Tarcísio plays soccer at the beach'.

Truthmakers of sentences have arguably to do with the reference of the sentences. According to [7], we must even identify the truth-values themselves (the True and the False) with the references of sentences. However, Frege's most important contribution to semantics was not the notion of reference. In his paper [5], he noticed that expressions of a language may have the same reference but diverge in their cognitive values. For explaining the notion of *cognitive value*, Frege famously introduced the notion of *sense*. But are Fregean senses just a new kind of makers?

4 Meaning: Identity and Difference-Maker

Fregean 'senses' or, as some prefer to say today, 'meanings', are certainly a kind of makers: they are at the same time identity- and difference-makers. Meanings are *identity*-makers because they ground the equivalence relation of synonymity: synonymous expressions 'have something in common', viz. the same meaning. In this way we can also explain the possibility and the nature of translation: ideally, a translation consists in the substitution of expressions (usually whole sentences) of a given language L1 by expressions of a different language L2 that express the same meaning. The identity of meaning is what makes translation possible. Meanings also make indirect discourse possible: a speaker may correctly reproduce someone's speech without using his own words: if Mary says 'John is a bachelor' we can correctly report her statement by saying that, according to her, 'John is not married'. Insofar we express the *same* cognitive value in our report, we are saying something truth. Since 'bachelor' and 'unmarried man' express the 'same' meaning, we are adequately reporting Mary's speech.

However, lets remember that [5] introduced senses not in order to ground identity, but difference of cognitive values, or, more exactly, to explain the puzzle of the non-triviality of true identity statements. Two different expressions 'a' and 'b' may have the same reference; therefore a statement of the form 'a is b' is true. After all, what we are saying is that the reference of 'a' (which is by hypothesis the reference of 'b') is identical to itself. But how can we explain that such a statement is not trivial given that it is necessarily and trivially true that everything is identical to itself? Well, explains Frege, the statement may not be trivial since 'a' and 'b' have different meanings. So, meanings are also *difference*-makers. They ground a difference in cognitive values of expressions which refer to the same entity.

Meanings seem therefore to be very valuable tools for understanding the nature of language. But do we really have to assume their existence, beyond the existence of words and references? Any adept of minimalism would certainly try a different route before postulating meanings. I don't know anyone who assures to fully understand the eight obscure paragraphs in [19] in which Russell explains why Frege's theory of meanings must be wrong. But the general lines of his own proposal, the theory of descriptions, give us a hint how to explain the way language touch reality without the intermediation of meanings. Given two expressions such as 'the star that sparkles in the morning' and 'the star that sparkles in the evening', we should analyse them in terms of different predicates or items with different references in a way that, once we archive the full analysed form of both singular terms we only have words with their corresponding references that pick out the same reference in different ways. But note, we do not have to reify 'ways of picking out' the same reference nor, as Frege claimed, different ways of presenting the same reference.

The result of this attitude is the development of a theory which experienced one of the most impressive stories of success in philosophy, viz., the theory of definite descriptions. Instead of assuming that any singular term has to be represented in formalization by a constant for individuals like 'a', we shall assume that singular terms, paradigmatically definite descriptions, codify information that can be better captured by means of the quantified form '$\exists x(Fx \wedge \forall y(Fy \to x = y))\ldots$'. With this theory we could solve some puzzles concerning bivalence of sentences of natural language and explain the non-trivial cognitive value of identity statements without appealing to Fregean senses among other things. Of course, we may, like [11], be critical on Russell's descriptivism about ordinary proper names, given their peculiar semantic behaviour. But with this theory we certainly advanced in the direction of a better understanding of the connections between the use of singular article in definite descriptions and its correlations with uniqueness and existence requirements. Notice further that new Kripkean orthodoxy concerning the semantics of proper names does not affect the minimalistic stance. Kripke's theory of proper names as rigid designators entails theoretical resources for explaining how an identity statement like 'a is b' can be true without being trivial even when 'a' and 'b' are proper names. In any case, the minimalist can happily conclude: meanings are not necessary.

So far we have been defending minimalism. But a caveat is important: the avidity for eliminating makers should not led the minimalist to carelessly discard a bunch of 'more or less' correlated entities altogether. Quine [16], for example, was so eager in eliminating all intensional entities that 'meaning' and 'necessity' were considered just relatives which could be eliminated in one and the same move. For him, modalities like necessity and possibility were just metaphysical counterparts of the meanings, all being equality intensional and, for this reason, being suspicious entities with dubious identity criteria and which we should abolished in an enlightened and rigorous naturalistic philosophy. Perhaps we do not need meanings, as we just saw. But are we justified to reject modalities in the same way?

5 Possible Worlds: Possibility Makers

I do not see good reasons for being optimistic about the prospect of eliminating modalities of our scientific and philosophical discourse. But something different could be defended concerning the entities usually appealed to for grounding modalities, viz. possible worlds.

Some philosophers already before Leibniz appealed to the notion of possible worlds [3]. But virtually no one nowadays dares talk of alethic modalities without defining them in terms of possible worlds. Something (a state of affairs, object, property or whatever) is possible if it happens (holds, exists or is instantiated) in at least one possible world, and something is necessary if it happens (holds, exists or is instantiated) in all possible worlds. Indeed, due to the notion of quantification over possible worlds we have a powerful semantics for modal logic and no one today can deny this nice formal development.

However, not everything that is formally neat and useful must be considered harmless from a metaphysical and from an epistemological point of view. And I think that possible worlds are indeed suspicious entities in many senses. As Fine [4] has pointed out, the usual definition of modalities in terms of quantification over possible worlds clearly invert the hierarchical structure of reality. Is p really possible *in virtue of* the existence of at least one possible world in which p is true? And is p necessary because p is true in all possible worlds? I am in Rio now, but I could have been in Fortaleza. So, do I really need the existence of a maximal consistent set of states of affairs with all its countless inhabitants, or the maximal mereological sum of all concrete entities counterparts of our world-mates, in order to ground the existence of the possibility that I could have been in Fortaleza now? This sounds quite odd to anyone with a robust sense of reality. Grounding the possibility of a single entity in the existence of one possible world in the infinite domain of all possible worlds would be equivalent as searching in Borge's Library of Babel[1] for finding if there is somewhere in some book the occurrence of the letter combination 'ptoupmdb' for grounding that it is a possible combination of letters. Similarly, from an epistemological point of view: Even if we *say* that p is possible *because* p obtains in at least one possible world, we certainly do not *know* or find out that p is possible because we search in the infinite domain of possible worlds until we find a world in which p occurs (except, of course, when we conclude the possibility from the actuality). And even more evident: we do certainly not conclude that p is necessary after we check all possible worlds and establish that in no one of them non-p is true.

So, even if we need a huge domain of all possible worlds in order to formalize our usual modal talk, we should keep clear about the fact that these worlds are not the ontological ground for our real possibilities and necessities. If it is true that I could possibly have been in Fortaleza right now, this is due to the nature of matter and its distribution in the realm of space-time . If it is really necessary that water is H_2O, than this is so not because in all worlds in which there is water, this liquid is H_2O, but rather because of the very chemical nature of water. The minimalist, instead of appealing to a realm of infinite many entities of enormous complexity, focus in the analysis of the nature of the involved entities and in the semantic behaviour of the relevant terms in order to conclude that some truths are necessary. Vertical deepness instead of horizontal amplitude is the best way.

6 Conclusion: Last words of a minimalist

In this paper I just tried to defend the minimalist stance by listing five examples on how to deal with philosophical puzzles without inflating unnecessarily our ontology. The list could go forever. But listing more than these five examples would witness against my own minimalism. So, let us stop here.

Minimalism in ontology is not a theory, but rather a methodological attitude. For the minimalist, the introduction of extraneous entities for playing roles like grounding similarity, grounding validity, or grounding possibility is an unjustified move in the scientific game. I do not think

---

[1]Borges is an Argentine writer who imagined a library consisting of infinitely many books that contain every possible ordering of just 25 basic characters (22 letters, the period, the comma, and the space). Of course, the vast majority of the books are pure gibberish. But the library must contain, somewhere, every coherent book ever written, or that might ever be written, and every possible permutation or slightly erroneous version of every one of those books.

there is a general rule, or a set of basic rules, for defining the minimalist method. The minimalist must devise in each new situation new ways to face the mythological creations of his opponent. For this reason, the minimalist does not see philosophy as a collection of doctrines, but as an activity—and no one understood this better than Tarcísio Pequeno.

References

[1] D. M. Armstrong. *Nominalism and Realism, Vol. I. A Theory of Universals*. Cambridge University Press, 1978.

[2] D. M. Armstrong. *Truth and Truthmakers*. Cambridge University Press, 2004.

[3] H. Burkhardt. "From Origen to Kripke". In: *Possible Worlds*. Ed. by Guido Imaguire and Dale Jacquette. Munich: Philosophia Verlag, 2010.

[4] K. Fine. "Essence and Modality". In: *Philosophical Perspectives* 8 (1994), pp. 1–16.

[5] G. Frege. "Sinn und Bedeutung". In: *Zeitschrift fur Philosophie und Philosophische Kritik* 100 (1892), pp. 25–50.

[6] G. Frege. *Grundlagen der Arithmetik: Studienausgabe mit dem Text der Centenarausgabe*. Meiner Verlag, 1988.

[7] G. Frege. "Der Gedanke. Eine logische untersuchung". In: *Wittgenstein Studien* 4.2 (1997).

[8] N. Goodman and W. V. O. Quine. "Steps Toward a Constructive Nominalism". In: *Journal of Symbolic Logic* 12 (1947), pp. 105–122.

[9] J. Heil. *From an Ontological Point of View*. Oxford University Press, 2003.

[10] G. Imaguire. "The Platonism vs. Nominalism Debate From a Metametaphysical". In: *Revista Portuguesa de Filosofia* 2-3 (2015), pp. 375–398.

[11] S. Kripke. *Naming and Necessity*. Harvard University Press, 1980.

[12] D. Lewis. "New Work for a Theory of Universals". In: *Australasian Journal of Philosophy* 4.61 (1983), pp. 343–377.

[13] D. Lewis. "Things qua truthmakers". In: *Real metaphysics: Essays in honour of D.H. Mellor* (2003). Ed. by H. Lillehammer and G. Rodriguez-Pereyra, pp. 25–38.

[14] T. Pequeno. "Pedras podem seguir regras?" In: *Colóquio Wittgenstein*. Ed. by Tarcísio Pequeno Gido Imaguire Maria Aparecida Montenegro. Fortaleza: Editora UFC, 2006.

[15] W. V. O. Quine. "On what there is". In: *The Review of Metaphysics* 2.1 (1948), pp. 21–38.

[16] W. V. O. Quine. "Two dogmas of empiricism". In: *Perspectives in the Philosophy of Language* (2000), pp. 189–210.

[17] G. Rodriguez-Pereyra. "Truthmakers". In: *Philosophy Compass* 1.2 (2006), pp. 186–200.

[18] Gonzalo Rodriguez-Pereyra. *Resemblance nominalism: A solution to the problem of universals*. Oxford University Press, 2002.

[19] B. Russell. "On denoting". In: *Mind* 14.56 (1905), pp. 479–493.

[20] L. Wittgenstein. *Philosophical investigations*. John Wiley & Sons, 2010.

How to Get Rid of Logic and Be Happy
Desidério Murcho

Consider the following argument:

> Mark is a novelist.
> Therefore, Samuel is a novelist.

This is not a valid argument. However, suppose that Mark is Samuel. In this case, it seems there is no way for the premise to be true and the conclusion false since, after all, it is the same person under different names. This sort of example was used by Dorothy Edgington in her 2004 paper 'Two Kinds of Possibility' [2] and it seems to suggest that an argument is deductively valid iff:

(1) It is not possible for its conclusion to be false if all its premises are true.

(2) One can know a priori that (1).

(1) is the metaphysical condition, (2) the epistemic one. Notice that this extends easily to logical truths. A statement is a logical truth iff:

(1) It is necessarily true.

(3) One can know a priori that (1).

This is the wide sense of 'logical truth,' under which 'No blue all-over particular is green' is deemed one. To capture the strict, formal sense just add another condition:

(4) Condition (2) is met using only formal methods of proof.

Thus 'Samuel is a novelist if he is a novelist' is a logical truth in this strict sense: it is necessarily true and one can know a priori that it is so using only formal methods of proof. Ditto for the strict, formal sense of deductive validity; the following argument is valid in this sense:

Copyright © 2019 by Desidério Murcho. All rights reserved.

Mark is a novelist.
Mark is Samuel.
Therefore, Samuel is a novelist.

This is a promising first step towards an explanation of deductive validity and logical truth in classical and other related logics, but fails regarding logical or quasi-logical truths like those of indexicals: 'I am here' is perhaps known a priori and it is a logical truth of sorts, but it is arguably a contingent rather than a necessary truth. Let us put aside these cases for the moment.

Apart from the logic of indexicals this proposal is compatible with classical and most non-classical logics. Furthermore, it is somewhat pre-theoretical in the sense that it does not depend on whether you assume that some sort of model theory or proof theory is the way to define or understand validity and logical truth. Thus, it is the sort of thing Hartry Field favours in his 2015 paper 'What is Logical Validity?' [3]: it is a tentative definition of validity that gives a common ground to logicians of different persuasions. Field believes that a model-theoretic or proof-theoretic account of validity 'couldn't possibly be what we mean by 'valid'' (p. 33) mainly because given any one account—classical, say—the rival logician will agree that a given inference is valid on that particular account, although she believes it not to be genuinely valid: 'even those who reject classical sentential logic will agree that the sentential inferences that the classical logician accepts are valid in *this* sense' (p. 33), that is, in the sense of being classically-valid. I doubt this sort of argument drives home his point because the proof-theoretic or model-theoretic logical relativist will insist that there is no such thing as validity outside any one given logical theory. Still, even if the account presented here does not attempt to provide for now any sort of argument against the proof-theoretic and model-theoretic relativist, it aims at the sort of theoretical neutrality Field supports:

> One way to try to explain the concept of validity is to define it in other (more familiar or more basic) terms. As we've seen, any attempt to use model theory or proof theory for this purpose would be hopeless; but there is a prominent alternative way of trying to define it. In its simplest form, validity is explained by saying that an inference (or argument) is valid iff it preserves truth by logical necessity. (p. 35)

The necessary preservation of truth in a valid argument is perhaps the same as the impossibility of having a false conclusion if all its premises are true; if that is so, Field's proposal is quite close to my reading of

Eddington's. He seems unaware, however, that without an epistemic clause he needs either to exclude the necessary a posteriori or declare that the argument 'Mark is a novelist; therefore, Samuel is a novelist' is valid after all.

Let us start with the first alternative: suppose you do not buy the idea that there are necessary a posteriori truths. In that case, you do not believe there are invalid arguments that meet condition (1) above. So far as it goes, this is a disagreement about the extension of the concepts of necessary truth and logical truth, not about the concepts themselves. You believe that those concepts are co-extensional (excluding perhaps the logic of indexicals and any such non-classical logics), and that amounts to the rejection of necessary a posteriori truths.

It may not seem that way, but if one distinguishes carefully the epistemic from the metaphysical components of logical truth this co-extensionality thesis is not easy to argue for. It certainly does not seem reasonable to just assume it. And why would the necessarily true be so finely tuned with what we are able to know a priori? Unless, of course, the claim is that 'necessary truth' just *means* 'knowable a priori, purely logically.' In that case however the claim is not just that the concepts of necessary truth and a priori knowledge are co-extensional; the claim is that it is the very same concept under different guises. This suggestion will be discussed in a moment. The point here is that rejecting the necessary a posteriori while accepting that the concepts of necessary truth and a priori knowledge are quite distinct is not very promising; if you want to reject the necessary a posteriori there is another way of doing it, to be presented shortly.

For now, assume that the concepts of necessary truth and a priori knowledge are quite distinct even if they are co-extensional. In this case, an eliminative reduction of the concepts of logical truth and deductive validity is at hand—as Field puts it, a definition 'in other (more familiar or more basic) terms.' A logical truth is nothing but a necessary truth that we are able to know a priori; a deductively valid argument is nothing but an impossibility—that it is impossible for the conclusion to be false and the premises true—that we are able to know a priori. Once you have a world with trees and atoms and all that, and cognitive agents capable of a priori knowledge, you have logical truths and deductively valid arguments. Perhaps there is also validity and logical truth in a realm of gods with formal languages and abstract objects, but we have an alternative, pedestrian way of understanding those concepts. The core logical relation of entailment, implication, the necessary preservation of truth or whatever you may want to use to talk about deductive validity, is thus nothing over and above the good old empirical world of trees and rivers and their relations plus a certain sort of cognitive workings of agents like us; ditto for logical truths. To put it differently: logical modality is reducible to metaphysical modality plus the a priori. There is no independent logical modality.

This raises the question, though, about the nature of necessary truths themselves. But why would this be so different from ordinary, non-necessary truths? Suppose that necessity really pertains to the world, and not to our knowledge, beliefs or other cognitive workings. In that case, the statement 'Water is necessarily H_2O'[1] is true if that is what water is necessarily, just like the statement 'Russell wrote *The Problems of Philosophy*' is true if he wrote that book. If the concept of necessity is taken as a bona fide metaphysical concept and not as a disguised epistemic concept, there is nothing special about at least some necessary truths; some of them have truth-makers just like some non-necessary truths.[2] Perhaps when a chemist examines a portion of water she does not see that it is necessarily H_2O. However, she does not see either that it is H_2O if by 'see' is meant anything direct and obvious. In order to *conclude* that that portion of water is H_2O the chemist goes through a complex set of tests and inferences. But suppose that the chemist could just look and see that that portion of water is H_2O, but could not see that it is necessarily so. How would that be a reason to believe that water is contingently H_2O? Unless we start by claiming that necessity is somewhat an epistemic concept, it is far from obvious that all necessary truths should be easily seen; why would they?

So, assuming that necessity is a bona fide metaphysical concept, pertaining to reality at large and not just to our ways of understanding it,

---

[1] Note that the chemist's claim that water is H_2O is not the claim that what we call 'water' in informal settings is *just* H_2O; it is rather the claim that there is a lot of H_2O in what we call 'water' in those settings and perhaps also that it is this chemical composition that explains lots of its chemical features. Of course that the water we drink has lots of other stuff, many important to our interests, and of course also that tomatoes and babies are mainly made of water but that is not the most relevant feature they have. (About this point, see Abbott's paper 'Water = H_2O.' [1]) Anyway, the metaphysician's claim is just that the chemist's claim, if true, is necessarily so, and this is compatible with the thought that chemists have it all wrong and water is really not H_2O. It would be somewhat surprising though to have a successful philosophical or linguistic argument against a well-researched result of chemistry.

[2] Suppose you believe that only the present is real and that necessary truths are true forever. In this case you believe that some necessary truths do not *have* truth-makers—even if they they *will* have them. This is a general point, though, regarding any truth about the future, and not specially about necessary truths: even if all truths are contingent or even simpliciter (in the sense of being neither metaphysically contingent nor necessary) this difficulty with truth-makers—if it is such—is still there.

we have a way of explaining logical truth and deductive validity. The world is such that makes it the case that there are necessary truths, and it makes it the case also that it is impossible for some statements to be false if some others are true; and then we have a way of knowing a priori some of those necessities and impossibilities. Both claims may need further clarification.[3] But neither is particularly egregious. Perhaps the first is just a primitive feature of the world, something we can explain but not in terms of other, more primitive aspects of the world. The statement 'Water is necessarily H_2O' can be formed from the simpler statement 'Water is H_2O' by the addition of the necessity operator, but that is not a good reason to believe that the world starts by having water that is simply H_2O and then some further feature—or possible world—makes it necessarily so. In general, it is not to be expected that the world is solicitous enough to mirror every aspect of our language—it is rather we that try to represent features of the world in our languages and we do not always do it in non-deceptive ways. As to the second claim, whatever you may think of a priori knowledge you have to accommodate the sort of justification mathematicians use and you have to accept that that is quite different from the chemists'. And this is enough to go on. Even if you reject the a priori as a philosophical term of art and believe that there is no clear and useful epistemic distinction here, you can accept this view because you surely accept that the standard methods of proof in mathematics are distinct, even if not a world apart, from that of the chemist's.

Did we get rid of logic, if we understand it this way? In a way, yes. There are no independent logical relations per se; there is just the world with its features, and cognitive agents capable of understanding some of them in a particular way. Logical truths are so just because they are metaphysical necessities that we are able to know a priori. It is not the case that metaphysical necessities are so because they are logical truths. Metaphysical modality and epistemic modality are primitive relative to logical modality; the former explain the latter, not the other way around. Notice that this does not mean that one is bound to reject logical modality as a primitive concept. Perhaps that concept is needed to do some further work, or perhaps there are other reasons not to get rid of logic as a primitive concept. The reasoning so far just expands a little a promising way of explaining logical truth using non-logical concepts.

What if you believe that in order to be a valid argument or a logical

---

[3]Namely, to tackle the worry that one may need negative facts. However, notice first that this is a difficulty also with non-modal truths and, secondly, that Eddington's metaphysical clause can be phrased without the worrying word 'impossible': necessarily, if the premisses of a valid argument are true, so is its conclusion.

truth, condition (1) above is quite enough? This is not a particularly promising view because 'Mark is Samuel' would turn out to be a logical truth and this certainly seems wrong. If you accept the necessary a posteriori, this statement is seen as necessarily true if true; but why would it be logically true also, and what are the inference rules of a logical theory according to which any statement of the form $a = b$ is a logical truth? And how do you distinguish those cases in which a statement of seemingly that very same logical form is false, like 'Quine is Kripke'? Thus, the belief that condition (1) is enough to explain logical truth and deductive validity is implausible unless one rejects the necessary a posteriori. The motivation for such a view is presumably the thought that logical truth and necessary truth walk hand in hand; the best way of going down that lane though is to reject the necessary a posteriori, thus rejecting that 'Mark is Samuel' is necessarily true if true, and rejecting also that the argument from 'Mark is a novelist' to 'Samuel is a novelist' is valid. This suggestion will be pursued in what follows, but let us pause a little.

What we have seen so far is that a reductive understanding of validity and logical truth is at least not completely egregious, as long as one takes the metaphysical concept of necessary truth seriously, together with some concept of a priory knowledge rich enough to account for mathematical knowledge. What we are about to see now is an alternative to this view for those that eschew metaphysical necessities. And one of the points of this paper is to motivate the thought that these are the only two promising views there are—seemingly rival views end up collapsing in one or the other.

Let us go back to indexicals like 'I am here.' Suppose you believe that this is a logical truth of sorts. Condition (1) above is seemingly not met: it is not necessary that I am here if I could have been elsewhere. And when condition (1) is not met, condition (2) is not met either (ditto for condition (3)). However, suppose you believe indexicals like that are logical truths. The obvious epistemic feature indexical statements like that have is that they are a priori: either by linguistic means alone or by reasoning alone I am able to know that I am here. So perhaps a more liberal view of the concept of logical truth demands only that we are able to know a priori that a statement is true. Perhaps in most cases necessity is also involved, but not in all.

This suggests a different reductive way, and quite minimalist, of understanding logical truth and deductive validity. Let us start with the former:

> A statement is a logical truth iff we are able to know a priori

that it is true.

Suppose now for a moment that this is taken seriously. In that case there is a way of giving something back to those who decry the necessary a posteriori: the concept of metaphysical necessity is not needed anymore.

Recall that one way to oppose the necessary a posteriori is phrased in such a way that the dispute is merely about the extension of the concepts. This is not plausible, among other reasons, because it leaves one with no plausible non-ad hoc reason to reject the necessary a posteriori. This difficulty however disappears if one eliminates the very concept of metaphysical necessity. It is not a matter of claiming that whatever we can only know a posteriori is contingent for quite mysterious reasons, but rather that philosophers have used all along the empty concepts of necessity and contingency: truth be told, nothing in the world responds to those concepts. It is not that water is contingently H_2O because we cannot know a priori that it is necessarily H_2O, but rather that water just is H_2O and that is it. There are no metaphysical modalities at all; there are only epistemic modalities.

A powerful reason to believe that there are necessary a posteriori truths emerges from the following simple logical result. Take any two names that denote the same particular, like 'Mark' and 'Samuel,' and form the sentence 'Mark is Samuel.' Add as a second premise the logical truth that Mark is necessarily Mark. Assume now an unrestricted form of Leibniz's law (the indiscernibility of identicals bit) that allows the substitution of 'Samuel' for 'Mark' in any non-epistemic context. The surprising conclusion is that Mark is necessarily Samuel. This conclusion is rock solid assuming that the concept of necessity here is metaphysical, not epistemic. In any non-epistemic context, Leibniz's law does allow unrestricted substitution of identicals for identicals. If the concept is not metaphysical, however, there is good reason to block this inference. After all, there is good reason to block the inference to 'Galen believes Samuel is a novelist' from the premises 'Mark is Samuel' and 'Galen believes Mark is a novelist.' So if one starts with the assumption that necessity is but an epistemic concept, there is no reason to accept the unrestricted substitution of identicals for identicals that one needs to prove the necessary a posteriori. From 'Mark is Samuel' and 'Mark is necessarily Mark' there is no valid route to 'Mark is necessarily Samuel' if 'necessarily' is not read metaphysically (or transparently if you prefer) because in that case the unrestricted use of Leibniz's law is invalid.

The funny thing is that even before we start exploring the thought that necessity is just an epistemic concept in disguise—the a priori, really—

we have an independent reason to suspect that the concept of logical truth is not purely metaphysical. For if it were the following inferences would both be indisputably valid, but the second certainly is not:

> It is true that Mark is a novelist.
> Mark is Samuel.
> Therefore, it is true that Samuel is a novelist.

> It is logically true that Mark is Mark.
> Mark is Samuel.
> Therefore, it is logically true that Samuel is Mark.

Either because one takes 'logically true' to be some sort of metalinguistic operator that requires quotes or for some other reason, the upshot is that if the restriction to the use of Leibniz's law is an indication that the concept involved is not purely metaphysical, then it seems that Edgington's suggestion is on the right track: the concept of logical truth, unlike that of truth simpliciter, may have epistemic content. The modal eliminativist just goes one (big) step further and proposes to eliminate altogether the concepts of logical truth and metaphysical necessity, reducing them to the a priori only. On this view, philosophers have all along been victims of a way of speaking, taking logical truth and necessary truth as concepts in their own right, whereas they are just different ways of talking about what we are able to know a priori. Notice that now there is no mystery regarding the fine tuning between the necessary and the a priori because one of the concepts just went out the window. It is not the case that there are two non-empty concepts—logical truth and metaphysical necessity—that happen to be mysteriously co-extensional; it is rather that there is only one concept—the a priori—under two different guises.

How promising is this? It is certainly quite minimalist. It takes two important concepts involved in a considerable volume of philosophical and formal work and proposes to throw them away, rereading them as just one epistemic concept under different guises. Will it work?

One initial worry is that the concept of necessary truth seems crucial to explain deductive validity. For take the following two arguments:

> Ruth and Dorothy are philosophers.
> Therefore, Ruth is a philosopher.

> Ruth or Dorothy are philosophers.
> Therefore, Ruth is a philosopher.

Suppose now that both Ruth and Dorothy are philosophers. Under this condition, both arguments have true premises and true conclusions. And yet, the first is valid, which the second is not. The difference of course is that in the first case it is not possible for its conclusion to be false if its premise is true, but this is exactly what is possible in the second case. So the modal concept of possibility plays a prominent role here. How would the eliminativist explain this difference?

Well, the difference is that in the first case, but not in the second, there are ways of knowing a priori that if the premise is true, so is the conclusion. Thus, on this view, the relation between the premises and the conclusion of a valid argument is explained as follows:

It is deductively valid to conclude B from A iff there are ways of knowing a priori that if A then B.[4]

On this view the difference between the two arguments above is epistemic, not metaphysical. And that seems to do the job.

The eliminativist thus proposes a double reduction. Metaphysical necessity is eliminated altogether in the sense that there is no such thing; there is only logical truth and deductive validity. The concepts of logical truth and deductive validity are then reduced to the a priori plus the good old concept of truth simpliciter, that is, truth without further metaphysical modalities. According to the modal eliminativist, there is no metaphysical difference between logical and non-logical truths; there is an important epistemic difference, though, and that is it.

However, is there some sort of metaphysical modality hidden here? For what does it mean to say that there are ways of knowing something or other a priori? Suppose I come to know a posteriori that Fermat's last theorem is true, by reading about it and even attending a lecture about it. I do not know how to actually prove this result though. So my knowledge seems to be a posteriori. Yet, there are ways of knowing a priori that Fermat's last theorem is true—by reconstructing the proof after studying it, say. Imagine, however, that I never actually do that. It seems then that saying that there are ways of knowing a priori that Fermat's last theorem is true just hides the modal counterfactual that I could have known it a priori.

How serious is this difficulty? Of course, if it proves to be the case that some sort of modal talk is hidden here in such a way that cannot

---

[4]Notice that technical and semi-technical terms like 'entailment,' 'implication' and so on are deliberately avoided here. Also, the classical material conditional is intended in 'if A then B;' perhaps the proposal works also with non-classical ways of understanding the conditional but if that is not the case then substitute 'not-A or B' as needed.

be eliminated, the proposed elimination of the concept of metaphysical necessity fails. However, the eliminativist has a ready answer. To say that I could have known something a priori is usually just a way of saying that others did know that a priori—so there are ways for us to know it a priori. In fact, if you look at the supposedly a priori knowledge one has of anything more demanding than a simple logical or mathematical truth, you will see how much this depends on a posteriori knowledge acquired from the work of others. So perhaps in most cases to say that I know something a posteriori that I could have known a priori actually means that somebody else did know that a priori. In a radical situation where a cognitive agent like us is all alone and without access to the work of others it is far from obvious that she could have known a priori something that she actually knows a posteriori. In this radical situation it seems that there is nothing she could have known a priori that she actually knows a posteriori.

The more important point, however, is that the demand being made to the modal eliminativist is perhaps unreasonable. For if one takes seriously the idea that modal concepts are after all epistemic, they are epistemic all the way down. In general, the eliminativist translates the boxes and diamonds of alethic modal logics thus:

$\Box A$: there are ways of knowing a priori that A.
$\Diamond A$: there are no ways of knowing a priori that not-A.

So, to say that I could have known a priori that p is translated as follows: it is possible that I know a priori that p, that is, there are no ways for me to know a priori that I do not know a priori that p. Yes, it is a bit of a tongue twister, but that is not a good reason to reject it.

So perhaps modal eliminativism has no immediate challenge on the epistemic front. What about the metaphysical one? From the eliminativist point of view, logical truths are just ordinary truths that we are able to know a priori. Ordinary truths, however, at least some of them, seem to have straightforward truth-makers: the statement 'Russell was born in England' is true because Russell was born in England, and it is this event that is responsible for its truth. Since the modal eliminativist claims that metaphysically there is no difference between logical and non-logical truths—the difference is merely epistemic—it seems that at least some logical truths have to have truth-makers, and furthermore they have to be just like ordinary truth-makers for ordinary, non-logical truths. How can that be?

The answer is that the truth-maker of a logical truth like 'Either Russell was born in England or he was not' is exactly the same aforementioned event of his birth in England. That is because given any statement of the form *p or q*, whatever is responsible for the truth of either *p* or *q* is also responsible for the truth of the larger statement.[5] A special realm of logical facts, formal features or abstract entities is not needed; the ordinary empirical world, and then linguistic agents like us that create languages with certain features, is enough. These linguistic features are not responsible by themselves for their truth; the world is responsible for that. But it is perhaps those features that allow us to know a priori that they are true without looking at the world. Again, the sharp distinction between what pertains to the world at large and what pertains to that part of the world that is us and our languages and beliefs and other cognitive workings is playing an important role here.

Notice that the modal eliminativist view is quite different in substance if not superficially in verba from the thought that logical truths are somehow mere linguistic conventions. Linguistic conventions do play a role in logical truths; but then again they play a role also in non-logical truths. After all, under different linguistic conventions the statement 'Russell was born in England' is false instead of true—a convention say, according to which 'born' means 'killed.' When it comes to language, convention *always* plays a role. However, the modal eliminativist is not saying that when it comes to logical truths linguistic conventions *alone* are responsible for their truth; the world is responsible for their truth, and linguistic conventions is at best responsible for our ability to know them a priori. The only sense in which linguistic conventions are responsible for logical truths is the very same sense in which they are responsible for non-logical truths: without linguistic conventions there would be no statements and if these are the primary truth-bearers there would be no truths whatsoever. This does not mean, however, that linguistic conventions are truth-makers; they are at best makers of truth-bearers.

Suppose now that someone argues that the world is not responsible for the truth of 'Either Russell was born in England or not' because this statement is true no matter what happens in the world. It is difficult though to make sense of this challenge. Given the linguistic conventions we have and the way the world is, the statement 'Russell was born in England' is true. The very truth-maker of this statement is also responsible for the truth of any statements like 'Russell was born in England or whales have no fins' and 'Russell was born in England or all cars are blue.' In general, given a truth-maker responsible for the truth of a statement *p*, it is also responsible for the truth of *p or q*. An important difference in the case of statements of the form *p or not-p* is that we know

---

[5]This sort of view is defended by Adrian Heathcote in his 2003 paper 'Truthmaking and the Alleged Need for Relevance.' [4]

a priori that they are true. Take a statement like 'There are extraterrestrials.' We do not know if this statement is true. We know a priori, however, that it is true if there are extraterrestrials and false otherwise. In the same vein, we know a priori that 'Either there are extraterrestrials or not' is true. In order for this statement to be true, however, it suffices that there are extraterrestrials, and it suffices also that there are no extraterrestrials. That is, whatever happens to be the case, it makes that statement true. How is this supposed to be a difficulty? Assume that everybody in the room is blond. The statement 'Someone in the room is blond' is true no matter what person I have in mind; what is the difficulty here? Do you need someone there not blond in order for that statement to be made true by those who *are* blond? That does not look promising. Yes, my statement is so to speak overdetermined: it is true no matter what person I have in mind. But that does not look like a serious difficulty.

So modal eliminativism does not seem to face any obvious difficulties. And it offers those who want to reject the necessary a posteriori—those who believe that if a statement is not a logical truth, then it is not a necessary truth—a promising way of doing it properly. For suppose you believe that necessary truths and logical truths are co-extensional and you did not get rid of the concept itself of necessary truth. In that case you do not seem to have a non-circular way of rejecting the unrestricted use of Leibniz's law that proves the existence of necessary truths that are not logical truths. On the other hand, you have to admit that everything that is logically contingent is really contingent—so that the success of science seems to become miraculous. Whenever a scientist predicts some outcome, anything might happen. Perhaps she drops a ball to prove a point about gravity, and the ball turns into a pink elephant. Why not? Everything that is not a logical contradiction is possible.

Suppose you react to this by saying that 'possible' just means 'logically possible' and that that is not *real* possibility. Well, do you still believe in the co-extensionality between logical truth and necessary truth, and that they are two different concepts? If so, notice that this entails that everything that is not a logical truth is contingent, and this does not mean that it is just logically contingent; it means it is *really* contingent. In contrast, modal eliminativism gets rid of metaphysical necessity and contingency altogether and so being logically contingent does not mean being contingent in any metaphysically relevant sense. That is why this looks like the best shot for someone who is suspicious of non-logical necessities.

A further worry, perhaps more promising, is that we are unable to explain the world without some sort of metaphysical modality—maybe because there is no non-modal promising way of understanding the laws of nature, say. That may be so, although I would doubt it until someone actually tried and failed to develop with the philosophical and logical tools we now have the old Humean thought that perhaps the world is just one damn thing after the other.[6] Still, suppose that the Humean hypothesis is hopeless. In that case, modal eliminativism is false. Notice, however, that the difficulty here is metaphysical, not conceptual. Conceptually, the reduction of logical concepts to the a priori (plus truth simpliciter) does not seem to face serious difficulties; it is just that if the world is not Humean, modal eliminativism is false.

It seems then that there are two non-equivalent promising ways of explaining logical truth and deductive validity, either by reducing them to the a priori plus metaphysical necessity, or by reducing them to the a priori (plus truth simpliciter). In both cases logical truth and deductive validity turn out to be non-primitive concepts, built on other more primitive concepts—the sort of thing Hartry Field is after. This looks promising for two reasons. First, as it was briefly seen, neither approach seems to face obvious difficulties from the start. Secondly, other approaches seem to be doing more or less the same—only not so obviously. Say for instance that you favour proof theory or model theory as the way to explain logical truth and deductive validity. It seems there are two ways of understanding any one of those. Either you believe that models model something in the world at large, or you want to cut all ties between them and the world at large. Ditto for proof theory: any step in a proof and any rule of inference can be seen as just formal devices cut loose from the world at large, or as representing garden-variety deductive validity.

Let us start with the thought that models and proofs are not about the world at large. Perhaps they are about an abstract, Platonic world? Then you are pushed to Edgington's view, substituting that abstract world for the world at large of rivers, galaxies and atoms; and you will have to include anyway our ability to know a priori what is going on in that Platonic world. Furthermore, it seems you need to accept that this Platonic world has some relation with our pedestrian one, for it is true that either Russell was born in England or not. So at the end of the day you will have to get back down to earth, the abstract world being just a temporary stopgap. Any plausible view that does not require that stopgap seems to have the upper hand here.

So perhaps things look brighter if one cuts any links between formal

---

[6]It is not being claimed that Hume thought that; and it is clear that he thought causality was crucial to understand the world. He thought however that we had a hard time trying to justify our causal beliefs.

languages and the world at large, rejecting also that those formal languages are about a Platonic world; they are not about anything, perhaps. In this case, the locus of logical truth is not garden-variety statements like that about Russell but rather formal statements like '$p \lor \neg p$,' and these are not about anything—neither a Platonic world nor the world at large. This is not particularly promising because it seems that a sequence of formulae like that represents an indefinite number of garden-variety statements that have that logical form. 'Russell was born in England' has truth value, but 'p' has no such thing: it is just a placeholder for different statements, some but not all of them true. And if 'p' has no truth value, '$p \lor \neg p$' is not true, let alone logically true. Still, suppose you insist that you want to cut all ties between formal languages and the everyday world. In that case, how are we supposed to understand your explanation of down-to-earth-seemingly-logical-truths? Even if they are not Real Logical Truths because they are not expressed using a formal language, they certainly demand an explanation. And you have provided none. Once you start providing one, though, again the world at large and our a priori faculties will seemingly play a role, and all that formal talk is again just a temporary stopgap.

So perhaps getting rid of logical modalities altogether is the way to go; the a priori plus truth—either necessary or simpliciter—seem to be able to explain by themselves logical truth and deductive validity. What about happiness? I leave that for another occasion. This was just the first logical step.[7]

References

[1] B. Abbott. "Water = H_2O". In: *Mind* 108.429 (1999), pp. 145–148.

[2] D. Edgington. "Two kinds of possibility". In: *Aristotelian Society Supplementary Volume* 78.1 (2004), pp. 1–22.

[3] H. Field. "What is logical validity?" In: *Foundations of logical consequence*. Ed. by Colin R. Caret and Ole T. Hjortland. Oxford: Oxford University Press, 2015, pp. 33–70.

[4] A. Heathcote. "Truthmaking and the alleged need for relevance". In: *Logique et Analyse* 183–184 (2003), pp. 345–364.

---

[7]Many thanks to Matheus Silva and Pedro Merlussi for generously providing me with constructive comments that allowed me to improve a previous version of this paper.

Necessity and Logic

Cícero Antônio Cavalcante Barroso

1 Introduction

In *Logical Forms*, Professor Oswaldo Chateaubriand, who was my thesis adviser, states that logic is "an ontological theory that is part of the theory of the most general and universal features of reality" [3](p.16). According to him, logic reveals necessary properties and relations existing in reality such as the property of self-identity and the relation of material implication. This view on logic has a long tradition in philosophy, and it has been supported by giants like Plato, Aristotle, Leibniz, and Frege. In fact, Chateaubriand's view is explicitly derived from Frege's view of logic, specifically the one advanced when he makes the claim at the beginning of *Der Gedanke*, that logic is an inquiry concerning the laws of true being [5](p.290). Professor Chateaubriand states this quite frankly in the following reasoning in favor of his point of view: "If logic has as at least part of its task the investigation of laws of truth, and if truth is an expression of reality, then it would seem that an aim of logic is the investigation of laws of being." [3](p.16). It is worth recalling, however, that logical truths are formal statements. This implies that a logic is not about particular contents, but about forms, i.e., logical forms.

This overtly metaphysical view on logic contrasts with two other views that were firstly presented to me by my once dissertation adviser and now dear friend Professor Tarcísio Pequeno: the vision of logic as engineering work and the vision of logic as a game. According to the first of these two perspectives, a particular logic is more like an engineering artifact. Logic does not tell us anything about the world, but it should be applicable to real problems or, in other words, it should serve us to solve problems that have a practical interest. The resemblance to engineering comes from the fact that the logician, like the engineer, has a certain freedom to develop his logic the way that pleases him most, but it is noteworthy that this logic should still meet a real demand. Effectively, alternative logics often arise from the need to formalize certain special

---

Copyright © 2019 by Cícero Antônio Cavalcante Barroso. All rights reserved.

types of reasoning such as reasoning for contexts with strict implication, reasoning of mathematics applied to quantum mechanics, and reasoning based on incomplete knowledge. The Logic of Epistemic Inconsistency, or LEI simply, one of the most influential works of Tarcísio, in partnership with Professor Arthur Buchsbaum, is effectively proposed from an analysis of the last type of reasoning aforementioned [6].

Nevertheless, there are logics that do not arise from a real demand, or at least do not from a previously known demand. They arise for different reasons such as the interest in studying the effects of certain variations on the semantic or inferential properties of logical connectives, the desire to do something that no one else ever did, or the wish to do something that is believed to be impossible to do. There are those who imagine that a wish of this kind was a major motivation for Professor Newton da Costa to create the family of C-logics [4]. These cases show that, when a logician creates a logic, he has a lot more degrees of freedom than the degrees of freedom of an engineer creating their projects. Consequently, it is better to draw an analogy with the design of games instead of engineering. A game is something that is not needed until it is invented[1]. This lack of prior use of the game is precisely what gives its creator greater autonomy than that of the engineer. Since some logics are created without specific purposes concerning ordinary needs, their creators can also create more freely, and it is in this sense that such logics are similar to games. In any case, it is to be supposed that some restriction still exists for the logician while performing his creative work because, in the end of day, logicians create logics and not games, at least not in a literal sense.

What can we say about these different views on logic? Would the view of logic as a product of engineering and the view of logic as a game be more consistent with our present scenario of logical plurality than the metaphysical view? In a certain sense, they would. After all, the metaphysical view of logic seems to be committed to the idea of an essential or real logic (rooted in reality), the one that unveils the 'laws of being'. If there are so many different logics, and so many logical principles that are valid in certain logics but not in others, how could anyone still argue that there is something like a real logic?

Per contra, Professor Chateaubriand does not defend a metaphysical view on logic by ignoring the plurality of logics that exist today. He thinks he has good reasons to maintain that a logic with ontological foundations must exist among various particular logics. And, indeed,

---

[1] Before Wittgenstein's comrades raise an objection here, I want to emphasize that I am using the word 'game' in its most prosaic sense, which does not apply to language or other complex human activities.

logical multiplicity does not seem to exclude this possibility. If the scenario is not only of plurality but also of diversity, it is possible that a real logic coexists with invented logics. Moreover, if logic tells us how we can convey the putative truth of the premises to the conclusion, whatever the set of axioms and rules of inference we adopt, it suggests to us how to relate certain putative facts to other inferred facts through a consequence relation[2]. At least in this sense, it seems reasonable to say that logic deals with facts.

As a former student of Professor Tarcísio Pequeno and Professor Oswaldo Chateaubriand, I was influenced by both the vision of invented logic and the vision of real (universal?) logic. Both, in my estimation, have qualities and, understandably, deficiencies. What I want to suggest then is a kind of eclectic view of logic. It seems to me that we must admit that the logician does the work of a creator, but at the same time, it is necessary to recognize that the logician does not create his logic *ex nihilo*. Indeed, it is not difficult to agree that this limitation also exists for engineers and game designers. Physics predates bridges, the human ability to follow conventional rules predates chess. In my opinion, the logician is like a builder who constructs his buildings on a foundation that he finds laid beforehand, in other words, he does his job by leaning on something that already existed. It is often difficult to probe the nature of this previous framework, but we can always come up with some assumptions. Professor Chateaubriand follows the trail of truth. As stated by him, the laws of truth refer us to the laws of being. In this article, I want to follow another clue (although I recognize that this clue leads me to cross the path of truth in some points).

There are several logics and, although they may be very different from the Aristotelian syllogism, we continue to call them logic. What makes these formal systems to be logical systems? What I want to suggest is that the *sine qua non* component is the 'follows from'. A logic is a formal system that has the primary function of demonstrating a particular relation between formulas, and this relation is expressed by the expression "...follows from...". It is important to say that such an expression does not indicate a trivial relation of inference, it indicates a necessary inference. By the 'follows from' of the logic we denote an inference relation that is notable for being necessary. In this way, a logic is a formal system that is founded on relations of necessity. And when I state that logic is founded on these relations, I mean that it does not invent them; indeed, I shall propose that it finds them in reality. In order to understand how

---

[2]Partisans of logical expressivism may not accept this suggestion, but it is offered anyway.

it occurs, I would like to examine the concept of logical necessity further, and to do so, we should begin with some considerations on the so-called 'necessary propositions'.

2 Necessary propositions and necessity

In general, the expression "necessary proposition" is used to designate necessarily true propositions, that is to say, propositions that can't but be true. Although propositions of the most varied types have already been considered necessary at some point, two types stand out. These are the logical truths and the mathematical truths. As it can be inferred from the previous discussion, my interest in this article rests on the former. My goal hereafter will be to show that, despite the present scenario of logical plurality to be suggestive of a relativization of the notion of logical necessity, there is a sense in which the notion of necessity can be maintained as strong as if the scenario were of logical uniqueness. If we can delimit this sense, we will have found the foundation upon which lies the whole variety of logics that proliferate in the world.

The propositions most widely accepted as necessary are the validities, that is, the propositions of logic, such as the statement "if everyone loves logic, there is no one who does not love logic." Why are propositions like this considered necessary? The standard answer is this: such propositions are necessary because every interpretation makes them true. That answer is enough for most people who study logic. But it instigates me to ask another question. How do we know that every interpretation makes validities true? Of course, it is not by testing the interpretations one by one. What we do is to examine the semantics, that is, the definition of truth that we adopt, and based on it we can say whether it is possible for a given proposition to be true or false. If we examine the semantics of classical propositional logic, for example, we can guarantee that the law of Duns Scotus is necessary. The application of semantics allows us to even construct a truth table to support our claim about the necessity of the proposition. On the other hand, if we wish to attest to the necessity of a proposition of the first-order logic, we have to use the semantics of the first-order logic; if we wish to attest to the necessity of the characteristic axiom of S4, we usually use a Kripke semantics with transitive accessibility relation, and so on. In general, if a necessary proposition φ belongs to a logic L, we have to use the proper semantics of L to attest to the necessity of φ. Based on these examples, we can see that when the problem is to attest to the necessity of a proposition, it is not enough to say that it is necessary because it is true for every interpretation. A so-called necessary proposition is always true for interpretations of a particular semantics, and therefore is only necessary in relation to that semantics. A proposition that is considered necessary according to a semantics S_1 might not be considered thus according to a semantics S_2. In short, if we understand that necessary propositions are true for any interpretation, we must admit that such propositions are only relatively necessary.

If we adopt a different approach, namely a syntactic approach, it is possible to define as logically necessary those propositions demonstrable from an empty set of premises. A more intuitive way of presenting this approach is by saying that the propositions will be considered necessary if they do not depend on any particular condition to be inferred. And if they are inferred unconditionally, they are inferred under any condition, inevitably. They are necessary just because they are unavoidable. Of course, when we demonstrate a logical theorem from an empty set, we construct a proof using as tools certain rules of inference and certain meta-theorems such as the deduction theorem. And it is also clear that the rules of inference that a logician is willing to use will not always be used by logicians who adopt a different conception of proof. Thus, also concerning this approach, the question of the necessity of the propositions of logic seems once again to refer to a kind of relativism. Only in relation to a given conception of demonstrability or proof can a proposition be logically necessary.

It is therefore clear that the necessity of the propositions of logic always depends on the conception of truth and the conception of proof that we adopt. In short, it depends on the logic that we adopt, and by 'logic' here I mean a formal system that encompasses a syntax and a semantics and presents a mechanism for the production of all valid inference relations of a given type, and only these. In this perspective, when a person asserts that a logical truth is necessary period, she seems to be naively presupposing absolute definitions of proof and truth. We are led to believe that the most reasonable thing to do when we want to assert the necessity of a proposition φ is to say that φ is necessary according to classical first-order logic, or according to intuitionistic propositional logic, or according to S5, or according to relevant logic etc.

All this leads us to ask whether it still makes sense to say that there are necessary propositions or not. The idea of propositions that have a relative necessity seems a distortion of our original intuition of the concept of necessity. Our original intuition of what is a necessary proposition is the intuition of a proposition that is always and unconditionally valid, it is the intuition of a proposition that asserts a universal and inexorable truth. A relatively necessary proposition is not taken as seriously as an unrestrictedly necessary one. For example, if one asserts that it is necessary for the phenomenon of pain to be identical to a type of neural activity, I believe that this statement is rather controversial, but if one asserts that it is necessary for the phenomenon of pain to be identical to a type of neural activity provided that we assume the truth of physicalism, I believe that this statement could hardly state something more innocuous. And, generally speaking, if something is necessary only under given conditions or assumptions, then it seems to me that such thing is no longer necessary.

But, pardon the pun, no relativism can be absolute. In physics, for instance, we know that time and space measurements are relative to the movement of the measuring instrument. However, there are fixed laws that determine how the measurement should be done. In ethics, there are many people who sympathize with the thesis that every list of moral values is relative and that the choice of a list always depends on how we conceive of certain aspects of human nature and the world, but it seems that when we assume a certain vision of the world, we are compelled to embrace certain values. In mathematics, we can relativize a result by saying that, to be demonstrated, it requires the admission of such and such methods of proof or of such and such notions. However, normally no one thinks of relativizing the statement that such a result is necessary provided that certain methods of proof and certain notions are adopted. In short, it seems that the relative settings are always to be printed on a fixed background without which no form could be delineated and no convention could be established.

Relativism as to what is or is not necessary also seems to be mounted against a metaphysical background of invariable colors. In fact, if we take L to represent a particular logic in the broad sense outlined above, and φ to represent a necessary proposition in L, we can formulate the following question: can the statement "φ is necessary in L" be considered necessary? In what system would it be necessary? That is the crucial question in this reflection. I dare to respond that the statement is necessary, and more than this, it is necessary regardless of any formal system. It is necessary because it expresses an invariant relation between φ and L.

Now observe that when I say that φ is necessary in L I am actually saying that in L there is a proof of φ from the empty set, that is, that the sequent $\emptyset \vdash \varphi$ is valid in L. If φ is an implication of type $\alpha \to \beta$, then I am saying that the sequent $\alpha \vdash \beta$ is valid in L. Following the same analysis, the statement "β follows from α" will be relatively necessary, but the statement "β follows from α in L" will be unrestrictedly necessary.

Now let us call the definitions of *proof* and *truth* included in L respectively *defP* and *defT*, and let us call the triad (Γ, defP, defT) a *frame*, where Γ is a set of formulas of the language of L. If, in a frame Q, Γ is a unitary set containing α, we can represent the first of the three elements of Q as a mere proposition. Hence we can say that the statement "β follows from the frame $(\alpha, \text{defP}, \text{defT})$" is equivalent to the statement "β follows from α in L", and that it is also absolutely necessary. Faced with this, we can say that there are two types of 'follows from'. There is a theoretical 'follows from' and a meta-theoretical 'follows from'. When we write, for example, that β follows from α, this 'follows from' is theoretical, it is a 'follows from' found in the framework of the particular logic L. But if we write that β follows from the triad $(\alpha, \text{defP}, \text{defT})$, then we are making use of a meta-theoretical 'follows from'. The last one is a 'follows from' that does not depend on any particular logical theory to be established.

Thus, it seems that it is in the meta-theoretical 'follows from' that lies at the core of genuine logical necessity. A relatively necessary proposition can be defined as that which expresses a theoretical 'follows from', whereas an unrestrictedly necessary proposition can be defined as that which expresses a meta-theoretical 'follows from'. My suggestion is that the meta-theoretical 'follows from' reveals the existence of an indissoluble nexus between certain states of affairs corresponding to frames and propositions. Although these nexuses are logical nexuses, they are not defined in any formal system; they are, so to speak, rooted in an extralogical reality. My opinion, indeed, is that this reality is not just an imaginary world created by logicians, it is the same reality in which we breathe and exist. This view, clearly, is sufficient to characterize my view on logic as a metaphysical view.

An objection can be made at this point. The existence of indestructible logical connections between frames and propositions may even be taken as evidence in favor of a conception of unrestricted or unconditional necessity, but the conclusion that such necessity is rooted on reality does not seem to be compelling. It seems obvious that definitions of truth, definitions of proof, and propositions must be understood as linguistic constructs, so that if there is any necessary relation between a frame and a proposition, this relation is also linguistic. The refined prose of Quine has been reminding us of this for a long time: "Necessity resides in the way we talk about things, not in the things we talk about." [7](p.174). Perhaps the necessity that is revealed in the meta-theoretical 'follows from' is a result of the structure of language. If so, one can understand to some extent why it does not depend on the specific theoretical constraints of a logic, since it depends on something much more general, namely, the language structure. However, at the same time one can interpret this as an indication that it does not inhabit a reality independent of language.

In order to answer this objection, I believe it is necessary to give a little more attention to the relationship between necessity and reality.

3 Necessity and reality

Let us consider the concept of a frame again. A frame can be seen as a structure composed of a base (the elements of Γ) and a set of construction rules (the elements of defP and defT). Construction rules, in the sense I want to use the term, are rules that allow us to construct certain arrangements of formulas, both syntactic arrangements (proofs) and semantic arrangements (truth tables, semantic trees, etc.). Such constructs show that there is a logical connection between a subset of Γ and some other subset of the universe of discourse. It should be noted, however, that the constructions do not enter into the composition of the frame, at least not in an actual way. What enters into the composition of the frame are the construction rules. In any case, it can be said that the possibilities of all constructions that are derivable from the frame are already given in the frame. In this sense, a frame is similar to a game, and a frame-derived construction is similar to a match of that game.

This exposition may give the impression that, instead of supporting my metaphysical view of logic, I am moving away from it. It seems that I am now closer to the vision of those who compare logic to a game. In a way, this is true, so I said in the introduction that I would advocate an eclectic view on logic. It seems to me that the work of the logician has many degrees of freedom (similar to the work of creating a game), but it also seems to me that this work has certain limitations (similar to the work of creating a game). Perhaps, when I speak of frames, the aspects of freedom appear more vividly. Now, I would like to highlight the limiting aspects.

A frame $Q = (\Gamma, \text{defP}, \text{defT})$ functions as the foundation of all logical constructs we can make from elements of Γ in a logic L. These constructs are syntactic and semantic proofs, paths that lead from Γ to other formulas of the universe of discourse. When the logician articulates a frame, he is creating a game, but when he builds constructions from there, he is playing the game. This is where the limitations begin to appear more evidently. Once the frame is complete, it is not possible to reach a certain outcome by playing at will. There are plays that lead to the result and plays that do not. Who determines this? Is it the logician? To some extent yes, but mostly not.

It seems to me that the best way to explain this is by postulating that there are two normative dimensions involved in logic, one denoted by the theoretical 'follows from' and another denoted by the meta-theoretical 'follows from'. The first is engendered by the logician when he creates the definitions of proof and truth, and the other is outside the jurisdiction of the logician, and, indeed, limits it. In this second dimension, which I will call the "*a priori* dimension of logic" (whether fairly or unfairly it will be judged at the end of this paper), it seems that what we have are second-order rules, rules that establish decrees of this type: "if you have such and such conditions, you have such and such consequences." Something similar to a perception of this *a priori* dimension of logic manifests itself precisely when we play a logical game and feel that our movements are blocked in certain directions and in others they are not. This happens, for example, when we realize that we can't force a logical connection between Q and a formula x working within a certain system of inference; the connection exists or does not exist independently of the will of the logician. If it exists, it can't be broken; if it does not exist, it can't be forced.

Could this second dimension be a consequence of the first one? I think I can imagine reasons that would lead someone to say 'yes'. An argument to support this answer might be as follows: the constructions we can make within a logic depend on the definitions of proof and truth; definitions of proof and truth, in turn, depend on the mood of the logician; therefore, the logical constructions depend on the mood of the logician. But this reasoning is misleading. Its fallacious component comes to light when we compare it with another similar reasoning: the trajectory of a projectile depends on how it was designed and programmed; how a projectile was designed and programmed depends on the mood of the engineers; so, the trajectory of a projectile depends on the mood of the engineers. The problem of the two reasonings is that 'depends on' denotes the relation between a result and a necessary condition, not between a result and a sufficient condition. It is true that the trajectory of the projectile depends on the decisions of the engineers, but it also depends on the laws of physics. In my opinion, the same thing happens with the constructions that we can make within a logic. They depend on the decisions of the logicians, but they also depend on certain laws that are already available before the logician starts to work. If so, the dimension of the meta-theoretical 'follows from' does not derive from the dimension controlled by the logician; on the contrary, it is logically previous; it is in this sense that it is *a priori*. It is like a plateau that at the same time supports and limits the movements of the mechanisms created by the logician.

How does this aprioristic logical dimension enable the work of the logician? I would say it does this by providing us with certain footholds. Let's suppose that a logician wants a logic that formalizes a type of reasoning including exception-sensitive inferences (e.g., "if it has a duck's beak, duck's feet, and sounds like 'quack, quack', conclude it's a duck")

3. What does this logician have to do to be able to develop such logic? He has to think about the kinds of consequences that will follow from the admission of non-monotonic rules of inference. In seeking such relations, he does not want to stipulate anything. He wants to find relations that anyone with common sense recognizes as inevitable. Where can he find such relations? My suggestion is that they are found in the *a priori* dimension of logic[4]. But how does he find them?

It is clear that the logician, as a rule, is not someone with a third eye that has special access to other dimensions. It seems that the only place where he can find indications of such relations is in his own cognitive apparatus. Indeed, in a recent work, I have argued that we are cognitively programmed to recognize the necessity of the meta-theoretical 'follows from'. This, however, does not mean that such relations of necessity exist only in the form of inference patterns congenitally imprinted on the circuitry of the cerebral cortex. My thesis was that the best explanation for the brain's implementation of such logical verification systems is that it works in this way to conform itself to reality [1]. Now what I am trying to do is to specify in a somewhat more detailed way what features of reality I have in mind. What I think is that reality includes laws that regulate the relations between sets of possibilities — possibilities represented by frames and possibilities represented by propositions. What the meta-theoretical 'follows from' denotes are these laws.

As I have said before, this *a priori* dimension of logic, the dimension of meta-theoretical 'follows from', at the same time enables and limits the behavior of logical artifacts. The previous paragraph intended to explain how it enables them, but, in a way, it also gave tips on how it limits them. The laws of the meta-theoretical 'follows from' tie the conditions selected by the logician to certain consequences, and the knot that they form can't be untied. So, what happens is that the logician is free to make certain choices, but he is not free to refuse the consequences that come with them. Consequently, what happens to these laws of the meta-theoretical 'follows from' is not much different from what happens to the laws of physics. If you want to convert matter into energy and want to obtain the amount x of energy, you must use the amount y of matter. There is a law that imposes this consequence. It is something we recognize, we do not create. We create the conditions that will bring forth a given consequence, but we do not create the law that determines that under such conditions we have such results. In my view, the laws

---

[3]Reiter [8] proposes such a logic.

[4]It is clear that in such a dimension a priori the logical will not seek reasonings about ducks, but the constraints that must constrain a class of exception-sensitive reasonings.

of meta-theoretical 'follows from' work in the same way.

At this point, it must have become clear that the necessity of the meta-theoretical 'follows from' does not originate in language. In fact, I am trying to argue that it does not even originate in human cognition. My claim is that such a necessity originates in reality. What we recognize as necessary, as mentioned above, are certain relations between possibilities, namely, possibilities represented by frames and possibilities represented by propositions. Here, however, a kind of skepticism about the relationship between the reality and the possibilities represented by frames and propositions may arise. Mere possibilities are not part of reality, how then is it possible to argue that the meta-theoretical necessity, which has to do with possibilities, is part of reality?

I can think of two answers to this question. First, it is true that mere possibility is not part of the world (if we mean by 'world' a 'maximal set of facts'), but my thesis is not that laws derived from the *a priori* dimension of logic articulate only mere possibilities; in my conception, they govern the relations between actualized possibilities as well. In such cases, the connections established by these laws are actually connections between real conjunctures and real states of affairs. But if we think that the totality of the facts of the world consists of actualized possibilities, then we can agree that the whole world is under the influence of meta-theoretical necessity. Second, we can think of the concept of reality as referring to something more comprehensive than the set of facts. In the Tractatus [9], Wittgenstein made a distinction between the world, which is the totality of facts, and reality, which is the occurrence or non-occurrence of states of affairs (TLP 1.1 and 2.06). It is common to interpret these tractarian concepts as referring respectively to the domain of the actual and to the domain of the possible, the latter encompassing the former. If we adopt this interpretation, then even more reasonably we can say that the meta-theoretical necessity is embedded in reality. Effectively, I believe that it would not be abusive to use the language of the Tractatus to declare that the laws that interweave the *a priori* dimension of logic are the very logical scaffolds of the world. In the light of this perspective, although each particular logic has the autonomy to define its own rules, their functioning is still guided by those laws erected in the outermost layers of reality.

Despite the characterization of the above paragraphs, I admit that there may still be several questions about these laws that I hold to be the source of the necessity expressed in the meta-theoretical 'follows from'. Where are they written? Who wrote them? What kind of ink was used? I confess that I can't give a short philosophical answer to these questions. But this is no reason to reject the idea of laws that determine what is necessary on a level that is independent of the will of the logician. The same questions could be addressed to those who postulate the existence of the laws of nature. To these questions, there are intelligent and well-founded answers that are offered by the proponents of scientific essentialism [2]. Nevertheless, even if there were no such answers, laws of nature could still be maintained because of their recognized explanatory power. Laws of necessity, such as scientific laws, can be admitted because they are good explanatory resources, but I believe the greatest evidence for them comes from the practice of logic, from the experience of designing frames that can't be separated from their consequences. This sense of insurmountability that we sometimes experience in relation to certain situations that we find in logical theories, it meets in the laws of the *a priori* dimension of logic a very plausible explanation.

The remarks in the previous paragraphs make clear that my view of logic is mongrel. On the one hand, it preserves the right of the logician to create their systems according to their tastes and concerns; but, on the other hand, it points to the limits of choices of the logician. His choices can't change what is established by the *a priori* laws of logic. One of the consequences of this view, therefore, is that it becomes possible to speak of an original logic again, of a logic behind the logics. In a strong sense, the logic can be seen as the description (or a mirror image) of a nomological structure that underlies all formal systems. Whenever a particular logic L establishes a result φ, this is a relative fact, a fact of L. But if we include in the same frame all the conditions that make it possible to derive φ from L, the law that determines the indissoluble connection between φ and such a frame is no longer a relative fact, it is a fact of a logic of background. This background logic has as its object these laws of the meta-theoretical necessity.

In metaphorical terms, if I may put it so, logic so conceived is like a labyrinth that limits and conditions our movements. In this labyrinth there are many entrances and many exits. You can enter wherever you want, but once entered you can't go anywhere you want. Building a proof is like following a path surrounded by walls and hedges. Our initial choices condition the outcome of our actions, and that is inexorable. Logic has to do with the mode of being of things, allowed modes and not allowed modes. Adopting certain rules is still not making logic, following these rules is what it is. By following them we discover ourselves walled, any movement we make in a direction not allowed will be blocked.

Of course, when you think about it, many questions come to the surface. Who built this labyrinth? Did it just come out of nowhere? Could things have happened any other way, in such a way that we could enter by $\alpha \to \neg\beta$ and get in $\alpha \wedge \beta$? It's strange to think so. Moreover, it is not clear why we can't renovate the labyrinth. Why can't we demolish the walls that keep us from going in and out wherever we want? It would be even easier to destroy the whole labyrinth, pull it all down and open up all possibilities. But if we did, we would destroy everything that is logical, and with that, everything that is thinkable.

References

[1] C. A. C. Barroso. "Uma explicação cognitiva do segue-se." In: *Revista Argumentos* 7.13 (2015), pp. 179–194.

[2] J. W. Carroll. "Laws of Nature". In: *The Stanford Encyclopedia of Philosophy*. Ed. by Edward N. Zalta. Fall 2016. Metaphysics Research Lab, Stanford University, 2016.

[3] O. Chateaubriand. *Logical forms: truth and description*. UNICAMP, Centro de Lógica, Epistemologia e História da Ciência, 2001.

[4] N. C. A. Da Costa. "On the theory of inconsistent formal systems." In: *Notre dame journal of formal logic* 15.4 (1974), pp. 497–510.

[5] G. Frege. "The thought: A logical inquiry". In: *Mind* 65.259 (1956), pp. 289–311.

[6] T. H. C. Pequeno and A. Buchsbaum. "The Logic of Epistemic Inconsistency". In: *Knowledge and Reasoning'91*. 1991, pp. 453–460.

[7] W. V. O. Quine. *The ways of paradox, and other essays*. Harvard University Press, 1976.

[8] R. Reiter. "A logic for default reasoning". In: *Artificial intelligence* 13.1-2 (1980), pp. 81–132.

[9] L. Wittgenstein. *Tractatus Logico-Philosophicus*. 1961.

Contextualism, Universal Pragmatics and Metaphysics

Manfredo Araújo de Oliveira

Translated from Portuguese by Fernando Alves Gomes.

1 Introduction

Philosophy, the knowledge of principles, more than any other form of knowledge has to lay its basis bare and make its fundamental categories explicit. The category "truth" is one of the oldest and most important categories of Western thought, so its not surprising that in times when the very structure and the specific proceedings of the philosophical activity are questioned, the concern about truth be shared by philosophers[1] just as much as it is by logicians, theoreticians of science and linguists. In other words, the problem of truth, its adequate treatment, and its presuppositions have implications for philosophy, logic, theory of language, theory of knowledge, and ontology[2]. In the systematic pursue of an enlightened "concept of truth"[3] I will confront Lima Vaz's metaphysical understanding of truth with two of the most remarkable trends of the theory of truth in contemporary philosophy.

Copyright © 2019 by Manfredo Araújo de Oliveira. All rights reserved.

A version of this paper has been previously published in Argumentos. Revista de Filosofia, ano 4, n. 7 (2012) 31.

[1] See Heckmann H-D., Was ist Wahrheit?. Eine systematisch-kritische Untersuchung philosophischer Wahrheitsmodelle. Heidelberg: Carl Winter. Universitätsverlag, 1981; Puntel L. B., Wahrheitstheorien in der neueren Philosophie, 2nd. ed., Darmstadt: Wiss. Buchgesellschaft, 1983; (ed.), Der Wahrheitsbegriff. Neue Erklärungsversuche. Wiss. Buchgesellschaft, 1987. Beck W., Wahrheit und sprachliche Handlung. Untersuchungen zur sprachphilosophischen Wahrheitstheorie, Freiburg/München: Verlag Karl Albert, 1988.

[2] See Puntel L. B., Grundlagen einer Theorie der Wahrheit, Berlin/New York: Walter de Gruyter, 1990, at 2.

[3] For L. B. Puntel, the expression "theory of truth" indicates a theory that consists of at least five parts: 1)Clarification of the concept of truth; 2) The problematics of the criteria of truth; 3)Typology of truth; 4) The paradoxical dimension of truth; 5) The place of the theory of truth in the horizon of the whole of scientifical and philosophical knowledge. Cf. *Puntel L. B.* (ed.), *Der Wahrheitsbegriff,* supra note 1, Einleitung, at 2-3.

A) Richard Rorty and the substitution of transcendence for solidarity: an anti-Platonic conception of truth.

The original intuition of metaphysics,[4] as it is conceived in the Menon (82b_85b), the programmatic writing of Plato's *Academy*, was that human knowledge cannot be reduced to the knowledge of experience, but that one can reach an objective type of knowledge through conceptual reasoning. The basis of truth, then, is not the empirical, material world, but the world of thoughts, where there is the rational structure of reality or its essence. Thus the correspondence between spirit and nature is believed to be an ontological one,[5] so that logical rules reflect the laws that govern reality itself, and reason is objectively embodied in the laws of nature. According to Rorty, classical philosophy intended to capture the form and the movement of nature and history with the use of concepts, and this ultimately led to the conviction that it is possible to find out how one can fix injustice[6] and hence human history.

In Rorty's view, this is precisely the disease that undermines Western thought. His intuition is exactly the opposite of that of classical metaphysics: there is no higher reality than daily reality[7] to provide human action with a horizon of reconciliation and salvation[8]. This is why his

---

[4] See Snell B., *Die Entdeckung des Geistes. Studien zur Entstehung des europäischen Denkens bei den Griechen*, Hamburg: Claassen Verlag, 1955, pp. 258 ff. Gigon O, *Der Ursprung der Griechischen Philosophie von Hesiod bis Parmenides*, 2nd. ed., Basel/Stuttgart: Schwabe & Co. Verlag, 1968. Hösle V., Hegels "Naturphilosophie" und Platons "Timaios" ein Sturkturvergleich, in: *Philosophiegeschichte und objektiver Idealismus*, München: Verlag C. H. Beck, 1996, pp. 39 ff.

[5] On the adequacy formula see *Aristotle: Met.* IV 7, 1011 b 26 ff. *Thomas Aquinas: De Veritate* I 1; S.Th. I q. 16 a 2 ad 2; I q. 21 a 2 ad 2.

[6] See Rorty R., *Objectivity, Relativism and Truth*, Philosophical Papers I, Cambridge Univ., 1991.

[7] In the case of the conception of truth this means to put it on the dependency of an instance presumptively autonomous from our beliefs, from our quests and from our use of language. The basic problem posed by pragmatism consists in knowing how to reconnect truth to intentions, to beliefs, to human desires. Thus Davidson, contrary to Nagel, who says that truth is conceived from nowhere, insists on the thesis that truth and falsehood have an essential relation to the existence of thinking creatures. See Nagel Th., *The View from Nowhere*, Oxford: Oxford Univ.Press, 1986. Davidson D., A Coherence Theory of Truth and Knowledge, in: Malachowski A, *Reading Rorty*, Oxford: Basil Blackwell, 1990; The Structure and Content of Truth, in: Journal of Philosophy, LXXXVII, n.6, 1990.

[8] According to J-P Cometti, American pragmatism, a philosophy ingrained in American culture, evolves around a philosophy of knowledge, but since the very beginning it drifted apart from conceptions that tended to prioritize the search for an absolute foundation or for a model of reason that would determine *a priori* the possibilities of inquiry and discovery. The idea according to which its a belief we have in the beginning and in the end of any research is the core of Peirce's objection to

proposal consists basically in ridding humanity from the Platonic disease, that is, metaphysics[9] – something that according to Richard Rorty will come as a consequence of the radicalization of the posture of an analytical philosophy that draws on the linguistic turn[10]. As the linguistic turn is taken to its last consequences as a pragmatic turn, there is the need to give up the premise that tacitly kept it connected to the western tradition of thought, namely, the idea that there are still philosophical truths to be discovered that can be grounded on arguments. Then the conclusion: the first thing to be done is to deconstruct metaphysics and unmask Platonism, and this means to demonstrate that even analytical philosophy remained tied to the metaphysical assumptions it fought, and that the western culture is rooted in the misunderstandings owed to Plato.

Analytical philosophy[11] meant, in the first place, a critique to the conceptual horizon in which modern philosophy of the subject[12] articulated

---

Descartes. See *Cometti J-P, Le Pragmatisme: de Peirce à Rorty,* in: *Meyer M.* (eds.), *La Philosophie anglo-saxonne,* Paris: PUF, 1994, at 396. Nevertheless, recent drawbacks towards pragmatism was made possible by a strong criticism against empiricism, especially in the Vienna Circle, whose influence spread like wildfire throughout the United States. Such criticism started with Quine's famous writing, Two Dogmas of Empiricism, and was carried on by N. Goodman, W. Sellars and D. Davidson. See *Cometti J-P.*, supra, pp. 446 ff.

[9] With a price to be paid. See Margutti Pinto P.R.,Pragmatismo, Ironismo e Ceticismo em Richard Rorty, in: Margutti Pinto P. R. et alii (eds.), Filosofia Analítica, Pragmatismo e Ciência, Belo Horizonte: Editora UFMG, 1998, at 34 : "Abandonar a terra firme metafísica significa embarcar num navio muito frágil e enfrentar águas turbulentas. Este não é um lugar fácil para se permanecer. O navio pode afundar a qualquer momento" ["To leave metaphysics firm ground means to embark in a very fragile ship and face troubled waters"] . See also in the same volume: Ramberg B. T., Rorty e os instrumentos da filosofia, pp. 81 ff.,. and Rajagopalan K.,O Radicalismo e os seus limites: Comentários sobre "Rorty e os instrumentos da filosofia, pp. 120 ff.

[10] See Rorty R., The linguistic Turn. Recent Essays in Philosophical Method, University of Chicago Press, Phoenix Edition, 1970.

[11] Which for Rorty actually has an antifundamentalist attitude, but notwithstanding does not assure its rupture with traditional theses. See *Rorty R., Pragmatismo, Filosofia Analítica e Ciência,* in: *Margutti Pinto P. R. et alii* (eds.), *Filosofia Analítica, Pragmatismo e Ciência,* Belo Horizonte: Editora UFMG, 1998, at 17: "Mas antifundacionalismo não diminui a força do seguinte argumento: uma vez que a verdade é uma noção absoluta, e consiste em correspondência, deve haver uma natureza absoluta, intrínseca, independente da descrição feita, à qual a verdade deve corresponder" [But antifundamentalism does not lessens the strength of the following argument: once truth is an absolute notion and consists in correspondence, there must be an intrinsic, absolute nature, detached from the description given and to which truth must correspond"].

[12] About the treatment of the question of truth in the horizon of the philosophy of consciousness see *Landim Filho R., Sobre a Verdade,* in: Síntese Nova Fase, v.20, n.63 (1993) 459-475.

itself – a critique that transformed the classical concept of objective rationality into subjective rationality but did not question its concept of reality. Reality, to be exact, was no longer the world in itself, but the world for us, the world of phenomenic objects represented by the knowing subject[13]. In this new conceptual context, the question of truth appears as a struggle to found genetically the agreement between representation and object, that is, to found it on the basis of self-evident experiences. For Rorty, what matters here is to show the insufficiency of the conception of knowledge in the "mentalist horizon", pictured as a product of the action of the individual's self-consciousness. The individual then ensures for himself a special range of readily accessible, absolutely certain experiences, though he does not have direct access to the objects, but to his own representations mediated by reflection. This has a fundamental implication that happens to be a distinguishing feature of modern philosophy: the subject / object dualism, the separation of spirit and object, and the modern type of skepticism[14] that follows it. The mentalist perspective, whose origins can be traced back to ancient philosophy, remained untouched in modern philosophy of consciousness. For mentalism, objectivity was guaranteed as long as the subject could relate to his object[15] correctly: concepts, a subjective reality, correspond to what is "outside", the objective world[16].

---

[13] Analytical philosophy can be considered a new form of primary philosophy that, as Tugendhat thinks, radicalized the "reflexive turn" in modern philosophy and therefore reflects upon basic universal questions from the perspective of language: What can be said? What can be thought?. See *Tugendhat E., Vorlesungen zur Einführung in die sprachanalytische Philosophie*, Frankfurt am Main: Suhrkamp, 1976. In Dummet's, Habermas's and mostly Tugendhat's steps: *Landim Filho R., Significado e Verdade*, in: Síntese Nova Fase 32, vol.X (1984) 33-47.

[14] Which for him is closely knit to the mentalist model.

[15] A conception, according to Dummett, simply unintelligible, and which must be replaced by one of a justified or verified assertion, some "verificationism", in Engel's view. Such perspective found inspiration in the intuitionist philosophy of mathematics and logic, according to which the conditions for the demonstration of a statement grant its truthfulness and that determine the existence of the objects to which it refers. See *Dummett M., What is a theory of meaning II*, in: *Evans G./Macdowell J.* (eds.), *Truth and Meaning*, Oxford: Clarendon Press, 1976; *Truth and Other Enigmas*, London: Duckworth, 1978; *The Logical Basis of Metaphysics*, Harvard Univ. Press, 1990.

[16] Its from this viewpoint that the debate between realism and anti-realism is seen in contemporary analytical phylosophy. In Cometti's definition (supra note 8, at 403), realism is that posture which establishes a connection between thought, or language, and an independent reality, transcendent or not, accessible or not, while anti-realism refuses to let thought, beliefs and language submit to the authority of a presumptively external reality. See *Putnam H., Reason, Truth and History*, Cambridge: Univ. Press, 1981; *Representation and Reality*, Cambridge Mass.: The Mit Press, 1988; *Realism with a human Face*, Cambridge Mass.: Harvard Univ. Press, 1990. *Engel P., Réalisme*

The linguistic turn consisted basically in demonstrating that the linguistic expression is the necessary mediation of all and any knowledge, the indispensable condition of our access to the world. Since language and reality interpenetrate within any comprehension of the situations and events of the world, the question about what can be known always implies the question about what can be said. Therefore, we never leave the world of language, and this implies an anti-fundamentalist concept of knowledge for, as Rorty says, one thing can only work as justification in relation to something else that we have previously accepted, which means that, outside our language, our beliefs, and our thoughts, we will never find criteria that are independent from the criteria of coherence of our claims[17]. Thus, according to the contextualist interpretation of the pragmatic turn, this means that there can be no access to reality either without the process of intersubjective understanding or outside the linguistic context of the lifeworlds of historically situated[18] communities.

For Rorty, such posture brings extremely important consequences for the concepts of reality and truth,[19] since it is not possible to think reality and truth without language[20]. The first consequence is that the knowledge of objects does not provide an adequate epistemic model. This is so

---

et anti-réalisme de Putnam, in: Meyer M.(ed.), supra note 8, pp. 363 ff. Dummett M., *Les origines de la philosophie analytique,* Paris: Gallimard, 1991. Engel P., *Réalisme et anti-réalisme semantiques: Davidson et Dummett,* in: Meyer M. (ed.), supra note 8, pp. 358 ff. Cometti J-P.(org.), *Lire Rorty,* Combas: L'Eclat, 1992.

[17] See Rorty R, *Der Spiegel der Natur,* Frankfurt am Main: Suhrkamp, 1981, pp. 199 ff.

[18] See Rorty R, *Der Spiegel,* Suhrkamp, supra note 17, at 191.

[19] Rorty seems not to be aware that Kant, in dealing with the problematic of the constitution of the objects of experience, had already strongly criticized the naive realist conception of the appropriateness rule: the object is not just simply "in-itself", for it is co-constituted by thought as long as the fundament of truth is found in the *a priori* of the laws of pure understanding, and such subjective conditions of thought have an objective value as they generate the object in accordance to the form. Therefore, there is no access to the objects of the world without the mediation of the laws of thought and understanding, i.e., without the mediation of consciousness. See *Höffe O, Anhang: Kritische Überlegungen zur Konsensustheorie der Wahrheit (Habermas),* in: *Ethik und Politik. Grundmodelle und –probleme der praktischen Philosophie,* Frankfurt am Main: Suhrkamp, 1979, at 258.

[20] Th. Kuhn understands the "reality" studied by sciences in a similar context. See Aguiar T. R. X de, *Realismo, Construtivismo e Progresso,* in: Margutti Pinto et alii (eds.), supra note 9, at 291: "O resultado é uma interessante forma de construtivismo: a realidade estudada pela ciência é _ de uma maneira que o próprio Kuhn tem, reconhecidamente, dificuldade de precisar _ determinada pelo paradigma corretamente adotado pela comunidade científica relevante" [The result is an interesting form of constructivism: the reality studied by science, in a way that Kuhn himself admittedly has difficulties to specify, is determined by the paradigm rightly adopted by the relevant scientific community"].

because what is important now is to apprehend states of affairs in view of their propositional articulations,[21] and also because truth is just a property of criticizable claims that may grounded on reasons. The consequence is that we move from the mere transference of "subject/object" relations to "claim/fact" relations that permit the whole linguistic turn to stick to the idea of *representation*, and therefore to a concept of truth as the "mirror of nature". When the linguistic turn is radicalized, the subjectivity of opinions no longer prevails due to its confrontation with an objective world, but because of the mediation of a community of public communication. The "objective world"[22] is not something that can be reproduced, but only the common reference point of a practice of reaching understanding among the members of specific communities of communication that agree over a subject matter. In this case objectivity is nothing more than the widest intersubjective agreement that is possible[23].

In one word, knowledge is no longer the mirror of nature, a correspondence between sentence and fact. As Rorty says, there is no contact previous to language that could allow us to point out what is an object in itself, in contrast with what it is in the light of the different descriptions that we make of it[24]. Knowledge is now understood as an intersubjective process of agreement in which language is no longer considered a mere grammatical form of presenting the world. Instead, one emphasizes the communicative dimension of language, and thus a world lived as an intersubjectively shared public space, in which interactions and traditions are structured as symbolic networks. Here, the dialogue among individuals (communicative model, intersubjective process of agreement) replaces the opposition between sentence and fact (representational model, language-world relationship[25]), leading to a radical rupture in the model

---

[21] See Habermas J.,Wahrheit und Rechtfertigung. Zu Richard Rortys pragmatischer Wende, in:Wahrheit und Rechtfertigung. Philosophische Aufsätze, Frankfurt am Main: Suhrkamp, 1999, pp. 236 ff.

[22] In Peirce's case, the association of belief as rule to action under the determination of habit allows pragmatism to avoid representation, by replacing the *external* relationship of idea and world, or of spirit and reality, for an internal relationship of belief and habit, i.e., of the *rule*. This is the origin of the criticism against a "contemplative theory of knowledge" also advanced by James and Dewey. See *Cometti J.P.*, supra note 8, at 398.

[23] See Rorty R. Solidarität oder Objektivität?, Stuttgart, 1988, pp. 14-15.

[24] See *Rorty R., Science et Solidarité*, L'Eclat, Combas, 1990, at 9.

[25] For Habermas, the vertical relationship of the world of statements about something surrenders to the horizontal relationship of being amid the participants in a community of communication. The intersubjectivity of the world lived holds back the objectivity a world opposed to that where the subject finds himself alone. See *Habermas J., Wahrheit*, supra note 21, at 241.

of knowledge as the mirror of nature. This reveals the uselessness of the traditional concept of truth as a correspondence between signifier and signified,[26] which is nothing but an illusion[27] since it implies that one can escape from language with language.

The epistemic authority thus moves from a subject that knows the objective world to a community of subjects that agree on the basis of a common ground of shared experiences with a world in view of which each subject justifies its conceptions. Knowledge is therefore that which is rationally accepted as according to the criteria of the praxis of a given community, and this means that the objectivity of experience is replaced by the intersubjectivity of comprehension – a comprehension that is mutable in principle, since the values and beliefs that constitute the symbolic universe of the subjects that reach an agreement over something could be completely different from what they actually are[28].

For Rorty, language is contingent, the self and the community are also contingent, and contingency is a basic rule of the thought. Therefore, the reference to absolute norms is just a foolish dream, given the finitude of the human condition. So, instead of transcendent norms[29] that we cannot access, all we have left is the solidarity of our common beliefs and values, of our preferences and choices in the common context of our form of life. Nevertheless, Rorty claims that this does not lead into relativism and its superseded representational model of knowledge, but brings as a result that any justification will always depend on different

---

[26] For Rorty, pragmatists in no way resort to a theory about the nature of reality, knowledge or man, asserting that nothing exists as truth or good. For him, this in no way implies a subjectivist or relativist theory of truth or the good, for what they wish is just to change the subject. See *Rorty R., Consequences of Pragmatism (Essays 1972-1980)*, Minneapolis: Univ. of Minnesota Press, 1982, pp. XIII ff. On Rorty's discussion with Davidson upon this topic see *Pereira A M., Lendo Rorty Lendo Davidson*, in: *Margutti Pinto P. R. et alii* (ed.), supra note 9, pp. 265-275.

[27] See *Rorty R., Pragmatismo*, supra note 9, at 16: "..a filosofia estará melhor servida sem as noções de "natureza intrínseca da realidade" e "correspondência com a realidade" do que com elas". [philosophy will be better off without such notions as "intrinsic nature of reality" than with them]

[28] See *Rorty R., Contingency, Irony and Solidarity*, Cambridge Univ. Press, Cambridge, 1989.

[29] Whose possession by philosophers in the past legitimated the specific supremacy of philosophy in the debate on humanities. If this access does not exist, pragmatism assigns a more modest function for philosophy, connected to the tasks of the daily live, with no privilege and no superiority over other forms of discussion. Philosophy, then, is available for other causes, the small everyday causes, and its task is at the same time critical and constructive. What will disappear in a post-philosophical culture – in the sense of a culture that abandons transcendental, foundational, and essentialist philosophical motives – is the caste of professional philosophers who protect their corporate interests. See *Rorty R., Consequences*, supra note 26.

criteria and vary according to the historical context. Consequently, we should stop worrying about objectivity and start feeling contented with intersubjectivity.[30]

The consequences to the problematics of truth are enormous: since knowledge is not the mirror of nature, but something that depends basically on dialogic practices and social context, the criticism of the different forms of social praxis loses its meaning, for we are bound to symbolic contexts, and any attempt to transcend them means a return to fundamentalist postures[31]. In Rorty's view, this new stance is where the history of western philosophy leads to, the final stage of a development that displays 3 basic, non-continuous paradigms: the metaphysical paradigm, in which the thought was centered on the "essence" of things (objectivity); the paradigm of modern theory of knowledge, revolving around the ideas constructed by a finite mind (subjectivity); and the paradigm of the philosophy of language, mainly concerned with linguistic expressions (intersubjectivity). It is in the latter that the epistemic authority moves from a knowing subject that engenders within itself the means to measure the objectivity of experience, to the practice of justification of a linguistic community[32], since the linguistic turn implies the primacy of the common language (once the explanatory power of the linguistic turn is based on the primacy of the common language). For this reason, the meaning of "intersubjectivity" here changes radically. It no longer signifies an observed convergence of thoughts and representations of different persons; instead, it incorporates the perspective of participants in

---

[30]See *Rorty R., Pragmatismo*, supra note 9, at 23.

[31]See Rorty R, Pragmatismo, supra note 9, p. "28: "Desistir da idéia de que a filosofia aproxima-se da Verdade e interpretá-la como fez Dewey é conceder primazia à imaginação sobre o intelecto argumentativo, e ao gênio sobre o profissionalismo" [To give up the idea that philosophy is close to truth and to interpret it as Dewey did is to give precedence to imagination over argumentative intellect, and to genius over professionalism.]. For Margutti, the skeptics are distinguished by their relinquishing of philosophical investigation in favor of practical affairs, "a field where there are no radical doubts". Margutti does not see any big difference between the Rorty's ironist and this type skepticism. See *Margutti Pinto P. R..Pragmatismo, Ironismo*, supra note 9, at 33.

[32]Putnam worked out the distinction between justification and truth. For him, rational acceptability based on our criteria does not allow us to go beyond justification, for the rejection of the existence of a coherent external perspective, of a theory that could be simply true in itself independently from other possible observations, does not legitimate the identification between truth and rational acceptability. Truth, not justification, is an inalienable property of propositions. If our knowledge is never definitive, even though we have reasons not to consider them arbitrary, we cannot see exclusively in those reasons the source of what we call the truth of a proposition. Truth is an idealization of rational acceptability. See *Putnam H, Reason, Truth and History*, supra note 14, p. 63 ff.

a common linguistic horizon in which the members of a community of communication are always already situated. This indicates the lack of meaning of skeptic views, for we as socialized individuals always already find ourselves against the linguistically interpreted horizon of the world we live in. This also implies the existence of a background of intersubjectively shared assumptions that are confirmed in practice, and that consequently makes it pointless to be totally doubtful about the possibility of us accessing the world. The modern dichotomy inside/outside just simply vanishes away.

The pragmatic turn eliminated the question about the existence of a world independent of our descriptions of it, but, on the other hand, accentuated the fallibility of any knowledge and its need for legitimacy in a context where the measure of the objectivity of knowledge is the public practice of justification. This raises the important question of whether it is still possible to separate truth from its context of justification. If we cannot transcend the linguistic horizon of justified opinions, how can we combine the basic thesis of the pragmatic turn with the intuition that true statements are adequate to the facts? The central problem of Rorty's contextualism is here: how can we relate truth and justification?

B) The interpretation of truth as unconditional validity: the universal pragmatic reading of J. Habermas

Habermas starts from the same conception of a reason embodied in language: we can only clarify what a fact is with the truth of a statement about a fact that is actually true[33]. Language therefore is an unavoidable condition of our access to the world,[34] so much so that we cannot consider experience and experimentation as means of control of knowledge absolutely detached from our linguistic system. It is language that provides the cognitive patterns of the objects of experience. In this sense, human knowledge cannot transcend the limits of language and reach the things "in themselves" because the relation with reality is built

---

[33] See Habermas J., Wahrheit und Rechtfertigung, supra note 21, at 246.

[34] See Costa, R. da, As vantagens de uma teoria consensual da verdade segundo Habermas, in: Cirne Lima C./Almeida C.(orgs.), Nós e o Absoluto. Festschrift em homenagem a Manfredo Araújo de Oliveira, São Paulo, 2001, 377: "Ora para ele, a teoria da verdade como correspondência não se apercebe de que ao termo < realidade > não podemos dar nenhum outro sentido do que o que vinculamos aos enunciados sobre fatos e de que a "realidade" só pode ser determinada mediante enunciados, ou seja, só no nível do discurso metalingüístico" [For him ,the theory of truth as correspondence does not take into consideration the fact that the term 'reality' cannot have any other meaning than the one we link to the statements about facts, and that the 'reality' can only be determined in view of statements, i.e., only in the level of metalinguistic discourse].

into language. As an immediate consequence, all philosophical problems necessarily entail a reflection of language on itself.

According to the discourse-theoretic stance we can only deal with the problem of truth by resorting to a universal pragmatics of the acts of speech[35]. We have to start from the practical contexts of communication and interaction among individuals in their historical worlds, as pragmatists do. Pragmatism eliminated, by means of the idea of universal linguistic mediation, the concept of truth as correspondence, since we cannot compare the linguistic expression with a piece of "naked", non-linguistically mediated truth, i.e., with something that is out of the realm of language. Nevertheless, one cannot escape from the questioning about the relationship between the linguistic system and the reality that it mediates, which means that there is at least one aspect of the theory of correspondence that persists in discourse theory. For Habermas, pragmatism has lost an essential aspect of the problematics of truth that was visible in the theory of correspondence, namely, the dimension of unconditioned validity.

This is why there is a fundamental distinction, in this context, between the communicative actions of the daily life and the discourse. In the sphere of our linguistic actions in the everyday life, we gather information about the objects of experience[36]. Furthermore, the validity claims[37] built in statements are implicitly presupposed and acknowledged

---

[35] Linguistic action in this sense has two basic aspects: the *performative moment* in which the type of action in question is expressed (a specific type of relationship between subjects), and the propositional content that expresses an objective state (knowledge about the object). See *Searle J. R., Sprechakte. Ein philosophischer Essay.* Frankfurt am Main: Suhrkamp, 1971. *Habermas J., Was heisst Universalpragmatik?,* in: *Vorstudien und Ergänzungen zur Theorie des kommunikativen Handelns,* 2nd ed., Frankfurt am Main: Suhrkamp, 1986., pp. 353-440. *Tugendhat E., Vorlesungen zur Einführung in die sprachanalytische Philosophie.* Frankfurt am Main: Suhrkamp, 1976. *Apel K-O, Zwei paradigmatischen Antworten auf die Frage nach der Logos –Auszeichnung der menschlichen Sprache,* in: H. Lützler (org.), *Kulturwissenschaften,* Bonn, pp. 31 ff.

[36] Habermas clearly distinguishes between the *objects* of experience about which we make statements, and the facts, which are what we state about the objects. Objects, thus, are something in the world, but not facts. See *Habermas J., Wahrheitstheorien,* in: *Vorstudien,* supra note 35.

[37] In considering the pragmatic dimension of language there is a move from well formed sentences to statements made in lifeworld contexts and for their underlying communicative competence. Instead of sentences, it is the acts of speech or linguistic actions which are taken into account. According to Habermas, for every linguistic system there are necessarily certain types of acts: communicative acts (which express the meaning of the speech as such), ascertaining acts (which express the propositional contents), representational acts (self-representation) and regulative acts (which express norms). Each one of these communicative acts is connected to specific forms of

without further problems, i.e., without having their validity questioned, once the human understanding cannot function unless the individuals relate to a single objective world, and thus stabilize an intersubjectively shared public space that can be clearly distinguished from everything that pertains only to subjectivity. Therefore, at the level where pragmatism is placed, the presupposition of an objective world[38] is the condition that enables all processes of cooperation and understanding among human beings[39]. In our statements we state facts about the objects in the world[40]. This means that we cannot apply the fallibilistic model of science to the lifeworld,[41] for this would make the actual, concrete life impossible.

Consequently, there is a "performative" need for certainty in the daily life that, in principle, makes any stock of truth unnecessary. The illocutionary force of statements is not the same in action and in discourse:

---

obligation, which means that *with each type of acts of speech we raise aspecific type of validity claim.* Thus, for instance, when I state something I take on the obligation of presenting the reasons of the *truthfulness* of the propositional contents of my statement, if I am asked to do so. See *Habermas J., Vorbereitende Bemerkungen zu einer Theorie der kommunikativen Kompetenz*, in: Habermas J./Luhmann N., *Theorie der Gesellschaft oder Sozialtechnologie - Was leistet die Systemforschung?*, Frankfurt am Main: Suhrkamp, 1975, at 111.

[38] Not every kind of pragmatism accepts this type of "realist" intuition about the everyday life. Rorty, for example, considers this posture to be a mistake and defends the re-education of the ordinary man through the substitution of the want for objectivity by the want for solidarity, and conceives truth as that which we, the members of liberal western societies, believe in. See *Rorty R., Is Truth a Goal of Enquiry? Davidson versus Wright*, in: Phil. Quart. 45 (1995)300.

[39] On the analysis of the concept of "world lived" see Habermas J., *Theorie des kommunikativen Handelns*, vol. 2, Frankfurt am Main: Suhrkamp, 1981, pp. 171 ff. Martini R.M.F., Uma arqueologia do conceito "mundo da vida" na teoria da ação comunicativa, in: De Boni L. A (org.) Finitude e Transcendência . Festschrift em homenagem a Ernildo J. Stein, Vozes/Edipucrs, Petrópolis, 1996, pp. 682-712.

[40] Habermas accuses transcendental theories of truth of mixing up the truth and the objectivity of experience, since they present the theory of the constitution of experience as a theory of truth. See *Habermas J.,Wahrheitstheorien*, pp. 151 ff. For a comparison between Kant and Habermas see *Heckmann H-D.*, supra note 1, pp. 41 ff. For the ambiguity, in this respect, of Habermas's theory of truth see *Apel K-O, C. S. Peirce and the Post-Tarskian Problem of an Adequate Explication of the Meaning of Truth: Towards a Transcendental-Pragmatik Theory of Truth*, in: Freeman E. (ed.), *The Relevance of Charles Peirce*, La Salle, Illi.,1983, pp. 208 ff. On a transcendental refutation of those ideas see *Höffe O*, supra note 19, pp. 257 ff.

[41] For Höffe, Habermas starts from the fact that linguistic systems produce a consensus in which we are always already immersed when we produce arguments. Habermas, in his turn, thinks that the linguistic systems are the result of the historical process of formation of the species, and that the cognitive patterns they provide are the result of the clash of personality systems, the societal system, and nature – i.e., derive from a non-communicative sphere. See *Höffe O*, supra note 19, at 264.

whereas the truth, in the context of the daily praxis, would destroy the certainty of action, at the level of discourse it constitutes the main reference point. This is so because in discourse we abandon the actual contexts of action to exchange reasons[42] about the validity of claims, i.e., we ask ourselves about the fulfillment of the conditions to satisfy those claims[43]. In short, the use of the adjective "true" explicitly shows that the claim was satisfied by being considered intersubjectively legitimate[44]. This is the reason why for Habermas the question of truth is not a problem of the contexts of action in the daily life, as pragmatism puts it. Truth has to do with the problematization of validity claims and its discursive retrieval, i.e., with the distinction between opinions (the rising of a validity claim for what is stated) and knowledge (the discursive solution of the problem raised with that claim, the demonstration of the reasons of its legitimacy), between being considered true and being true.

This presupposes a fundamental distinction that disappears in pragmatism: the difference between the experience of certainty in action, that is, the experience of self-evident facts of the everyday life which are deeply rooted in our vast background of intersubjectively shared convictions (or, in other words, the objectivity of experience), and the discursively legitimated possibility of making true statements (i.e., truth)[45].

---

[42] On the concept of argumentation see Toulmin S., Der Gebrauch von Argumenten, Kronberg, 1975; Kambartel F., Überlegungen zum pragmatischen und argumentativen Fundament der Logik, in: Lorenz K. (org.), Konstruktionen versus Positionen, Berlin/New York, vol. I., at 221; Kopperschmidt J., Argumentation, Sprache und Vernunft. Parte II, Stuttgart, 1980, pp. 85 ff. In this respect, Habermas follows Toulmin, who uses the model of jurisprudence, not of mathematics, when explaining the logic of scientific inferences and the speech in ordinary, everyday life. On this issue and in the same sense see Perelman Ch/Olbrechts-Tyteca L., Traité de l'argumentation. La nouvelle rhétorique, Éditions de l 'Université de Bruxelles, qui. Ed., 1988. For Höffe, the pattern of argumentation with which Habermas works is neither something new, since it corresponds to the standard model of the Hempel-Oppenheim scheme of explanation of natural sciences plus an inductive support of legal hypothesis, nor is related to a genuine process of unification. See Höffe O, supra note 19, at 263.

[43] These conditions are not on the level of the contents of empirical statements, for they are universal formal conditions subjacent to any statement about an object of experience, whose validity is always already presupposed and recognized. See *Becker W.*, supra note 1, at 30.

[44] See *Becker W.*, supra note 1, at 29.

[45] For Höffe, Habermas focus only upon this aspect of truth and excludes the multiple senses of the use of the word 'truth' which can be found both in everyday life and in philosophical tradition. His question is whether an encompassing theory of truth should not at least inquire if this is not the case of an analogous use of the word, and how we could face it – a question that Habermas does not pose. As a consequence, Habermas narrows down the problematic of truth and leaves outside his theory some important aspects, such as the ontological dimension. See *Höffe O*, supra note 19,

Hence truth is a validity claim that we connect to statements while we make them,[46] that is, it is a quality of statements[47] that exists only when generated by an act of speech and inasmuch as a claim is raised whose validity may be retrieved by means of arguments. This means that truthfulness is something that concerns the acts of speech in argumentation – acts through which participants in rational discourses engage in clarifying cognitively solvable validity claims by the exchange of reasons (namely, an obligation of justification) so as to reach a legitimate consensus.[48] Discourse is intended to make a factual, contingent agreement evolve into a rational one – that is, one that is reached through an argumentative procedure; thus discourse finds its place in a larger tradition of western thought by tying up true knowledge to justification.[49]

Truth, then, is composed of three[50] basic elements: a validity claim;[51]

---

pp. 254-255. For some examples of the multiplicity of senses of truth in tradition see *Aristotle: Nic. Et.*, VI; *Thomas Aquinas: De Veritate* I.

[46] See Habermas J., Wahrheitstheorien, supra note 35 at129.

[47] On the discussion about the bearer of truth in modern analytical philosophy see *Heckmann H-D.*, supra note 1, pp. 23 ff.

[48] For Höffe, the different consensual theories of truth have some features in common: a) They refer exclusively or primarily to the truth of statements; b) They define truth as a procedure for the investigation of true statements; c) This procedure is based on a dialogical situation, and is oriented to the reaching of an agreement; d) Agreement does not mean, in the last analysis, a historical-factual event, but should be understood in a normative way as true consensus or objective agreement; e) What stands for true consensus is the potential agreement of everyone, or everyone who is linguistically competent. See *Höffe O*, supra note 19, at 252.

[49] See *Höffe O*, supra note 19, at 261.

[50] Ever since Descartes there is a trend of modern philosophers who include a fourth element: evidence. See *Heckmann H-D.*, supra note 1, pp. 148 ff. *Tugendhat E., Der Wahrheitsbegriff bei Husserl und Heidegger*, Berlin: Walter de Gruyter, 1967.*Landim Filho R., Evidência e Verdade no Sistema Cartesiano*, Loyola, São Paulo, 1992.

[51] Landim Filho, with Searle, Tugendhat, and Habermas, clearly separates the propositional content from its claim of truthfulness. See *Landim Filho R., Significado e Verdade*, supra note 13, at 41: "Como o ato único do falante é decomposto em dois momentos – o que é dito e a pretensão de verdade, a compreensão pelo ouvinte da asserção é a compreensão da possibilidade de uma tomada de posição: o conteúdo proposicional pode ser negado ou reafirmado, e então a resposta do ouvinte é uma asserção; pode também ser questionado e a resposta então é uma pergunta, etc." [As the single act of the speaker is divided into two moments – what is said and its claim of truthfulness – the listener's understanding of the statement is also the understanding of the possibility of adopting a posture: the propositional content can be either denied or reaffirmed, then the listener's answer is an assertion, or it can be questioned, then the answer is a question etc.].

its discursive retrieval,[52] and its relationship with objects.[53] This implies a passage from action to discourse which causes the naive reliance on what one takes as the truth that characterizes the experiencing of certainties on the level of the contexts of action to disappear, and which also produces the transformation of preexisting statements into hypothesis whose validity must be proven through an argumentative process.[54] In short, the certainties of the contexts of action that have been unsettled for different reasons become disputable validity claims of hypothetical statements on the level of discourse. Those claims need to be substantiated and once so legitimated they can be reintroduced in contexts of action,[55] thus making it possible that a collective form of life be based on truth, i.e., on a rational consensus.[56] But the sharp distinction between Rorty's pragmatism and universal pragmatics is exactly in the role of an argumentation which clearly separates the proceedings of justification in actual contexts of action from truth and its unconditional, context-independent meaning, whose intrinsic universality claim[57] cannot be associated with specific peoples, groups or epochs.

In an argumentative process, the force that generates consensus has to do with the consistency between language and the corresponding conceptual system. One can only speak of a satisfactory argument when

---

[52]See *Landim Filho R., Significado e Verdade*, supra note 13, at 45: "Uma asserção é verdadeira se a sua pretensão de verdade é justificada" [A statement is true if its claim of truthfulness is justified]. At 42 we read: "A Expressão "é verdade" remete assim o conteúdo proposicional às razões e revela por isto a natureza do ato como uma tomada de posição, e como uma interpelação que é um desafio" [The expression 'it's true' links the propositional content to reasons and thus reveals the nature of the act as an adoption of posture and a challenging inquiry].

[53]Habermas focus his systematic works on the two first and then forgets about what E. Tugendhat called 'the objective component' of his theory of truth. See *Puntel L. B.,Wahrheitstheorien in der neueren Philosophie*, 2nd ed., Darmstadt: Wissenschaftliche Buchgesellschaft, 1983, at 149. For Becker, Habermas tries to incorporate elements of coherence and correspondence theories of truth into his own theory. Nevertheless, the incorporation of elements from correspondence theories regarding the relationship between language and reality mediated by experience, as reference points for the judgement of the adequacy of a linguistic system, is incompatible with other central tenets of discourse theory. See *Becker W.*, supra note 1, at .321.

[54]On the logic of this argumentative process see *Habermas J., Wahrheitstheorien*, supra note 35, pp. 159-174; *Heckmann H-D*, supra note 1, pp. 37 ff.

[55]Habermas speaks of a circular process involving action and discourse. See *Habermas J., Wahrheit und Rechtfertigung*, supra note 21, at 254.

[56]Which cannot be explained by merely logical motives or by the evidence of experiences, but through the formal-logical properties of discourse. The logic of the discourse, in Habermas's view, has three distinctive features: a) A specific scheme of argumentation; b) Substantial criticism of language and self-reflection; c) The conditions of the ideal situation of speech. See *Höffe O*, supra note 19, at 262.

[57]See Habermas J., Vorbereitende Bemerkungen, supra note 37, at 124.

all parts of that argument belong to the same language, since the basic concepts of a legitimating procedure are predicated upon the linguistic system in such a way that a statement can only function as justification when it is part of a linguistic system. From the vantage point of discourse theory, justification has nothing to do with the relationship between single statements and reality, but first and foremost with the consistency of sentences in a linguistic system. To put it in a nutshell, the consensus among the parties to an argumentative process can be effectual only insofar as one manages to demonstrate through argumentative procedures that what one states is consistent with other statements in the linguistic system.[58]

The argumentative answer to validity claims must lead to consensus, but a justified form consensus, otherwise we would still be moving amidst the unquestionable truths of the everyday life. The most important question in this context is: how can one distinguish the factual, variable consensus common to contexts of action from a justified consensus that claims immutability?[59] What are the formal qualities of a justified consensus?[60] How can one tell true statements from false ones? The first task of a theory of truth in the sense of universal pragmatics is to thematize the formal requirements of all substantive truth. For Habermas, a genuine consensus can only be reached under the counterfactual conditions of an ideal[61] situation of communication,[62] which means that the unavoidable requirement of true statements is that they could be accepted by everyone.[63] This unavoidable presupposition associated with consensus has four major features,[64] all of them related to equality of chances: a) equality of all participants in discourse in the use of communicative acts of speech; b) equality in the thematization and critique

---

[58] See *Habermas J., Wahrheitstheorien*, supra note 53, at 165 ff. For Becker, this means adopting elements from coherence theories into discourse theory. See *Becker W.*, supra note 1, at 318 ff. Höffe in turn maintains that the logic of discourse cannot go without some element from coherence theory in the explanation of validity claims. See *Höffe O*, supra note 19, at 264.

[59] See, on this 'eternity' claim and the ontological problematic there implied, *Heckmann H-D,* supra note 1, at 206, note 17.

[60] According to Becker, Habermas reverbs Peirce's pragmatic theory of truth here. See *Becker W.*, supra note 1, at 315, note 47.

[61] This is what Habermas calls the epistemic concept of truth seen from a pragmatic perspective. He identifies this view in Putnam (utterances are true when they can be justified under ideal epistemic conditions) and in Apel (truth springs from an argumentative consensus in an ideal community of communication). See *Habermas J., Wahrhiet und Rechfertigung*, supra note 21 , at 256.

[62] See *Habermas J., Wahrheitstheorien*, supra note 35, at 174 ff.

[63] See Habermas J., *Wahrheitstheorien*, supra note 35, at 137.

[64] See *Heckman H-D,* supra note 1, at 21.

of preceding opinions; c) equality in the use of representational acts of speech; d) equality in the use of regulative acts of speech. The ideal situation of speech as a necessary reference is neither a phenomenon nor sheer construction, but an essential, unavoidable assumption that provides the normative foundations of linguistic comprehension. It is always necessarily given in advance in speech and as such it operates in human life.

The basic question[65] here is whether under the conditions of an ideal situation of speech we can obtain a satisfactory formal basis for the rational agreement of all those involved in discourse, and thus, as Habermas believes, a satisfactory criterion for determining truth. First of all, according to Becker, it is obvious that a participant in discourse cannot turn either to an argumentatively reached agreement or to the existence of symmetrical relations so as to make his agreement intelligible from a formal perspective.[66] This is so because when one makes use of consensus one cannot avoid the question about what provides the basis for the agreement of all other participants in discourse, just as in the case of symmetrical relations one cannot avoid the question about which symmetrical conditions may give rise to an agreement about the legitimization of a statement. In both cases we have to make use of a basis of agreement that is not sufficiently determined by purely consensual reasons, i.e., that does not derive from the structure of consensus itself. Consequently, we have to look for ways to transcend the integrative force of the criterion of truth of a consensus theory and reach the justification of a rational agreement.[67] As a result we have that the conditions of an ideal situation of speech constitute the universal premises of argumentation and the requisite conditions for the participation in discourse, but they do not provide a sound basis for the judgment of the legitimacy of a claim.[68] This means that the potential agreement of all possibly affected people is not the deciding formal requirement,[69] and therefore

---

[65]See *Becker W.*, supra note 1, pp. 321 ff.

[66]Which means consensual theory of truth cannot provide a criterion for truth., for an ultimate rational consensus cannot be anticipated, once we don't even know whether it will actually take place. See Hösle V., Die Krise, supra note 4, at 198.

[67]For Hösle, a consensus is only rational when it meets certain criteria that precede consensus, *evidence* and *coherence*. See *Hösle V., Die Krise der Gegenwart und die Verantwortung der Philosophie. Transzendentalpragmatik, Letztbegründung, Ethik*, München: Beck, 1990, at 199.

[68]For Alexy, the conditions of an ideal situation of speech can be fulfilled in part. They are an instrument of criticism, and provide at least a negative criterion for correctness and truth. See *Alexy R., Theorie der juristischen Argumentation. Die Theorie des rationalen Diskurses als Theorie der juristischen Begründung*, Frankfurt am Main: Suhrkamp, 1983, at 170.

[69]For Puntel, Habermas's formulations express an extraordinary global intuition

that consensus cannot be the criterion of truth[70] or the foundation of the justification of validity claims. Grounded consensus, the rationally legitimated agreement of all affected people is the form in which the result of argumentation is expressed, the manifestation of a rational judgment of the justification of validity claims. The thesis according to which truth is what could be acknowledged as such by all rational beings[71] in an ultimate rational consensus is correct, but it does not provide a formal requirement of truth modeled on the characteristics of an ideal situation of speech.[72]

---

that Habermas does not work out rationally. When one intends to determine more concretely what consensus means as a criterion for truth, one has to say that it is a manifestation of the 'essence' of truth in the pragmatic-intersubjective dimension. This is certainly a primary dimension, though not the only one. To deal with the revelation of truth in this sphere means that it has a concrete meaning. The corresponding abstract meaning is the idea of coherence. In this context, it is possible to say that a statement is true when we can reach a universal consensus about it. In other words: a statement is true when it can be integrated in the total system of statements that can be coherently demonstrated. See *Puntel L. B., Wahrheitstheorien*, supra note 1, at 162-163. On his coherence concept of truth see *Puntel L. B., Grundlagen einer Theorie der Wahrheit*, Berlin/New York: Walter de Gruyter, 1990; *Uma versão forte do princípio do contexto (Frege)*, in: *De Boni L. A* (ed.), *Finitude e Transcendência. Festschrift em homenagem a Ernildo J. Stein*, Petrópolis: Vozes/Edipucrs, 1996, at 371-387. For Hösle coherence theory of truth is correct in saying that there is a reality independent from all theories and all experience, but we can only say that a theory expresses such reality if it manages to integrate all data collected from reality into a coherent whole. See *Hösle V., Die Krise*, supra note 67, at 198.

[70] For Landim Filho this is not a criterion, but rather a consequence. See Landim Filho R, Significado e Verdade, supra nota 13, at 45: "Ao invés de procurar através da noção de consenso fundamentado um critério de verdade, poder-se-ia modificar a investigação e pensar este consenso como consequência de uma asserção verdadeira" [Instead of seeking for a criterion for truth through the concept of grounded consensus, we could change our inquiry and conceive such consensus as a consequence of a true statement"] In the same sense Hösle says that one could see an equivalence between truth and recognition through an unlimited community of communication when one says that truth does not depend on such recognition, but rather that such recognition presupposes criteria for truth that represent more than the mere factual agreement. See. Hösle V., Die Krise, note 4 supra, p. 198.

[71] See *Hösle V., Die Krise,* note 4 supra, p. 198.

[72] For Höffe, the ideal situation of speech produces only what one could call the "political bypassing conditions" so that argumentation is possible and we can reach an agreement without much pressure. The elimination obstacles to communication is not the same as providing a basis for the legitimation of a truth claim. In his view, Habermas actually takes the moments of coherence and correspondence in his theory for criteria for truth. See *Höffe O.,* note 4 supra, p. 270-271. Against the idea that the conditions of the ideal situation of speech can provide solid foundations for validity claims see *Skirbeck G., Rationaler Konsens und ideale Sprechsituation als Geltungsgrund? Über Recht und Grenze eines transzendental-pragmatischen Geltungskonzepts*, in: *Kuhlmann W./Böhler D.*(orgs.), *Kommunikation und Reflexion. Zur Diskussion der Transzendentalpragmatik. Antworten auf Karl-Otto Apel,* Frankfurt am Main:

C) An ontological-metaphysical concept of truth: Henrique Cláudio de Lima Vaz.

1. *The ontological concept of reason: ontological premises of the question of truth*

Lima Vaz's thought belongs to the great Western metaphysical tradition[73] as contemplation or theory of the totality of the being, of the reasons of the beings, their principle and their concept of truth as correspondence.[74] The main purpose of his theory – at least in its metaphysical version – is not so much to give an account of the nature of that correspondence, but rather to thematize its *deepest foundations*. Lima Vaz's effort is twofold: first, he makes the ontological dimension[75] explicit by thematizing the being to which human reason is fundamentally open;[76] second, he demonstrates the existence of ultimate metaphysical foundations by thematizing the absolute being as the ultimate basis of truth, i.e., as the primordial truth.[77] He addresses the problematics of ontology

---

Suhrkamp, 1982, p. 63.

[73] For Lima Vaz the many, mostly anti-metaphysical revolutions of the 20th century ended up by showing the unavoidable advent of metaphysical problems. The theory of knowledge was the field where metaphysics was destroyed, but was also the province where metaphysical questions reappeared. See *Lima Vaz, H. C. de, Tópicos para uma metafísica do conhecimento*, in: *Ulmann R. A* (ed.), *Consecratio Mundi. Festschrift em homenagem a Urbano Zilles*, Porto Alegre: Edipucrs, 1998, p. 431. On the current negation and the future of metaphysics see *Lima Vaz H. C. de, Esquecimento e memória do ser: sobre o futuro da metafísica*, in: Síntese – Rev. de Filosofia, v. 27. n. 88 (2000), p.149-163.

[74] For Lima Vaz, one cannot face the problematics of truth outside the ontological horizon and, in a final analysis, outside the metaphysical horizon, as the Greeks had already understood. This means that the ontological dimension cannot be considered as something outside the concept of truth. See *Lima Vaz, H. C. de, A dialética das Idéias no Sofista*, in: *Ontologia e História*, São Paulo: Duas Cidades, 1968, pp. 15-66.

[75] On a different conception of the relationship between the problematics of truth and ontology in view of the opposition between the "principle of semantic-sentential compositionality" and the "context principle". See *Puntel L. B., Uma versão forte do Princípio do Contexto (Frege)*,in: *De Boni L. A* (ed.) *Finitude e Transcendência*. Festschrift em homenagem a Ernildo J. Stein, Vozes/Edipucrs, Petróplis, 1996, p.371-387; *Truth, Sentential Non-Compositionality and Ontology*, in: Synthese, vol. 106 (2001), p. 221-259.

[76] Which means that the metaphysical tradition has placed knowledge in the symbiotic convergence of being and knowing, and thus as the interpenetration of noiesis and metaphysics. Modern nominalist philosophy, on the contrary, has thought of this problematics from based on the opposition between being and representation, and this implied the separation of metaphysics and the theory of knowledge. See *Lima Vaz, H. C. de, Tópicos*, note 73 supra, pp.431-432.

[77] See *Lima Vaz H. C. de, Esquecimento*, note 73 supra, p. 154 : "... p*ensar e enunciar o Ser* na sua amplitude inteligível só é possível como discurso sobre as

in a manner similar to that of J. Maréchal[78] and J.B. Lotz,[79] i.e., through a reflexive analysis[80] of the act of judging in Aquinas' philosophy.[81] This reflexive analysis conduces him to demonstrating the absoluteness of the principle of contradiction with the mediation of counterargumentation.

According to Lima Vaz this procedure is developed in three steps: a) The specificity of humans in their relationship with the world appears above all in their intellectual[82] capacity of elevating what is in the perceptual level to the universalistic level of meaning – where meaning is to be understood as a necessary intelligibility that is at the basis of the understanding of perceptions. In a final analysis, this is the same as introducing the given in the sphere of the being to produce an intelligible order of beings in such a way that the universe of ideas coincides with the totality of what is intelligible. But the whole of Western metaphysical thought springs from Plato's discovery of a realm of Ideas[83] where

---

razões do Ser que tem um ponto de partida, ou seja a sua intuição (noeîn) do Ser como absolutamente inteligível ou como o Absoluto *pensado* (Parmênides) e a intenção voltada para um ponto de chegada, ou para o Absoluto *conhecido* como Absoluto *real*"..... ["... to think and to *enunciate* the *Being* in its full intelligibility is only possible as a discourse on the *reasons of the Being* if there is a starting point, that is, an insight (noeîn) of the Being as absolutely intelligible or as Absolute thought (Parmenides), and also if the intentions are directed towards an end, towards the Absolute *known* as *real* Absolute.]

[78]See *Maréchal, J., Le point de départ de la Métaphysique*, Cahier V, 2nd ed., L 'Édition Universelle Desclée? Bruxelles/Paris,1949.

[79]See *Lotz J. B., Das Urteil und das Sein*, 2nd ed., Pullach bei München: Verlag Berchmanskolleg, 1957; *Sein und Existenz. Kritische Studien in systematischer Absicht*, Freiburg/Basel/Wien: Herder, 1965.

[80]Which as philosophical reflection is rooted in the very original reflection of the intellect upon its own action as intentionally bound to the object. See *Thomas Aquinas, De Veritate*, q. 1 a 9. Lotz J. B., *Transzendentale Erfahrung*, Freiburg: Herder, 1978.

[81]See *Aquino, M. F. de., Experiência e sentido II*, in: Síntese Nova Fase , v. XVII, n.50 (1990), pp. 31-54.

[82]This is what Lima Vaz calls "spiritual intelligence". See *Lima Vaz, H. C. de, Tomás de Aquino: pensar a Metafísica na aurora de um novo século*, in: Síntese Nova Fase, v. 23, n. 73 (1996), p. 173: "A aventura da Metafísica ocidental....teve como protagonista uma certa concepção da *inteligência* (ou nous na terminologia grega) cujo exercício permitiu ao filósofo, seguindo a rota da segundo navegação platônica, estender sua inquirição além do horizonte do sensível...A inteligência metafísica...é propriamente uma *inteligência espiritual*. [The adventure of Western metaphysics (...) was performed by a certain concept of intellect (or nous in greek) whose usage allowed the philosopher, by following Plato's route, to extend his inquiry beyond the sensorial level (...) Metaphysical intelligence (...) is most suitably qualified as a spiritual intelligence.]

[83]See *Lima Vaz, H. C. de, A metafísica da idéia em Tomás de Aquino*,in: Síntese – Rev. de Filosofia, v. 28, n. 90 (2001), p. 7. For Lima Vaz, the discovery of the world of ideas means that Plato found out what he calls the "ideonomic paradigm", which is distinguished precisely by its grounding rational knowledge in the world of ideas,

we can find the true essence of all beings,[84] as well as the articulation of a new form of language intended to express in discourse an insight into the new reality,[85] which is *"transnatural*, epistemologically *transempirical* and ontologically *transcendent"*.[86] The discursive knowledge of ideas consists in the negation of the qualities of perception so that the Idea appears as singular (by the negation of plurality), immutable (by the negation of change), simple (by the negation of composition), incorporeal and indivisible etc.; b) This knowledge takes place in judgement through a process of unification of the plurality of the formal determinations of the object by the mediation of the synthesizing activity of an intellect that thematizes the basic ontological core within substance, that is, the essential configuration of reality expressed in judgments by the predicate, whereas the subject expresses the singular substance,[87] the fundamental unity with which mental activity is primarily concerned in this synthesizing process; c) The being reveals itself in judgment firstly within that process of synthesizing unity and multiplicity as essence, i.e., as the organized structure of reality. Nevertheless, judgement is characterized by the fact that it goes beyond the level of mere logic and beyond the level of the epistemological composition and division of "quiddities" to reach the affirmation of the being.

For Lima Vaz[88] what makes Aquinas unique in Western philosophy

---

where universal, transempirical concepts are to be found that relate to intelligibility as such. See *Lima Vaz, H. C. de, Escritos de Filosofia V. Introdução à Ética Filosófica 2*, São Paulo: Loyola, 2000, p. 27.

[84]See *Lima Vaz, H. C. de, Antropologia Filosófica II*, São Paulo: Loyola, 1992, p. 111.

[85]See *Lima Vaz, H. C. de, Transcendência: Experiência Histórica e Interpretação Filosófico-Teológica*, in: Síntese Nova Fase, v. 19, n. 59 (1992), p. 444: "...a primeira acepção do termo transcendência.....pretende designar um aparentemente incoercível movimento intencional pelo qual o homem transgride os limites da sua situação no Mundo e na História e se lança na direção de uma suposta realidade transmundana e trans-histórica que se eleva como cimo do sistema simbólico através do qual as sociedades exprimem suas razões de ser". [... the first meaning of transcendence (...) is intended to designate an apparently unstoppable intentional movement by which man oversteps the bounds of his position in the world and history and moves towards a supposedly trans-worldly and trans-historical reality which represents the highest point of the symbolic system societies use to express the reasons for their beings.]

[86]See *Lima Vaz, H. C. de, Escritos de Filosofia V*, note 83 supra, p. 98.

[87]For L. B. Puntel, this stance presupposes an ontology whose basic elements are the objects (individuals) and their attributes (properties and relations), i.e., *an ontology of substance*. In his view, this type of ontology should not be accepted. On the arguments supporting an alternative ontology whose basic elements are not substance, but "behaviors" (Verhalte) see *Puntel, L. B., Grundlagen*, note 2 supra, p. 220 ff.; *The Context Principle, Universals and Primary States of Affairs*, in: American Philosophical Quartely 30 (1993)123-125.

[88]Following E. Gilson's interpretation. See *Gilson E., L'être et l'essence*. Vrin,

and sets him apart from the essentialist tradition of Greek metaphysics is that for him the being ultimately reveals itself as the action of existence under the form of "the first, constitutive act of the reality-in-itself of the being",[89] the actuality of all actions and the perfection of perfections (De Pot. Q. 7, a 2 ad 9m),[90] and whence as absolute perfection – so that the judgement no longer concerns essence, but existence.[91] Here is the starting point of metaphysics, for its object, the being as being or as universal being, appears in each act of judgement[92] and in the self-affirmation of the being, which is the substitute for the intellectual intuition of the being in pure intellect. The affirmation of the being[93] which is precisely the

---

Paris, 1948. In the same sense see *Siewerth G, Das Schicksal der Metaphysik von Thomas bis Heidegger*, Einsiedeln: Johannes Verlag, 1959. *Lima Vaz H. C. de, Presença de Tomás de Aquino no horizonte filosófico do século XXI*, in: Síntese Nova Fase v. 25, n. 80 (1998), p. 38 ff.

[89]See *Lima Vaz H. C. de, Presença,* note 88 supra, p. 38.

[90]On the difference between this metaphysics of the act of existing and the hermeneutical ontology that puts the being as condition of the possibility of understanding the ens see *Müller M., Existenzphilosophie im geistigen Leben der Gegenwart*, 3rd ed., Heidelberg: F. H. Kerle Verlag, 1964. *MacDowell J. A, A Gênese da ontologia fundamental de Martin Heidegger,* São Paulo: Loyola, 1993. For Lima Vaz, modernity is characterized by the precedence of the subject over the being, and this is precisely what the so called anthropocentric turn of the thought consists in. Lima Vaz's thought is clearly part of the metaphysical tradition, not the anthropocentric, as A. J. Severino believes. See *Severino A J., A Filosofia Contemporânea no Brasil. Conhecimento. Política e Educação,* 2nd ed., Petrópolis: Vozes, 1999, pp. 134-139 (but particularly p. 138).

[91]Puntel would certainly ask if the centrality of the act of existing in the thought of Aquinas – even though we acknowledge its originality in view of the essentialism of Greek philosophy – really puts in question the traditional ontology of substance or is just a new version of that same ontology in spite of its novelty. Is not the object as a central category preserved in Aquinas? Does he break with the object-centered ontology of tradition?

[92]Which means that, for Vaz, the insertion of the human being in the historical contexts of its ordinary life does not deny, as for Rorty, the existence of its metaphysical dimension. On the contrary, it necessarily points to it. See *Lima Vaz, H. C. de, Antropologia II,* note 84 supra, at 112: "Com efeito, não há nenhuma possibilidade de se pensar qualquer experiência humana sem que a contingência e o efêmero acontecer da experiência não sejam atravessados pelas questões propriamente transcendentais em torno da unidade, da verdade, da bondade, da beleza e, finalmente, do *ser* dessa realidade sempre fugidia que está em jogo na experiência". [In fact, there is no possibility of thinking any human experience without its contingency and ephemerality being pervaded by specifically transcendental questions about the unity, truthfulness, goodness, beauty, and, last but not least, the *being* of the always-fleeting reality of experience.]

[93]See *Lima Vaz, H. C. de, Presença,* note 88 supra, p. 39: "Com toda ênfase, sobretudo na citada passagem do comentário ao *De Trinitate* de Boécio, Santo Tomás aponta, na síntese judicativa e na afirmação do esse (existir) no juízo, o lugar inteligível do encontro entre a *inteligência* e o *ser* na sua plenitude existencial, de tal sorte que

passage from representation to the being is irrefutably established by the counterargumentation directed against skepticism:[94] the negation cannot grasp the being since it would deny itself by denying the being.[95] We could say that "as a being-of-language and bearer of the logos the human being necessarily experiences the infiniteness of the being and therefore its transcendence of all finite beings".[96] Here, "absoluteness as *form* (the absolute truth of the contradiction principle) and as *act* (the absolute necessity of existence) emerges a necessary condition for the stating of judgement to have an ontological scope as well as for the metaphysical discourse to be possible."[97]

The force of the argument[98] stems from the nature of the relationship

esse encontro venha a operar a *identidade*, na ordem *intencional*, entre o *sujeito* cognoscente e o *objeto* real conhecido". [Aquinas, specially in the aforementioned passage of the commentaries on Boecio's *De Trinitate*, emphasizes that the synthesis of judgment and the affirmation of the *esse* (existence) in judgement provide the intelligible site for the intersection of *intellect* and *being* in their existential plenitude, so that such intersection operates the *identity*, on the *intentional* level, between the knowing *subject* and the real *object* known.]

[94] See *Aristotle, Met.* IV, 4, 1006 a 13-28.

[95] For Lima Vaz, this is the core of the counterargument against the attempts to introduce contradiction into the straightforward affirmation of the being. Negation always takes place inside the original affirmation of the being, which entails negation but remains untouched by it. Inside affirmation, negation exercises its power. See *Lima Vaz, H. C. de, Antropologia II*, note 84 supra, at 133, note 93. See on this problematics is transcendental tomism: *Holz, H., Transzendentalphilosophie und Metaphysik. Studie über Tendenzen in der heutigen philosophischen Grundlagenproblematik,* Mainz: Mathias-Grünewald-Verlag, 1966. And in the thought of K-O Apel: Oliveira, M. A de, *Sobre a Fundamentação*, 2nd ed., Porto Alegre: Edipucrs, 1997, pp. 57 ff.

[96] See *Lima Vaz, H. C. de, Antropologia Filosófica II*, São Paulo: Loyola, 1992, at 111.

[97] See *Lima Vaz, H. C. de, Tomás de Aquino*, note 82 supra, at 186.

[98] Lima Vaz always refers to counterargumentation as the argument that legitimates the necessary affirmation of the being, and consequently provides the basis, in a first moment, for metaphysical discourse. Lima Vaz's thought is analogous and in close proximity to transcendental tomism, particularly to Maréchal's e Lotz's thought, but he makes no reference to the need of using the transcendental method to reach and develop metaphysical discourse, a need that Lotz, Rahner and Coreth explicitly accept (the question about whether he would accept Maréchal's view, according to which the transcendental method is not necessary, except in a cultural sense, due to its influence in contemporary philosophical culture, remains unanswered). As a consequence, there is no further explanation of the semantic and epistemological presuppositions of metaphysical discourse but for its foundation on counterargumentation. In this case, it is difficult for him to establish a dialogue with the theories of truth marked by the linguistic turn, even because he understandably rejects those philosophies that, as the ones mentioned above, limit themselves to the analysis of the semantic and logical forms, as incapable of thematizing the ultimate question about the being. Since this question is irrefutable due to the inevitable affirmation of the being, the metaphysical

that weds the finite human intellect to the being as such[99] - a relationship marked by the finitude of an intellect that needs to act so as to move from the possibility to the activity of knowing. Thus it has to identify intentionally with the being, implying a necessary minimal determination of its object – something *is*! This makes the skeptical contradict himself when he means and says that nothing is. In short, finite intellect needs to act in order to know, which in turn leads to a minimal determination of its object and to a meeting with elementary determinations of the being. For Vaz the originality of Thomas Aquinas' thought "resides in this ingenious intuition according to which the very object of metaphysics is not to be found in the end of an abstractive process of the intellect as a universal notion of the being (*ens generalissimum ut nomen*), but appears in the dynamic intentionality of an act of judgment as dialectical identity between the *form* of judgment (est) and the supreme act or perfection (existing, *esse*)".[100]

This is exactly what the metaphysical structure of judgment and consequently the structure of human knowledge consist in: judgment firstly takes its object to the level of formal universality of the being. This means that the being asserted in a judgment (the predicative being)[101] is sublated in the transcendental being, i.e., in the horizon of the being as such,[102] which is the same as to detect formal absoluteness in the

---

horizon is the inescapable horizon of the ultimate philosophical questions. Nevertheless, one can and must, in principle, present a metaphysical theory of truth as an explanation of the ultimate dimension of truth, one that is capable of comprising all other aspects of the truth that have been worked out by other theories, once the discussion itself demonstrates the complex character of the truth, and the metaphysical horizon is the horizon of totality.

[99] See *Lima Vaz, H. C. de, Transcendência*, note 85 supra, at 448: "....a evidente finitude do nosso espírito, situado na contingência do Mundo e da História, só pode comportar-se com a sua também evidente infinitude intencional, atestada no pensamento do Ser, se aceitarmos obedecer à exigência lógica e existencial de afirmar o Transcendente como Absoluto do ser". [... the evident finiteness of our spirit, which placed in the contingency of the world and history, can only be harmonized with the also evident infiniteness of our intentionality, which is asserted in the thought of the Being, if we obey the logical and existential requirement that we affirm Transcendence as the Absolute of the being.]

[100] See *Lima Vaz, H. C. de, Tomás de Aquino*, note 82 supra, at 181.

[101] See *Lima Vaz, H. C. de, Tomás de Aquino*, note 82 supra, at 184, note 111.

[102] For Lima Vaz, what distinguishes contemporary philosophy is its effort to prevent any passage to transcendence and to reduce metaphysics to anthropology. See *Lima Vaz, H. C. de, Antropologia II*, note 84 supra,. at 121: "As filosofias da história, da cultura, da existência de um lado e, de outro, as diversas versões do positivismo, bem como, mais recentemente, as filosofias da linguagem e, caso exemplar, o programa heideggeriano de um retorno aquém de Platão em busca do 'impensado' do Ser no universo pré-socrático atestam as variadas formas desse imenso rito de exorcismo do Absoluto no qual se transformou a filosofia contemporânea". [The philosophies of

immanence of a free (freedom, will), intelligent subject (rationality, intellect) who thinks it.[103] But the being is in itself and subsists in itself, "which means that it possesses intrinsic intelligibility or *ontological reflexiveness* (the being is understood on the basis of its self, for there is nothing that precedes it logically that has any explanatory power upon it)".[104] As being-for-self it consists in a primary identity with thought,[105] or, in other words, being and truth are equivalent and thus they make up transcendental truth,[106] which grounds logical truth. Transcendental truth is the truth of human spirit, the intended identity of intellect and object,[107] for it is structurally underpinned by the universality of the being as such – an intended universality that is coextensive to the totality of being and ontic truth, the truth of things which is due to their participation in the being.

In view of its being subordinate to the being as such, human spirit faces an open horizon of unlimited intelligibility within which each and every particular entity may be cognized. Pure being then presents itself as truth (identity between truth and being) and through counterargumen-

---

history, culture, and existence, on one hand, and the different versions of positivism, on the other, as well as the more recent philosophies of language, and the exemplar case of Heidegger's project of moving back beyond Plato in search of the 'unthought' of the Being in the pre-socratic universe, testify to the several different forms of this huge rite of exorcism of the Absolute which contemporary philosophy has turned out to be.]

[103] This is a proof that metaphysics is not the alienation of human life, but a deep plunge into the "raízes do nosso ser, onde o Absoluto está presente como princípio fontal" [roots of our being, where the Absolute appears as its principle and source]. See *Lima Vaz, H. C de, Transcendência,* note 85 supra, at 447.

[104] See *Lima Vaz H. C. de, Antropologia II,* note 84 supra, at 111.

[105] See *Lima Vaz, H. C. de, Esquecimento,* note 73 supra, at 154: "O inteligível, na sua primeira manifestação ao pensamento, só pode ser pensado como Ser absoluto: o absolutamente *um*, o que significa imediatamente a sua *identidade* com o próprio pensamento, segundo a célebre proposição parmenidiana: *Com efeito, é o mesmo o pensar e o ser* (D-K 28, B, 3)". [Intelligibility, in its first manifestation to thought, can only be conceived as absolute Being: the absolute *one*, that which means an immediate identity with thought itself, according to Parmenide's famous proposition: *Veritably, thinking and being are the very same.*

[106] Which Lima Vaz also calls *noetic experience* of truth. See *Lima Vaz, H. C. de, Antropologia II,* note 84 supra, pp. 102 ff. In E. Coreth's version, being is primarily and properly knowing itself, being aware of itself before itself (ein wissendes Bei-sich-Sein), and thus *identity of being and knowing.* See *Coreth, E. Metaphysik. Ein methodisch-systematische Grundlegung,* 2nd ed., Innsbruck/Wien/München: Tyrolia Verlag, 1964, at 354. See also *Oliveira, M. A de, Ética e Justiça num mundo globalizado,* in: Veritas, v.45, n. 4 (2000), pp. 561 ff. The main feature of modernity, in Vaz's view, is the passage of the sense of transcendence from ontology to epistemology, and its transformation into a subjective condition of the possibility of the objects of human experience. See *Lima Vaz, Antropologia II,* note 84 supra, pp. 104-105.

[107] See *Lima Vaz H. C. de, Ética Filosófica 2,* note 83 supra, at 60.

tation it appears as perfect unity and perfect goodness – transcendental concepts[108] that are logically equivalent to that of being, that are identical to it in reality though presenting formal differences. Those provide the metaphysical foundations of human thought so that the core of philosophy is the discourse about the singleness, truthfulness and goodness of the being.

Human intellect, inasmuch as it correlates with the being as such, also correlates with the unity, the truth (the object of theoretical reasoning), and the goodness (the object of practical reasoning) of the being. That is why the basic structure of human beings can be said to be a synthesis of unity, truth, and goodness or liberty: "From the perspective of the intellect, human beings as spiritual beings must be defined as beings-for-truth; from the perspective of liberty, beings-for-the -good. These two intents of man's spirit or of the man as an intelligent and free spirit intersect in the unity of the spiritual movement, for truth is a good of the intellect and goodness is the truth of liberty. Such is the intersection of the finite spirit which in the infinite spirit is the absolute identity of truth and goodness."[109] Truth in the sphere of human life is the reception of the intelligible form or the perfection of the act of the ens into the spirit. And the received ens is also said to be true since as long as it takes part in the being it also takes part in truth and is thus open to the intellect, and through the act of cognizing it then becomes intentionally present in spirit. The ens is subordinate to intellect, and intellect is subordinate to the ens: such is the correspondence that constitutes the (logical) truth of intellect and the (ontological) truth of ens.

There is a reciprocal relationship here that is essential for both ens and intellect, a relationship that Lima Vaz calls "categorical truth". This, in turn, is firstly founded on the truth of the being as such: the truth of intellect and the truth of ens are rooted in the truth of the being, i.e., in the "transcendental truth".[110] That is to say, truth is a relationship from man to the being: the relationship of transcendence is the ontological constituent of humans as spiritual beings and thus irreducible to

---

[108] See on the 'doctrine of the transcendental' Lima Vaz, H. C. de, *Antropologia II*, nota 84 supra, at 129, note 42; *Transcendência*, note 85 supra, pp. 455-456.

[109] See Lima Vaz, H.C. de, *Antropologia Filosófica I*, São Paulo: Loyola, 1991, at 213.

[110] See Lima Vaz, H. C. de, *Antropologia II*, note 84 supra, at 128, note 40: "O problema lógico da verdade, que acabou prevalecendo na literatura filosófica sobre o tema, pressupõe, em última instância, a *estrutura ontológica do ser verdadeiro*, tal como foi primeiramente estabelecida por Platão". [The logical problem of truth, which ended up prevailing in the philosophical literature on the subject, presupposes, in the final analysis, the *ontological structure of the true being*, as first established by Plato.]

the somatic and psychic dimensions. Since the human being is spirit, its primary determinateness is transcendental in the sense of a radical openness to the being and of a formal identity with the being as truth and liberty.[111] On the other side, as that human spirit is finite, it moves within the sphere of categorical truth, i.e., within the opening to the finite and contingent inner and external worlds, where truth is made relative in the multiplicity and fluidity of things. This is what makes man be a basic inner tension between transcendental openness and categorical openness, and thus a tension between transcendental truth and categorical truth: in other words, "the paradox of a subject *placed* in the externality of the world and history, but innerly *open* to the universality of the being".[112] That is why human spirit is characterized by a dual structure: it is understanding (categorical openness) and reason (transcendental openness).

2. *Metaphysical truth as ultimate foundation of transcendental truth and categorical truth (logical and ontological).*

To demonstrate the real Absolute – the primal truth, the ultimate foundation of all truth – is the end and fulfillment of a metaphysical discourse that starts, in a first stage, from the original understanding of the Absolute found in the affirmation of the being that occurs in each act of judgment. Metaphysical discourse starts precisely with a reflexive analysis of the acts of judgement intended to show the conditions for the possibility of demonstrating the Absolute in a formal manner,[113] which

---

[111]See. Lima Vaz, H. C. de, *Antropologia I*, note 109 supra, at 223: "O homem é espírito significa,pois, a abertura transcendental do homem à universalidade do ser segundo o duplo movimento do acolhimento e do dom, da razão e da liberdade". [That man is spirit then means the transcendental openness of man to the universality of the being, according to the double movement of the receiveing and the gift, reason and liberty.] This has ethical consequences. On page 224 we read: "...a abertura transcendental ao horizonte universal do ser (como Verdade e Bem) impõe ao homem a tarefa de sua auto-realização segundo as normas dessa universalidade". [... the transcendental openness to the universal horizon of the being (as Truth and Good) burdens man with the task of self-realization, in accordance with the norms of such universality.]

[112]See *Lima Vaz, H. C. de, Transcendência*, note 85 supra, at 446.

[113]Those conditions are the predominance of existence on the level of metaphysical intelligibility and the dynamics of the assertion of judgement oriented toward absolute Existence. See *Lima Vaz, H. C. de, Presença*, nota 88 supra, at 40. Such stance, in which the formal demonstration of the real Absolute takes place a posteriori, is a consequence of the non-acceptance of the validity of the ontological proof, although Lima Vaz states that "essa argumentação redargüitiva ou de retorsão (obrigando a negação a negar-se a si mesma) pode ser entendida em analogia com o chamado 'argumento ontológico'" [such counterargumentation, which forces negation to negate

are to be found *a priori* in the very structure of human intellect. In short, this is a question of showing that the dynamic nature of judgment is the condition for the possibility of demonstrating the existence of the real Absolute, since, given the finite and measurable character of the intellect, subordination to the formal Absolute (Unity, Truth, Goodness) implies subordination to the real Absolute.[114]

Lima Vaz, following Aquinas' steps in his metaphysical discourse, states that "the positioning of the being in judgment reflexively denotes: a) the *separatio* of the existent (esse) as supreme perfection of the being and the impossibility of identifying *esse* and *essentia* in the finite being which is structurally subjected to the *eidetic* limitation that occurs in the formation of concepts; b) the reference to the Absolute in the dynamics of the spirit that moves the judging activity of the intellect."[115]

The real Absolute as subsisting being or Absolute-of-existence is the absolute One from which all unity in the level of the beings originates. It is expressed as intellectual intuition due to the identity between intellect and intelligibility as actual identity and formal difference – thus as absolute intuition of itself and thinking of thinking. As absolute intelligence and primary truth, it is the radical principle of all intelligibility, i.e., the reciprocal relationship of human intellect to its objects is based on the truth of the very being, which in its turn is grounded on the subsisting truth, which is the ultimate foundation of the discourse of the intelligibility of everything, namely, metaphysics, which culminates in the passage from transcendental truth to infinite truth. The Infinite Spirit therefore

---

itself, may be understood in comparison with the so called 'ontological argument'] (*Antropologia Filosófica II*, nota 84 supra, at 126, note 16). See also *Henrich, D, Der ontologische Gottesbeweis*, 2nd ed., Tübingen: J.C.B.Mohr (Paul Siebeck), 1967.

[114]See *Lima Vaz, H. C. de, Presença*, note 88 supra, at 39: "...essa teoria oferece-nos a possibilidade de articularmos intrinsecamente a tese da primazia do *existir* e suas conseqüências metafísicas. Com efeito, a afirmação do *esse* (existir) no juízo vai além necessariamente, no seu dinamismo intencional, da *limitação eidética* dos objetos *finitos*, a que se aplica e, em virtude da *ilimitação tética* do próprio ato de afirmação *põe* ineluctavelmente, como horizonte último, não contemplado, mas dialeticamente implicado, o Existir subsistente *infinito* na sua absoluta transcendência, cuja existência, no âmbito da inteligibilidade analógica, será formalmente demonstrada nas provas clássicas da existência de Deus". [... such theory offers the possibility of an intrinsic articulation between the thesis of the predominance of existence and its metaphysical consequences. In fact, the affirmation of the *esse* (existence) in judgement necessarily goes, in its intentional dynamics, beyond the *eidetic limitation* of the *finite* objects to which it applies, and, in virtue of the *thetical unlimitedness* of the very act of affirming, unavoidably establishes, as an ultimate, non-contemplated but dialectically implied horizon, the infinite subsistent Existence in its absolute transcendence, whose existence, on the level of analogical intelligibility, is formally demonstrated in the classical proofs of the existence of God.]

[115]See *Lima Vaz, H. C. de, Tomás de Aquino,* note 82 supra, at 187.

reveals itself as absolute identity of Truth and Good, Intelligence and Love, as the basis of transcendental truth and categorical truth. To put it in a nutshell, "as the act of existing is absolute perfection in itself, its primacy in the dialectical structure of the spirit leads us to affirm the absolute Spirit as absolute Existence (*Ipsum esse subsistens*), in which the difference returns as *absolute identity* of intellect and intelligibility (Truth) and of liberty and loveliness (Good), so that intellection and love interpenetrate in the infinite totality of existence".[116] When one locates the theory of truth in metaphysical horizon as a theory of correspondence it then necessarily incorporates core elements of coherence theories, once the relationship of correspondence is introduced in a global metaphysical constellation, i.e., in the horizon of the totality of beings.[117] In such horizon the human being and his truthfulness and goodness understands himself as a result of the "superabundance and infinite ontological generosity of the Absolute",[118] so that the transcendental relationship of man and being reveals here its deepest foundations. "Man is because the Absolute is: as First Cause, Infinite Perfection and End";[119] the human being as finite subject, as unity of truth and liberty, is constitutively referred to the Absolute, i.e., it is a being-for-the-Absolute, for absolute Truth and absolute Goodness.

---

[116]See *Lima Vaz, H. C. de, Antropologia I*, note 109 supra, at 221.

[117]On an alternative interpretation of totality see *Puntel, L. B., Grundlagen*, note 2 supra, pp. 251 ff.

[118]See *Lima Vaz H. C. de, Antropologia II*, note 84 supra, at 122.

[119]See *Lima Vaz H. C. de, Antropologia II*, note 84 supra, at 124.

Sense, Reference, and Connotation
OSWALDO CHATEAUBRIAND

1 Formulations

My purpose in this paper is to combine a fregean account of senses with a kripkean account of reference and a descriptivist account of connotation. The main ideas derive from my book *Logical Forms* [1, 2], and the paper complements Chateaubriand [3] where my account of senses, thoughts, and truth is developed in some detail. I am very pleased to dedicate it to Tarcísio Cavalcante Pequeno on his 70th birthday.

I interpret Frege's idea that a sense is (or contains) a manner of presentation in terms of the notion of identifying property, where an identifying property is a property whose logical content is such that if the property applies to anything at all, it applies uniquely to that thing. Identifying properties are properties denoted by descriptive predicates of the form

$$x \text{ is the } F, \tag{1}$$

or properties necessarily equivalent to such properties. The logical structure of a descriptive predicate is

$$Fx \wedge \forall y (Fy \rightarrow y = x), \tag{2}$$

which can also be expressed in predicative notation as

$$[Fx \wedge \forall y (Fy \rightarrow y = x)](x), \tag{3}$$

and which I abbreviate as

$$[!xFx](x). \tag{4}$$

It is part of the logical content of an identifying property that if it applies to anything at all, it applies uniquely to that thing. Thus, in this interpretation, *senses are manners of presentation that may or may not present something*.

About names Frege says [4, p. 31]:

Copyright © 2019 by Oswaldo Chateaubriand. All rights reserved.

> A proper name (word, sign, sign combination, expression) expresses its sense, stands for or designates its reference. By means of a sign we express its sense and designate its reference.

Since senses are manners of presentation of entities, the relation of sign to reference (if any) goes via the sense. If the sense is an identifying property, and this property does identify an entity, then this entity is designated by (or is the denotation of) the sign. What is not made explicit in Frege's account is the relation between sign and sense. How does a sign express a sense?

For Frege a sign is not a mere syntactic string—a mark, say—but is something that "already" contains a sense. In his argument about identity at the beginning of "On Sense and Reference" he says [4, p. 26]:

> If the sign 'a' is distinguished from the sign 'b' only as object (here, by means of its shape), not as sign (i.e. not by the manner in which it designates something), ...

But, then, what is the relation between the syntactic string (mark, sound, etc.) and the sign? In other words, how does a syntactic string become a sign?

My view is that a syntactic string becomes a sign by *coding* a sense, so that a sign expresses the sense coded by the syntactic string, and designates the entity (if any) identified by the sense. Let me illustrate with a mathematical example.

Consider the proper name '$\emptyset$', of the empty set in set theory. The sense coded by '$\emptyset$' is the property denoted by the descriptive predicate

$$x \text{ is the set that has no elements,} \qquad (5)$$

and in various set theories we proceed to show that there is such a unique set. Hence, the sense identifies an object, the empty set, which is the denotation of the sign. This procedure is typical of the introduction of signs in mathematical theories, although some signs are introduced by means of axioms rather than by definition. Does this idea apply to proper names in general? It is in this connection that I appeal to Kripke's views on naming.

One of Kripke's main ideas about the reference of proper names and natural kind terms is that they are introduced into the language by means of baptismal acts. These baptismal acts may involve ostension, or description, or both, and may be explicit or implicit. According to Kripke by means of such baptismal acts one *fixes the reference* of the proper name or natural kind term. Although I find this idea of the introduction of names very persuasive, I do not agree that one should describe the introduction of names and natural kind terms in terms of *fixing a reference*. My main reason for this is that in many cases *there is no reference*, for the baptismal act may fail to identify something. This is especially clear in cases where the baptismal act is by description. I may think that an unknown kind of bacteria is the cause of a disease afflicting a certain population, and "name" the alleged kind of bacteria K, and yet there may be no such kind. Similarly, I might think that an unknown planet is causing perturbations in the orbit of certain other planets, and "name" this alleged planet N, and yet there may be no such planet. In fact, not only we may fail to fix a reference because of some such failures, but there is also the important case of fictional names, where names are introduced in a way that *there cannot be a reference*. Nevertheless, these names without reference are introduced into the language and transmitted from speakers to speakers. But if we are not fixing a reference, then what are we doing?

My view is that what is fixed when a name or natural kind term is introduced is a *sense*, not a reference. If that sense identifies an entity, then a reference is also fixed. In fact, what we are doing is introducing *signs*, in essentially the same way in which we introduce a sign such as '∅' in a mathematical theory. And although we may believe that the sign we introduced has a reference, we may be wrong.

The last notion I introduce is the notion of connotation. I define the *connotation of a proper name for a speaker at a given time as the totality of properties the speaker associates with that name at that time*. Thus, the connotation of the proper name 'Aristotle' for me at this time includes the properties of being human, being a philosopher, being Greek, being a disciple of Plato, etc. Some of these properties may be senses as well, for example, the property of being the disciple of Plato who was a teacher of Alexander the Great, or the property of being the foremost logician of antiquity. According to various descriptive theories of reference, it is the connotation of a name for a speaker at a given time that determines the reference of the name for the speaker at that time. Although I agree with Kripke's rejection of this general thesis, in its various formulations, I hold that this notion of connotation is an important aspect of our *use* of signs. For it is obvious that if a name has no connotation for a speaker at a time, then it is not possible for this speaker to use the name significantly at that time. A similar notion of connotation may be introduced for natural kind terms and for other terms.

2 Senses as Identifying Properties

Although Frege never developed a theory of senses in his published works, three characteristics of senses emerge quite clearly from his remarks.

One, already mentioned, is that senses are, or contain, manners of presentation. Thus, he says [4, p. 26]:

> It is natural, now, to think of there being connected with a sign (name, combination of words, letter), besides that to which the sign refers, which may be called the reference of the sign, also what I should like to call the *sense* of the sign, wherein the mode of presentation is contained.

A second characteristic, also mentioned above, is that a sense may not present an entity [4, p. 28]):

> It may perhaps be granted that every grammatically well-formed expression representing a proper name always has a sense. But this is not to say that to the sense there also corresponds a reference. The words 'the celestial body most distant from the Earth' have a sense, but it is very doubtful if they also have a reference. The expression 'the least rapidly convergent series' has a sense but demonstrably has no reference, since for any given convergent series, another convergent, but less rapidly convergent, series can be found. In grasping a sense, one is not certainly assured of a reference.

A third characteristic of senses is their *objectivity*. A sense is not a subjective idea, but "is the common property of many and therefore is not a part or a mode of the individual mind" [4, p. 29]. The relation between an individual mind and an objective sense is one of *grasping* [4, p. 27].

My interpretation of senses as identifying properties clearly conforms to the first two characteristics, and since I take properties to be objective entities we can grasp, or apprehend, it is also in agreement with Frege's third characteristic.

What does not seem to agree with Frege's views about senses is the interpretation of senses as *properties*, for in his posthumously published manuscript "Introduction to Logic" [5, p. 192] Frege maintains that the sense expressed by a proper name is an *object* (saturated) whereas the sense expressed by a predicate is a *property* (unsaturated). This view is never developed, however, and there is no indication in any of Frege's works of what *kind* of object is in question here. Therefore, it seems to me more natural to take all senses to be identifying properties, which may be manners of presentation of objects or of properties.

3 Sign, Sense, and Reference

Since I take senses to be objective entities I do not see them as necessarily associated to language. Insofar as a *sense* (identifying property) identifies something, we may say that it is a sense of that thing—be it an object or a property. Thus, objects and properties *have* senses independently of these senses being expressed in language or being grasped by anyone. It is a purely ontological relation between a sense and an entity identified by that sense. Nevertheless, senses are *expressed* by signs.

But how does a sign express a specific sense? By a convention, not necessarily explicit, that introduces a syntactic expression as code for that sense. It is easy to see how we do this in our theories, and I gave earlier the example of the expression '$\emptyset$' as a code for the sense denoted by the descriptive predicate 'x is the set which has no elements'. One of the merits of Kripke's views on naming is the realization that names are generally introduced in this way. Thus, if I buy a dog and decide to name him 'Fred', I am introducing the expression 'Fred' as a sign that codes some such sense as denoted by the descriptive predicate 'x is the dog I just bought', or 'x is the dog in front of me now', etc. Given that in most cases of this kind the coded sense does identify something, it is quite natural to say that by this baptismal convention I am *fixing the reference* of the name 'Fred'. What I maintain is that this fixing of the reference is indirect, via the sense.

Suppose I see an ad in the paper offering a dog for sale. I contact the seller, who says his name is 'Burns', and negotiate the purchase of the dog. He asks me what name I intend to give the dog so he can process a pedigree, and I say that I am calling the dog 'Fred'. The sense coded by the name may be the property denoted by the descriptive predicate 'x is the dog I am acquiring from Mr. Burns'. Everything may be above board and I have fixed the reference of 'Fred' to the dog that Mr. Burns is selling. It may happen, however, that I have been taken in, and there is no dog. Nevertheless, I have introduced the name 'Fred', which I can now use and transmit to other speakers. In this case, it is *incorrect* to say that I have fixed the reference of the name 'Fred', although I have introduced it as a new sign in the language.

The inadequacy of describing this procedure as a fixing of a reference is also illustrated by the case of fictional names. The previous paragraph is a little story I just made up. In making up my story I decided to introduce the name 'Fred' for the fictional dog-character in my story, and the name 'Burns' for the fictional dog seller-character in my story. These names are now signs, and it is part of the very sense they express that they do not—and cannot—designate anything. This is akin to Kripke's conclusion [6, p. 158] that a name such as 'Sherlock Holmes', which is introduced in a fictional story, not only does not designate anything in the actual world, but does not designate anything in any possible world (or counterfactual situation).

I conclude, therefore, that the introduction of signs into the language—be they proper names, natural kind terms, or other terms—fixes a sense coded by the sign and only indirectly may fix a reference for the sign.

References

[1] O. Chateaubriand. *Logical forms. Part I—Truth and Description*. Coleção CLE. UNICAMP, 2001.

[2] O. Chateaubriand. *Logical forms. Part II—Logic, Language, and Knowledge*. Coleção CLE. Campinas: UNICAMP, 2005.

[3] O. Chateaubriand. "The truth of thoughts: variations on Fregean themes". In: *Grazer Philosophische Studien* 75.1 (2007), pp. 199–215.

[4] G. Frege. "Sense and Reference". In: *Translations from the Philosophical Writings of Gottlob Frege*. Ed. by Peter Geach and Max Black. Oxford: Blackwell, 1960.

[5] G. Frege. "Introduction to Logic". In: *Posthumous Writings*. Ed. by Hans Hermes, Friedrich Kambartel, and Friedrich Kaulbach. Trans. by Peter Long and Roger White. University of Chicago Press, 1979.

[6] S. Kripke. *Naming and Necessity*. Cambridge, MA: Harvard University Press, 1980.

Reflections on Frege's Platonism

MATTHIAS SCHIRN

1 Introduction: Arithmetical platonism in *Die Grundlagen der Arithmetik* (1884)

In the current literature on Frege, it is not undisputed that he was a full-fledged platonist. Independently of the dispute, I think that putting his platonism in the right perspective is of considerable importance for appropriately assessing his overall philosophy of arithmetic. In my view, Frege's logicism goes hand in hand with his endorsement of an arithmetical version of ontological platonism. At least before Russell discovered the inconsistency of the logical system of [2, 3], Frege was convinced that all numbers are logical objects which exist independently of human minds. In particular, his ontological platonism in the period 1893—1903 was meant to apply to logical objects of a fundamental and irreducible kind, namely to value-ranges of functions, which include extensions of concepts and extensions of relations as special cases. According to Frege's logicist credo, all numbers are to be identified with logical objects of this prototype (cf. [2], §9).[1]

We find the first clear expression of Frege's arithmetical platonism in his philosophical masterpiece [1] when he comes to reject Hankel's formal theory of arithmetic. There he says that even the mathematician cannot create something arbitrarily, any more than the geographer; "he too can only discover what is there and name it" ([1], pp. 107f.). In the Preface to [2] (p. XIII), Frege argues exactly in the same vein.

The *epistemological* thesis that we derive our knowledge concerning numbers exclusively from the logical source of knowledge qua reason obviously does not imply the *ontological* claim that they do not exist

---

Copyright © 2019 by Matthias Schirn. All rights reserved.

[1]Both in *Grundlagen* and in *Grundgesetze* Frege identified the cardinal numbers with special extensions of concepts, with equivalence classes of equinumerosity. In [3], he intended to define the real numbers as Relations of Relations but due to Russell's Paradox his project of laying the logical foundations of real analysis remained a fragment; cf. [10, 11]. A Relation is the extension of a relation. On the role of extensions of concepts in Frege's logicist programme see [5, 6, 7, 11].

independently of human reason, let alone the stronger claim that they are creations of human reason. (Roughly speaking, in *Grundlagen* Frege seems to understand by *reason* the capacity of grasping thoughts, of making judgements, that is, of acknowledging thoughts as true, and of drawing inferences, that is, of acknowledging a thought (the conclusion) as true on the basis of thoughts which have already been accepted as true (the premises).)[2] In Frege's opinion, the number 2, for instance, exists independently of our conceiving or grasping it through our reason. At the same time, the number 2 *is what it is*—the sum of 1 and 1, the smallest prime number, the immediate successor of 1 in the natural series of numbers, the first Fibonacci prime, etc. not only independently of our sensation, intuition, imagination and our ideas, but independently of our mental activity in general including our arithmetical judgements about the number 2. Yet to answer the question of what the number 2 is independently of human reason would be to judge without judging (cf. [1], §26). Furthermore, several fragments written by Frege during the period of *Grundlagen* make it clear that in his view judgeable contents in general and those expressed by the laws of arithmetic in particular are not the product of any mental process. On the contrary, the principles of arithmetic are not only true independently of our recognizing them to be so, but they are also independent of our thinking as such. Yet if this is so, then according to Frege the numbers for which they hold good must likewise exist independently of our mental life.

In what follows, I shall first briefly set the stage for my critical discussion of two problems regarding Frege's platonism which we come across in [2], §10 and later in [3], §§146-147. While in [2], §10, Frege seems to ignore his platonism altogether—we shall see, however, that this does not apply to his argument in the long footnote to §10—in [3], §§146-147, he seems to be vacillating between the acceptance of a full-blooded platonism with respect to value-ranges (and hence with respect to numbers of all kinds) and the view according to which his introduction of value-ranges in [2], §3 via a second-order abstraction principle might be called a creation.

In the course of carrying out the exposition of his logical theory in the first volume of *Grundgesetze*, Frege encounters a variant of his old Julius Caesar problem from *Grundlagen*, §66, now clad in formal garb.[3]

---

[2] I assume that Frege uses the term "source of knowledge" to refer to a cognitive faculty of the human mind, and in doing so he is following deliberately in Kant's footsteps. In a late fragment, he characterizes a source of knowledge as that which *justifies* the acknowledgement of truth, the judgement (cf. [4], p. 286). On Frege's sources of knowledge see [9, 10, 14].

[3] On Frege's Caesar or indeterminacy problem both in *Grundlagen* and in *Grundge-*

The problem arises here from his semantic stipulation in §3, which is intended to govern the identity conditions of value-ranges. This stipulation concerning the metalinguistic analogue of the value-range operator "ἐφ(ε)" (which is one of the two primitive monadic function-names of second level in *Grundgesetze*) reads as follows:

> I use the words 'the function $\Phi(\xi)$ has the same value-range as the function $\Psi(\xi)$' generally as coreferential [*gleichbedeutend*] with the words 'the functions $\Phi(\xi)$ and $\Psi(\xi)$ always have the same value for the same argument'.

Henceforth I call this stipulation *the contextual stipulation* because it is immediately reminiscent of the tentative contextual definition of the direction operator "the direction of line a" in *Grundlagen*, §65, which in fact reads very similarly. The contextual stipulation, which is later encoded in the formal version of Basic Law V: $\vdash (\grave{\varepsilon}f(\varepsilon) = \grave{\alpha}g(\alpha)) = (-\mathfrak{a}-f(\mathfrak{a}) = g(\mathfrak{a}))$ (cf. [2], §§9, 20), is non-standard because, unlike the elucidations of the other primitive function-names of Frege's logical system, it does not directly assign a reference (and a sense) to the name of the value-range function by stating the values that this function receives for appropriate arguments, in this case for monadic first-level functions as arguments. It is rather designed to fix at least partially the reference of the name of that function by licensing the mutual transition from one mode of speaking which involves that name to another which does not (cf. [3], §146). Note that in contrast to the contextual stipulation, Basic Law V *asserts* something: the value-range of f is identical with the value-range of g if and only if f and g are coextensive. In the case of the contextual stipulation as well as in the case of Basic Law V we have an abstraction principle of the form "$Q(\alpha) = Q(\beta) = R_{eq}(\alpha, \beta)$". Here "$Q$" is a singular term-forming operator, α and β are free variables of the appropriate type, ranging over the members of a given domain, and "R_{eq}" is the sign for an equivalence relation holding between the values of α and β. Like Hume's Principle "$N_x F(x) = N_x G(x) \leftrightarrow Eq_x(F(x), G(x))$"[4] which both in *Grundlagen* and in *Grundgesetze* is

---

setze see [5, 7, 10, 13].

[4] In words: The number that belongs to the concept F is equal to the number that belongs to the concept G if and only if F and G are equinumerous. Frege defines the relation of equinumerosity in second-order logic in terms of one-to-one correlation (cf. [1], §72). Note that the above formulation of Hume's Principle is a schematic one; here its two sides are (closed) sentences, that is, "F" and "G" are schematic letters for monadic first-level predicates, not variables for first-level concepts. By contrast, in "$\forall F \forall G(N_x F(x) = N_x G(x) \leftrightarrow Eq_x(F(x), G(x)))$" "$F$" and "$G$" are variables for first-level concepts; here we have the universal closure of the open sentence "$N_x F(x) = N_x G(x) \leftrightarrow Eq_x(F(x), G(x))$".

the linchpin for the proofs of the basic laws of cardinal arithmetic, Basic Law V is a second-order abstraction principle and hence the structural analogue of the former. The equivalence relation on the right-hand side of Basic Law holds between monadic first-level functions and is therefore of second level.[5] So much for my introductory remarks. Let us now turn to [2], §10.

2   *Grundgesetze der Arithmetik*, vol. I, §10

In [2], §9, Frege introduces the value-range notation and begins §10 by pointing out that that the contextual stipulation does not yet by any means fix the reference of a name such as "$\dot{\varepsilon}\Phi(\varepsilon)$" completely.[6] In what follows, he draws attention to a "dual undecidability". (U1): "we cannot decide yet [that is, by appeal to the contextual stipulation] whether an object that is not given to us as value-range is a value-range or which function it may belong to". (U2): we cannot "decide in general whether a given value-range has a given property if we do not know that this property is connected with a property of the corresponding function." After having stated (U1) and (U2), Frege presents an argument for his previous claim that the contextual stipulation fails to determine the reference of a name such as "$\dot{\varepsilon}\Phi(\varepsilon)$" completely. Here I skip the argument. Frege then proceeds to characterize what I call *the piecemeal strategy* and presents it as the appropriate means of resolving the indeterminacy mentioned at the outset of §10. At the stage of §10, the proposed solution, namely the piecemeal determination of the values of the primitive first-level functions for value-ranges as arguments, "just as for all other arguments", eventually boils down to the case of $\xi = \zeta$, and "all other arguments" are reduced to the True and the False. However, the restriction seems to go against the use of the phrase "all other arguments", if this phrase is meant to include all the objects there are, not only the logical objects whose existence is required by the axioms of the formal theory of *Grundgesetze*: the two truth-values and value-ranges of functions.[7]

---

[5]For a detailed discussion of the nature and role of second-order abstraction in Frege's foundational project see [5, 7, 12].

[6]Once the second-level function-name "$\dot{\varepsilon}\varphi(\varepsilon)$" is available in the formal language and we are entitled to apply the rule of insertion as one of the two permissible modes of forming new names from the primitive function-names, we may regard any term that results from the insertion of a monadic first-level function-name into the argument-place of "$\dot{\varepsilon}\varphi(\varepsilon)$" as a *canonical value-range name*.

[7]In [13], I argue against the claim, recently defended by several Frege scholars, that the first-order domain in *Grundgesetze* is restricted to value-ranges (including the truth-values), but conclude that there is an irresolvable tension in Frege's view. The tension has a direct impact on the semantics of the concept-script, not least on the semantics of value-range names.

Having arrived at this point in §10, Frege sets out his permutation argument—which not least for reasons of space I shall not present and analyze here—and goes on to state the identifiability thesis: "Thus, without contradicting our equating '$\dot{\varepsilon}\Phi(\varepsilon) = \dot{\alpha}\Psi(\alpha)$' with '—$\overset{a}{\smile}$— $\Phi(\mathfrak{a}) = \Psi(\mathfrak{a})$', it is always possible to determine that an arbitrary value-range be the True and another arbitrary value-range be the False." By invoking the latter, he goes on to identify the True and the False with their own unit classes. Thanks to this dual transsortal identification, the intended complete determination of the references of canonical value-range names is now supposed to come within reach. Nevertheless, at the end of §10 Frege points out that a little more work has still to be done in subsequent sections (§§11–12) in order to achieve this goal.

In a long footnote to [2], §10, Frege examines the possibility of generalizing the twin stipulations just mentioned so that all objects whatsoever, including those that are referred to by canonical value-range names, are identified with their unit classes. He rejects the suggestion on the ground that it may be inconsistent with the criterion of identity governing value-ranges if the object to be identified with its unit class is already given to us as a value-range. At the same time, Frege dismisses the less general, but intuitively more appealing proposal of identifying with their unit classes all and only those objects which are given to us independently of value-ranges. He does so by using basically the same argument that he advances in *Grundlagen*, §67, saying that the mode of designation of an object must not be regarded as its invariant property, since the same object may be given in different ways.

Despite appearances, Frege's argument in the long footnote to §10 concerning the possibility of generalizing the dual stipulation in §10 does not undermine or invalidate his identification of the True and the False with their unit classes. By invoking the permutation argument, we can say that the dual stipulation is logically consistent with the contextual stipulation and, hence, legitimate on logical grounds.

Nonetheless, the sceptic might object that the twin stipulations in §10 fly in the face of Frege's platonism. He or she might wish to argue that for Frege it should be an objective fact whether, say, the True, is a value-range, and if it is one, which one it is. From the point of view of Frege's platonism, this has to be fixed once and for all in the mind-independent universe of logical objects and, hence, can never be a matter of arbitrary stipulation. On the one hand, I do not think that we are entitled to claim, by appealing to Frege's practice of making certain stipulations in the course of constructing his logical theory in *Grundgesetze*, that he was not a platonist or only the shadow of a platonist. On the other hand, it is hard to deny that *prima facie* there is a tension between Frege's platonism and certain stipulations that he makes in [2]. To be sure, we have no evidence that he was fully aware of this conflict; nor do we know whether he thought that he could lightly pass over it, insisting that he was at liberty to make certain stipulations—consistent with the set of assumptions underlying his logical theory—in order to secure a unique reference for every well-formed expression of his formal language. In my view, the right way to look at the problem is probably as follows: Platonist concerns are completely ignored in §10, but they apparently play a certain role in the long footnote to §10. In §10, Frege has a very pragmatic attitude, one that seems to contravene the strong platonism which we find in other places of his work including *Grundgesetze*. More specifically, in §10, he probably does not care at all about the question of whether there is some fact of the matter about whether, say, the True is a value-range and, if so, which one it is or which function it may belong to. Seen in this way, we may understand why it is not mandatory for him to explicitly rule out, prior to the twin stipulations in §10, that the True and the False are classes containing more than one object or no object at all.[8]

Suppose that Frege had been asked: If from the point of view of your platonist stance you had to answer the question "Is there is a fact of the matter in the mind-independent universe of (Fregean) logical objects whether, say, the True is a value-range, and if so, which one?", what would you say? I think that Frege would probably have answered with "yes". In particular, he might have pointed out that mind-independent and epistemically inaccessible facts about the truth-values and value-ranges do not undermine his dual stipulation in §10, since its legitimacy—its consistency with the constraints that he had previously imposed on value-ranges and the truth-values[9]—rests entirely on

---

[8] If there were a (metaphysical) fact of the matter that the True is identical with $\dot{\varepsilon}(—\varepsilon)$ —a case which the contextual stipulation does not rule out—then the first of the twin stipulations in §10 would amount to identifying the True with itself. From a metaphysical or platonist point of view, it would then be an unerring stipulation. By contrast, if there were a fact of the matter that the True is identical with $\dot{\varepsilon}(\varepsilon = \underset{a}{\frown} a = a)$ —a case which the contextual stipulation does not exclude either—then the second stipulation in §10 would amount to identifying the False with the True. From a platonist point of view, this stipulation would then miss its mark. To be sure, if the first-order domain in *Grundgesetze* is taken to be all-encompassing, then Frege could not even rule out, prior to §10, that $\dot{\varepsilon}(—\varepsilon)$ and $\dot{\varepsilon}(\varepsilon = \underset{a}{\frown} a = a)$ coincide with two distinct celestial bodies, say, with Jupiter and Mars.

[9] The only constraint imposed on value-ranges up to §10 is the contextual stipulation in §3. From a logical point of view, this is the only relevant thing that we know about value-ranges up to §10. Regarding the truth-values, I consider the constraints to be their characteristic marks which Frege mentions in [2], §2, in particular this one: a declarative sentence refers either to the True or the False, and the True and

the permutation argument and the identifiability thesis both of which are grounded in logical considerations. In his response to the hypothetical question above, Frege might have added that in general it is not metaphysical, but logical constraints that have a formative influence on the semantics of his concept-script. Plainly, the reference of, say, "$\acute{\epsilon}(-\epsilon) = -\!\!\stackrel{a}{\smile}\!\!- \mathfrak{a}$" has not been settled by any stipulation prior to the stipulation in §10. The contextual stipulation neither confirms nor rules out that a given value-range name corefers, say, with a concept-script name of the False, and that is one important point at issue in §10, although it is not the only one. Thus, I presume that from the point of view of his platonism Frege would have said that "$\acute{\epsilon}(-\epsilon) = -\!\!\stackrel{a}{\smile}\!\!- \mathfrak{a}$" has a determinate reference (truth-value) independently of our judgements and assertions. Yet in the same breath, he would probably have insisted that within the framework of his stipulations—and this is obviously the only thing that matters when he deals with the semantics of value-range names in §10 and later in §31 in the course of carying out his proof of referentiality for all primitive and non-primitive concept-script expressions— "$\acute{\epsilon}(-\epsilon) = -\!\!\stackrel{a}{\smile}\!\!- \mathfrak{a}$" lacks a determinate reference prior to the twofold stipulation that he is going to make in §10. Claiming the opposite would be at odds with what Frege says in the first half of §10.

3 *Grundgesetze der Arithmetik*, vol. II, §§146-147

In [3], Frege takes up the topic of the creation of new numbers through definition or abstraction which he had already discussed in earlier work. His demonstration in §143 that every attempt to create mathematical objects through definition is bound to fail, I take to be convincing. A mathematician who, in the spirit of Frege's contemporary Stolz, wishes to introduce new mathematical objects through a creative act must, prior to performing this act, prove that the properties he intends to assign to an object, which is initially devoid of properties, do not contradict each other. The consistency of these properties cannot be proved save by establishing the existence of an object that possesses all of them. Thus, the claim that the properties of the set $\{F_1,...,F_n\}$ do not contradict each other requires the proof that for at least one object a it holds: $F_1(a) \wedge ... \wedge F_n(a)$. Yet if one can prove this, then there is no need to create such an object through definition.

A few sections later, in [3], §§146–147, Frege considers the issue of how we have cognitive access to the objects of arithmetic. Surprisingly, he does not give a clear-cut answer to the question of whether the step of logical abstraction from right to left in Basic Law V could reasonably be

---

the False are distinct (logical) objects.

called a creation. In §147, he writes:

> If there are logical objects at all—and the objects of arithmetic are such objects—then there must also be a means of conceiving, of recognizing, them. And to this end serves us that basic law of logic that permits the transformation of the generality of an equality into an equality. Without such a means a scientific foundation of arithmetic would be impossible. For us it serves towards the ends that other mathematicians intend to attain by creating new numbers ... Can our procedure be called a creation? Discussion of this question can easily degenerate into a quarrel over words. At any rate, our creation (if you like to call it that) is not boundless and arbitrary; the mode of performing it, and its admissibility, are established once and for all.

In this passage, Frege does not justify his inference from the presupposed existence of logical objects to the necessary existence of a means of conceiving, of recognizing, them. In any event, he seems to identify a scientific foundation of arithmetic with a logical foundation, and the success of the latter depends, in his view, essentially upon a methodologically sound introduction of logical objects. For Frege, the primordial problem of arithmetic is first and foremost an epistemological, not an ontological one. In his view, then, it is precisely with the answer to the question 'How do we grasp logical objects, in particular the numbers?' that the scientific foundation of arithmetic stands or falls.

Despite the vagueness of the quoted passage, three points at least seem clear to me.

Firstly, there is ample evidence that in [3] Frege's platonism did not undergo any significant change.

Secondly, by his own lights, Frege should never have conceded that discussing the question of whether his introduction of value-ranges via logical abstraction can be called a creation, may easily degenerate into a quarrel over words. To my mind, he should have avoided raising this issue at all in [3], §146 instead of backing himself into a corner by leaving it undecided. Plainly, once the issue was brought up, Frege should have given a definite answer, not an evasive one. In the light of the available evidence about the status and the role that he assigns to Basic Law V in his logical system, a coherent answer would have been one along these lines: Basic Law V is designed to function as the appropriate means of coming into epistemic contact with value-ranges, of grasping them; it is not intended to call them into being. Like any explicit definition that meets the requirements of eliminability and non-creativity it would be powerless to achieve this anyway. It is true that, unlike explicit definitions which are immediately turned into epistemically trivial declarative sentences once the *definiendum* has been defined, Frege considers basic laws or axioms to contain real knowledge. Nonetheless, nowhere does he unambiguously claim that they have any creative potential. When he raises the epistemological key question of his logicist project: "How do we grasp logical objects?", he presupposes that they exist prior to their apprehension by us. The answer to the question, though not exactly in Frege's words, is as follows: they are grasped by means of logical abstraction.[10]

Thirdly, in the light of his overt fondness of ontological platonism, Frege would have been well-advised to refrain from conceding that one might perhaps call the procedure of logical abstraction a *creation*, if creation is meant in a rigid sense, implying a barrier to its executability. For even if a creation of logical objects (if it were possible at all) were to proceed in a regulated, non-arbitrary fashion and thus within sharp boundaries, it would nevertheless be a creation and as such clash with the platonist aspirations that Frege manifests in several places of his writings, not least in *Grundgesetze*. In other words: prohibiting or condemning any arbitrary and boundless creation of mathematical objects and, in the same breath, licensing in certain cases a creation of such objects, if the mode of carrying it out and its admissibility are established once and for all (cf. [3], p. 149), marks a position that Frege could not consistently maintain, quite apart from the fact that he fails to spell out what "admissibility" is to mean here precisely and how it could be established. In short, he could not have accepted any creation of mathematical or logical objects, no matter how it were performed.[11]

In my view, another issue raised in §147 would require clarification. When in §147 Frege asks whether "our procedure can be called a creation" and responds by saying that this question may easily degenerate to a quarrel over words, it is not quite clear what he means by "our procedure". Considering the entire context of his remarks, I presume that he appeals to the step of logical abstraction inherent in Basic Law V, that is, to the transformation of the generality of an equality between function-values into a value-range equality. Yet I do not wish to vouch for this option. (Frege explains that in carrying out this transformation we

---

[10] I call a Fregean abstraction abstraction principle *logical* if the equivalence relation on its right-hand side can be defined in second- or higher-order logic.

[11] Concerning Frege's attack on the practice of bringing numbers into existence by means of definition, which was apparently widespread among his fellow mathematicians see [8], pp. 156ff.

acknowledge something in common to the two functions—namely their value-range; this is how he characterizes the move of abstraction in Basic Law V in [3], §146; cf. also [4], p. 198). It is true that in the third passage and especially in the first half of the fourth and concluding passage of §147 Frege focuses entirely on the alleged virtues of Basic V: (1) that it is the appropriate cognitive means of grasping logical objects, if there are such objects at all; (2) that it is scientifically indispensable, in particular, that without it a scientific foundation of arithmetic would be impossible; (3) that it serves the same ends that other mathematicians seek to attain by creating new numbers. (Concerning (3), Frege had already made it clear that the transformation in Basic Law V differs fundamentally from the unregulated and arbitrary creation of numbers by other mathematicians.) He goes on to say: "We thus hope to be able to develop the whole wealth of objects and functions treated of in mathematics out of the eight functions whose names are listed in I, §31, as from a seed. Can our procedure be called a creation?" I find this transition irritating, especially since Frege spares himself the trouble of explaining it to his readers. In particular, I fail to see why and how Frege's hope of being able to develop all the objects and functions dealt with in mathematics out of the primitive functions of the system of *Grundgesetze* should derive from the virtues that he claims for Basic Law V. Admittedly, one might perhaps say that value-ranges are developed out of the primitive function $\dot{\varepsilon}\varphi(\varepsilon)$ via Basic Law V. And in a sense, Basic Law V is designed to "yield" (not to create) all objects dealt with in arithmetic. Recall that according to Frege's logicist credo all numbers are to be defined as value-ranges. Nonetheless, at least the way in which the functions that occur in arithmetic sprout from the seed of the primitive, logically simple functions is a matter quite distinct from logical abstraction embodied in Basic Law V.

In any event, at this point of Frege's exposition it seems that we cannot definitely rule out that with "our procedure" he intends to refer quite generally to the development of the objects and functions dealt with in arithmetic out of the primitive functions and not exclusively to the transformation of the coextensiveness of two monadic first-level functions into a value-range identity as represented by Basic Law V. However, the content of the last sentence of §147, vague at it is, appears to speak again in favour of my presumption that with the use of the phrase "our procedure" a few sentences earlier Frege intends to refer only to Basic Law V. In this sentence: "And with this, all the difficulties and doubts that otherwise call into question the logical possibility of creation disappear, and we may hope that with our value-ranges we achieve everything what has been missed by following those other paths", he mentions explicitly value-ranges and thus appeals implicitly to Basic Law V. Be that as it may, the question as to how the development of the objects and functions dealt with in mathematics out of the primitive functions is to proceed is passed over in silence by Frege. This is unfortunate because he missed the chance of dispelling any remaining doubt about what he meant by "our procedure" in the relevant context. As to the development of objects and functions out of the primitive functions, I conjecture that what Frege had in mind was the construction of logically complex function-names and object-names by iterated application of the formation rules of his system, which are the extraction of function-names from more complex names via "gap formation"[12] and the "insertion"[13] of admissible argument-expressions into the argument-place(s) of function-names. In this way, Frege does indeed obtain special functions (for example, the relation of an object falling within the extension of a concept, the single-valuedness of a relation, the following [succession] of an object after an object in the series of a relation) and likewise special objects (for example, equivalence classes of equinumerosity, extensions of relations (= Relations), Relations of Relations) that are required for laying the foundations of cardinal arithmetic and real analysis—and he is able to define them via constructive definitions. [14]

4 Conclusion

The upshot of the preceding discussion is as follows. Regarding arithmetic and its logical foundation, Frege was undeniably a platonist at least during the period 1884–1902.[15] However, we have seen that in [2],

---

[12]The rules of gap formation (as I call them) govern the formation of (i) monadic first-level function-names, (ii) dyadic first-level function-names and (iii) monadic second-level function-names with an argument-place of the second or of the third kind by removing some or all occurrences of a proper name either from a more complex proper name—case (i) in the standard form—or from a first-level monadic function-name—case (ii)—or by removing some or all occurrences of a monadic first-level function-name from a proper name—case (iii)—and by marking the resulting gap(s) as an argument-place of the appropriate kind (cf. [2], §26).

[13]Frege calls insertion the first way of forming a name and gap formation the second way of forming a name (cf. [2], §30).

[14]See the table of definitions in [2], pp. 240 f.; see also [3], for example, §§167, 173, 175, 193.

[15]As far as Frege's late idea of providing a geometrical foundation of arithmetic is concerned, it was, to my mind, not remotely a new awakening but only a desperate move, indeed a non-starter. And I trust that he had at least an inkling that proposing such a sea change in his philosophy of arithmetic and his foundational outlook in general did not carry an awful lot of conviction. I fail to see that the geometrical source of knowledge qua spatial intuition, which for Frege is far more restricted in scope than the logical source of knowledge, could persuasively account for the distinguishing marks setting arithmetic apart from intuition-based geometry.

§10, when he comes to diagnose the referential indeterminacy of value-range terms arising from his contextual stipulation in §3 and suggests to resolve it by carrying out what I called the piecemeal strategy, he completely disregards his commitment to platonism (and his closely related "mode of being given" caveat in the footnote to §10), although otherwise this commitment is a common theme throughout his logicism. Frege's chief concern both in §10 and the closely connected §31 is providing a satisfactory semantics for his formal language. This is why I suggested that regarding the determinacy or indeterminacy of the reference of a mixed equation of the form "$\grave{\varepsilon}\Phi(\varepsilon) = \Delta$"[16] at the stage of §10 but prior to the identification of the truth-values with their unit classes, he would most likely have responded in distinct ways from the point of view of his platonism on the one hand and the point of view of the semantics of the concept-script on the other. In section 3, we have seen that in [3] Frege seems to take a wavering attitude towards the introduction of value-ranges via the contextual stipulation: Can it be called a creation of logical objects or not? I have argued that Frege would have been well-advised to refrain from conceding that one might perhaps call the procedure of logical abstraction inherent in the contextual stipulation or Basic Law V a creation, if creation is meant in a strict, rule-governed sense. For even if a creation of logical objects were carried out in a regulated manner (assuming, for the sake of argument, that such an act of creation would in principle be possible), it would nevertheless be a creation and as such it would clash with the platonist credo that Frege manifests in several places of his writings, not least in *Grundgesetze*.

References

[1] G. Frege. *Die Grundlagen der Arithmetik. Eine logisch mathematische Untersuchung über den Begriff der Zahl*. Breslau: W. Koebner, 1884.

[2] G. Frege. *Grundgesetze der Arithmetik. Begriffsschriftlich abgeleitet*. Vol. I. Jena: H. Pohle, 1893.

[3] G. Frege. *Grundgesetze der Arithmetik. Begriffsschriftlich abgeleitet*. Vol. II. Jena: H. Pohle, 1903.

[4] G. Frege. *Nachgelassene Schriften*. Ed. by H. Hermes, F. Kambartel, and F. Meiner F. Kaulbach. Hamburg, 1969.

---

[16]The auxiliary name "Δ" is here supposed to represent a truth-value name (sentence) of the concept-script.

[5] M. Schirn. "Fregean Abstraction, Referential Indeterminacy and the Logical Foundations of Arithmetic". In: *Erkenntnis* 59 (2003), pp. 203–232.

[6] M. Schirn. "Concepts, Extensions, and Frege's Logicist Project". In: *Mind* 115 (2006), pp. 983–1005.

[7] M. Schirn. "Hume's Principle and Axiom V Reconsidered: Critical Reflections on Frege and His Interpreters". In: *Synthese* 148 (2006), pp. 171–227.

[8] M. Schirn. "Consistency, Models, and Soundness". In: *Axiomathes* 20 (2010), pp. 153–207.

[9] M. Schirn. "Frege's Approach to the Foundations of Analysis (1873-1903)". In: *History and Philosophy of Logic* 34 (2013), pp. 266–292.

[10] M. Schirn. "Frege on Quantities and Real Numbers in Consideration of the Theories of Cantor, Russell and Others". In: *Formalism and Beyond. On the Nature of Mathematical Discourse*. Ed. by G. Link. Boston and Berlin: Walter de Gruyter, 2014, pp. 25–95.

[11] M. Schirn. "Frege's Logicism and the Neo-Fregean Project". In: *Axiomathes* 24 (2014), pp. 207–243.

[12] M. Schirn. "Second-Order Abstraction Before and After Russell's Paradox". In: *Essays on Frege's Basic Laws of Arithmetic*. Ed. by P. Ebert and M. Rossberg. Oxford University Press, 2018.

[13] M. Schirn. "The Semantics of Value-Range Names and Frege's Proof of Referentiality". In: *The Review of Symbolic Logic* 11 (2018).

[14] M. Schirn. "Frege's Philosophy of Geometry". In: *Synthese*, published online (August 2017). DOI: 10.1007/s11229-017-1489-6.

The Need for Indexical *Sinn*
João Branquinho

1 Notational Variance Arguments

One might summarize as follows the main sort of criticism developed by Millian theorists.[1] It is argued that neo-Fregean theories about *de re* modes of presentation[2] for indexical expressions are bound to face the following dilemma. Either they can be reconstructed as notational variants of direct reference theories, *de re* indexical senses having no clear explanatory function and being thus wholly dispensable in favour of a Millian semantics for indexicals; or they yield results which are unacceptable in the light of our intuitions about the use of indexicals in the ascription of attitudes. In what follows, I concentrate on the first horn of this putative dilemma, even though I also consider some issues related to the second. [3]

The Millian theorist typically argues for the semantic redundancy of indexical modes of presentation by claiming that *de re* senses are not needed to explain apparent failures of substitutivity of co-referential indexicals in attitude-ascriptions, or to block certain apparently problematic results involving attitudes.[4] If sound this claim would constitute a serious objection to any Fregean account of indexicality, since what is taken to be the privileged role of senses, and what is often proposed as the crucial rationale for their introduction, consists precisely in their status as theoretical entities postulated to explain why co-referential singular terms are not in general interchangeable salva veritate when occurring in the embedded sentences of propositional-attitude constructions. The

---

Copyright © 2019 by João Branquinho. All rights reserved.

[1] See [15, 16, 17]. See also [20, 19, 21].

[2] Concerning the general notion of a *de re* sense, see [10]. A criticism of the general notion is available in [18].

[3] The following papers critically deal with other (related) aspects of Millian notational variance arguments: [1, 3]. For a critique of the approach developed in the latter paper and for a different view, see [11].

[4] See [21, pp. 154–5]. Although Soames's arguments are mainly directed against Evans's particular version of Fregeanism, they could be easily generalized to other neo-Fregean approaches.

anti-Fregean argument for the above claim runs as follows. Clearly, a necessary condition for inferences falling under the general pattern **x V's that S(i), i=i'. Ergo, x V's that S(i')**[5] to be rated as invalid by the proponent of a Fregean theory is that such a theory must provide us with a criterion for sameness of indexical sense; that is, it should state clearly under what conditions an indexical **i** used in a context **c** has the same sense as an indexical **i'** used in a context **c'**. And, since sameness of reference is thought of as being necessary for sameness of sense, one should expect such a test to be given in particular for the case in which the referent of **i** in **c** is identical with the referent of **i'** in **c'**. Yet, the Fregean theory does not contain a uniform criterion for the sameness of indexical sense, i.e. a means of decision capable of being applied to the different categories of indexicals, such as personal pronouns like 'I' and 'he', demonstratives like 'this' and 'that', temporal indexicals like 'now' and 'today', etc. Therefore, it is in general unclear how an appeal to senses might even account for failures of substitutivity (assuming for the sake of argument the anti-Millian thesis that co-referential indexicals are not interchangeable salva veritate in attitude contexts). The Millian critic would discern a certain tension in the neo-Fregean account, a tension which reflects the alleged absence of a clear and uniform means of individuating indexical senses. On the one hand, the Fregean treatment of temporal indexicals, spatial indexicals, and perceptual demonstratives allows utterances of sentences containing different but co-referential indexicals of these kinds, as used in distinct contexts, to express the same (token) Fregean thought; hence, it allows the possibility of the same particular mode of presentation being associated with different indexicals in different contexts of use. As a result, substitutivity and other problematic results about attitude-ascriptions would apparently be forthcoming in a neo-Fregean account of such categories of indexicals. On the other hand, the Fregean treatment of personal pronouns precludes utterances of sentences containing distinct but co-referential indexicals (used in possibly different contexts) from expressing the same (token) Fregean thought; hence, it disallows the possibility of the same particular sense being attached to different indexicals of that sort (in possibly different contexts). As a result, substitutivity and other problematic results about attitude-ascriptions would be blocked in a neo-Fregean account of such a category of indexicals. The consequences the Millian theorist urges us to draw from the adoption of such allegedly disparate verdicts on same-

---

[5] Here the letter **x** is replaceable by a designator of a subject, **V** by a propositional-attitude verb, **S(i)** by a sentence containing an indexical term **i**, and **S(i')** by a sentence that results from **S(i)** by replacing at least one occurrence of **i** with an indexical term **i'**.

ness of indexical sense are as follows. If indexical expressions are treated along the lines suggested above for temporal indexicals, etc., then the resulting theory will no longer be Fregean in nature; it will be simply a notational variant of a direct reference theory, redundant *de re* indexical senses being eliminable and the referents of indexicals in given contexts doing all the relevant semantic work. If, on the other hand, indexicals are to be treated on the model of personal pronouns, then the resulting theory, though presumably Fregean in nature, will be implausible since some of its consequences are incompatible with the way we intuitively use indexicals in attitude-ascriptions. The implication is, of course, that we should generalize in the former direction, i.e. from temporal indexicals to other indexicals, in which case the Millian Notational Variance Claim would be warranted.

Before assessing the claim, let us check the details involved therein.

According to the brand of neo-Fregeanism put forward by philosophers like Gareth Evans, Christopher Peacocke and others,[6] the following kind of result holds with respect to indexical expressions such as perceptual demonstratives and temporal and spatial indexicals. There are circumstances in which two (syntactically simple) indexicals **i** and **i'** of those types, taken as used in distinct contexts **c** and **c'** where they turn out to be co-referential, are to be seen as expressing the same particular sense, or as being associated in **c**, **c'** with the same particular way of thinking of their common denotation. As a result, sentences **S(i)** and **S(i')** uttered in **c**, **c'** are assigned the same propositional content, i.e. the same Fregean thought, with respect to **c**, **c'**.

Take the case of spatial indexicals. Suppose that on a certain occasion I am at a certain place **p**, e.g. a certain corner of my living-room, and that I utter a token of the sentence

(1) It is cold here.

I then move to a different place **p'**, e.g. the opposite corner of my living-room, and utter a token of the sentence

(2) It is cold there

(while pointing to **p**). Then, on the neo-Fregean view, the co-referential indexicals 'here' and 'there' in **(1)** and **(2)** have the same sense with respect to the contexts in question: in both cases I am entertaining the

---

[6]Frege is "neo-Fregean" on this score. See [9]. See also the following publications defending neo-Fregeanism about indexicals: [5]; [6], especially 291-321; [7]; [8]; [12]; [13]. For a more recent and rather sophisticated neo-Fregean approach, see [14], especially Chapter 6.

same way of thinking of a place (viz. **p**); and with my utterances of **(1)** and **(2)** I am expressing the very same Fregean thought[7].

Or take the case of perceptual demonstratives. Suppose that I am faced with a set of briefcases. I hold one of them and assert

(3) This briefcase is heavy.

A few moments later, having managed in some way to track the briefcase in question, I point at it from a distance and assert

(4) That briefcase is heavy.

Again, according to the neo-Fregean account, I attach to the demonstratives 'this' and 'that' in **(3)** and **(4)** the same mode of presentation of an object; in spite of there being superficial differences between the tactual way of thinking of it I employ in **(3)** and the visual way of thinking of it I employ in **(4)**, I am expressing the same particular Fregean thought on both occasions.

Now a familiar anti-Fregean notational variance claim is that there would be no substantive way by means of which one would be able to distinguish the above sort of account from a direct reference theory of indexicals. It is alleged that results which are quite similar to the ones outlined are forthcoming in such a theory, and that the differences between the two kinds of account might be counted as being minor (simply terminological) ones. In effect, given the same set of starting assumptions about indexicals **i** and **i'** in contexts **c** and **c'**, and given that (syntactically simple) indexicals are construed as directly referential expressions, it follows that **i** and **i'** in **c, c'** make exactly the same contribution to propositional content; indeed, on the Millian view, they just contribute their common denotation with respect to the given contexts **c, c'**. Hence, assuming compositionality, the propositions expressed by sentences **S(i)** and **S(i')** in **c, c'** are one and the same, viz. a certain neo-Russellian proposition. Thus, in our examples, the tokens of 'here' and 'there' in **(1)** and **(2)** (respectively the tokens of 'this' and 'that' in **(3)** and **(4)**) make the same contribution to propositional content with respect to the contexts in question: they contribute **p** (respectively the briefcase referred to); and with my utterances of **(1)** and **(2)** (respectively **(3)** and **(4)**) I am expressing the same proposition, viz. the ordered pair of **p** and Coldness (respectively the ordered pair of the briefcase and the property of being heavy).

---

[7] I ignore the difference in time of 'it is'.

2 Indexical Sense and Attitude-reports

I turn now to the Millian arguments concerning attitude-attributions. Recall that one of the claims here is that neo-Fregeanism of the sort described earlier on would entail the *prima facie* unFregean consequence that certain transitions (see below) between attitude-reports containing occurrences of co-referential indexicals (of the envisaged kinds) within the 'that'-clauses are to be rated as legitimate; thus, in this respect there would be again no difference between such a Fregean theory and a Millian one, from which the consequence in question is in general acknowledged to be derivable. And a different (but related) kind of claim is that the neo-Fregean would apparently put a Fregean believer in the same sort of position as Salmon's Millian believer Elmer,[8] who believes—at a given time or, without changing his mind, on different occasions— a pair of mutually inconsistent propositions while failing to recognize the same proposition in both cases and hence without being illogical; thus, likewise, there would allegedly be Fregean thoughts which are not completely transparent to their thinkers.

Let us focus on temporal indexicals (having in mind that the results obtained might easily be made to apply to spatial indexicals and perceptual demonstratives). Suppose that Jones, a logical thinker and a fully competent speaker of English, sincerely and reflectively assents at a certain time **t** on a certain day **d** to a token of the sentence-type "Today is fine". Accordingly, one would expect the belief-ascription

(5) **Jones believes that today is fine**,

as uttered by someone (say Ralph) at a certain time **t'** on **d**, to be true of Jones (here **t'**> or = **t**, and if **t'**>**t** one would have also to suppose that Jones does not change his mind about the weather on **d** at any time between **t** and **t'**). Now the sort of neo-Fregean account subscribed to e.g. by Evans would entail, given certain additional assumptions, the consequence that the belief-report

(6) **Jones believes that yesterday was fine**,

as uttered by Ralph at a certain time **t"** on **d+1**, is also true of Jones. And the assumptions in question are: (i)- that Jones keeps track of the days from **d** to **d+1** (and one would have also to assume that the ascriber, i.e. Ralph, does not mistrack time either, otherwise there might be a possible divergence between the senses attached by him and by Jones to 'yesterday'); and (ii)- that on **d+1** Jones has not changed his beliefs about the weather on **d**.

---
[8]See [15, p. 92].

Therefore, it seems that the neo-Fregean theory validates transitions such as the one from (5) to (6). In general, such transitions might be characterized as consisting in carrying out the following two steps: (i)- interchanging certain pairs of co-referential indexicals within the (semantic) scope of psychological verbs in propositional-attitude constructions, e.g. replacing in (5) 'today' by 'yesterday'; and (ii)- readjusting in an appropriate way the times at which the attitudes are held, e.g. changing the time **t'** (or the day **d**) of Jones's belief to **t"** (or to **d+1**). Notice that, in virtue of step (ii), the transitions in question are obviously not cases of substitutivity salva veritate; thus, there is a clear contrast between moves of the above sort and moves such as e.g. the one from 'I believe at **t** that I am ugly' to 'I believe at **t** that he is ugly', which, assuming that I am the male demonstrated at **t**, is a valid move according to a Millian theory of indexical belief but an invalid one on a Fregean view.

However, contrary to the Millian claim, the sort of consideration employed by the neo-Fregean theorist to ensure the legitimacy of a transition such as the one from (5) to (6) (under the given circumstances) has nothing to do with a mere identity of indexical reference. Indeed, according to his proposed individuation of temporal modes of presentation in terms of ways of tracking times, the neo-Fregean theorist is appealing rather to identity of indexical sense here. The sense referred to by 'today' in (5) is judged to be the same as the sense referred to by 'yesterday' in (6), the same particular way of keeping track of a day (viz. **d**) being employed by Jones on both occasions. Hence, denoting such a common mode of presentation of **d** by '**MP$_d$**', the belief-reports (5) and (6) might be (respectively) given, with respect to the contexts in question, the following sort of representations under the envisaged neo-Fregean account:

(5)' $B_{t'}$ [Jones, ⟨**MP$_d$**, Fineness⟩]

(6)' $B_{t''}$ [Jones, ⟨**MP$_d$**, Fineness⟩]

(the Fregean thoughts believed by Jones at different times being thus one and the same); here '**B$_t$**' stands for the binary Belief-relation as relativized to a certain time **t**.[9]

I want now to argue with a view to establishing the following two points, which taken together provide us with a refutation of the Millian arguments for the dispensability of *de re* indexical senses. First, on the

---

[9]I employ the usual notation of ordered pairs only for reasons of simplicity; in fact, it sounds strange to say that believing a Fregean thought is something like standing in a certain relation to a set.

above sort of neo-Fregean view, it turns out that transitions of the form mentioned earlier on *may* fail to obtain; and such a possibility, which is presumably unavailable under a Millian account, is also explained by means of an appeal to indexical senses. Secondly, contrary to appearances, the Millian critic is definitely wrong when he holds that in the end one would not be able to differentiate between Fregean thinkers and Millian thinkers in respect to the possibility of unknowingly believing contradictory thoughts. The upshot of my discussion is that the main *rationale* for the introduction of senses in a semantic theory, viz. that of blocking problematic results involving attitudes, is still available in this area of indexicality.

As to the first point, it is indeed a feature of the neo-Fregean account, a feature which is not usually displayed by a Millian semantics for e.g. ascriptions of temporal beliefs, that transitions sharing the form of the one from **(5)** to **(6)** are not always legitimate under it; i.e., there are circumstances in which, although such an account would count a token of a sentence sharing the form of **(5)** (used on **d**) as holding, it would not count a token of a sentence sharing the form of **(6)** (used on **d+1**) as holding at all. In effect, consider the case in which, after assenting on **d** to a token of "Today is fine" uttered by John, I take a 24-hour "nap" and unknowingly lose track of the days. Thus, John's belief-ascription on **d**

(7) He believes that today is fine

(where 'he' refers to me) would be true. Yet, my assenting on **d+1** to a token of "Yesterday was fine" uttered by John does not put him in a position to make on **d+1** the following belief-ascription:

(8) He believes that yesterday was fine.

Assume that John reports all his beliefs according to the neo-Fregean theory, that he keeps track of the days correctly, that he remembers the weather on the previous day, and that he is aware of my situation. Then, since when I give my assent to "Yesterday was fine" I am not actually entertaining any way of thinking of **d**, I am not entertaining then any Fregean thought about **d**, and hence I am not having any belief whatsoever about **d**.[10] Therefore, **(8)** does not come out as true under the neo-Fregean account (with respect to the given context), but rather as a **false** belief-ascription. Hence, the following report

---

[10] Here one would have to rule out the possibility that **(8)** is true because of some mode of presentation of **d** under which I have the belief and which I do not associate with the word 'yesterday', e.g. a memory-based mode of presentation of **d**.

(9) He lacks the belief that yesterday was fine

(as uttered by John on **d+1**) would come out as true (assuming that **(9)** is the negation of **(8)**).[11]

Obviously, this kind of result would not constitute any problem for neo-Fregeanism since the thought expressed by **(7)** would not be inconsistent with the one expressed by **(9)**; and hence the ascriber (John) would not be contradicting himself, the Fregean regimentations for such belief-reports being:

(7)' B_d [J.B., $\langle MP_d,$ Fineness$\rangle$]

(9)' $\neg B_{d+1}$ [J.B., $\langle MP_d,$ Fineness$\rangle$].

At most, the neo-Fregean account would allow a thinker to hold *at different times* conflicting attitudes towards the same proposition.

From the preceding reflection I think that one is entitled to draw the conclusion that, from the standpoint of the neo-Fregean account of indexicality, there are in fact **illegitimate** transitions involving co-referential indexicals of the envisaged types in attitude-ascriptions, and that it is in terms of indexical *Sinne* that such an illegitimacy is to be accounted for.

It could be replied that there is a sense in which the transitions in question might also be deemed illegitimate on a Millian view. Suppose that one supplements a directly referential account of temporal indexicals, spatial indexicals and perceptual demonstratives with some epistemic notion of tracking an object over time and/or space. And suppose that it is possible to do it in such a way that the following sort of general condition would obtain: there exists a guise (or other suitable Millian construction) under which a given subject stands in the Belief-relation to a neo-Russellian proposition containing an indexically presented time,

---

[11]Note that in place of **(9)** one might have used

(10) He does not believe that yesterday was fine

(which is—at least syntactically—a straightforward negation of **(8)**). The problem with a report such as **(10)**—and the reason why I avoid employing it—is that it is ambiguous between **(9)** and

(11) He disbelieves that yesterday was fine,

or

(12) He believes that yesterday was not fine

(which I construe as having the same meaning as **(11)**). If **(10)** were read in the sense of **(11)** (or of **(12)**), then it would surely come out as false with respect to the case discussed; thus, a confusion between **(10)** and **(11)** (or **(12)**) would help generate the wrong conclusion that **(10)** is false and hence that **(8)** is true in our story (by means of the wrong premise that **(10)**—assimilated to **(11)** (or to **(12)**)—is the negation of **(8)**).

or place, or spatio-temporal item, only if the subject is somehow able to track the time, or the place, or the spatio-temporal item, in question. Then Millianism could presumably be made to yield the same verdicts as neo-Fregeanism on the truth-values of certain belief-ascriptions containing indexicals of the above kinds. For instance, reports like **(7)** and **(9)** would both come out as true, and **(8)** as false, under the extended Millian account (relative to the given contexts); so there are after all moves of the sort discussed which would not be validated by Millianism either. However, assuming that such a notion of tracking could be harmoniously incorporated into a Millian theory, it is obvious that it would have to be located at the *pre-semantic* level. As a result, the Millian and neo-Fregean analyses of belief-ascriptions of that kind, in spite of being materially equivalent, would not be logically equivalent to each other; a report like **(9)**, for example, would be in each case assigned substantially different truth-conditions and meanings. And this would provide us with sufficient grounds on which to reject the Millian claim about notational variance. On the other hand, if a notion of tracking is to be in the end acknowledged as theoretically relevant, then one might always raise the question concerning the overall advantages, for explanatory purposes, of taking it as semantically relevant as well. Furthermore, cases might be introduced where the same object is tracked separately by hand and eye by a given subject and where she does not know that the touched object is the seen object. Concerning such cases, it is very likely that even the extended Millian account would yield different verdicts as the neo-Fregean account on the truth-values of belief-reports such as e.g. 'She believes that *that* (the touched object) is **F**' and 'She believes that *that* (the seen object) is **F**'.

As to the issue about the possibility of a (rational) subject's believing contradictory Fregean thoughts, consider the following sort of case.[12] Suppose now that Jones sincerely and reflectively assents to a token of "Today is fine" on **d** at 11:58 p.m., so that **(5)** is then true of him; and that, without taking the trouble to look at his watch, three minutes later (i.e. at 00:01 a.m. on **d+1**) he sincerely and reflectively dissents from a token of "Yesterday was fine" (thinking, of course, that he is referring to **d−1**). Assume further that the ascriber is as before, i.e. that he does not mistrack time, that he is aware of Jones's situation, etc. And one might also assume that, on the later occasion, Jones has not changed his mind about his previous belief (he remembers what the weather was like on **d**).

---

[12]This happens to be the kind of case used by Soames in support of his notational variance arguments against neo-Fregeanism; see [21, pp. 154–155].

The question I want to take up, and to which I shall eventually give a negative answer, is this. Does it follow that the ascription

(13) Jones believes that yesterday was not fine,

taken as uttered on **d+1**, holds of Jones? If so then a consequence of the neo-Fregean account would be that Jones, *ex hypothesi* a rational thinker, comes to believe a pair of contradictory thoughts on different occasions without apparently having meanwhile changed his mind (I think it would be manifestly implausible to construe **(13)** as implying that a change of mind has taken place). In effect, the thoughts referred to by the 'that'-clauses in **(5)** and **(13)** would clearly contradict each other, the regimentations for **(5)** and **(13)** being:

(5)" $B_{\text{11:58 p.m.},d}$ [Jones, $\langle MP_d, \text{Fineness}\rangle$]

(13)' $B_{\text{00:01 a.m.},d+1}$ [Jones, $\langle MP_d, \neg\text{Fineness}\rangle$].[13]

A positive answer to the above question would thus give us the result that neo-Fregeanism is committed to reporting Jones's doxastic states in a way which is strikingly similar to the way in which a Millian theorist would report them. And an additional problem for neo-Fregeanism would be that it does not seem to contain a notion designed to fulfil a role similar to that of the Millian notion of a **guise** or appearance under which one may be acquainted with a proposition (such a role being mainly that of rendering given propositions opaque to the thinker's awareness, so that in some cases one may be prevented from re-identifying a proposition previously entertained). Take Salmon's Millian believers, for instance. They may find themselves in a situation in which they believe inconsistent propositions, at the same or at different times, but (if rational) they would necessarily do it under different guises; thus, Salmon's analyses for **(5)** and **(13)** (taken with respect to the given contexts) would be as follows:

(5)''' $(\exists g)[G_{\text{11:58pm},d}(\text{Jones},w,g)$ &
$\text{BEL}_{\text{11:58pm},d}(\text{Jones},w,g)]$

(13)'' $(\exists h)[G_{\text{00:01am},d+1}(\text{Jones},w,h)$ &
$\text{BEL}_{\text{00:01am},d+1}(\text{Jones},\neg w,h)]$

---

[13] It is assumed that the negation of a thought consisting of a certain mode of presentation of an object and a property is the thought consisting of that mode of presentation and the negation of that property.

(where 'w' stands for the Russellian proposition ⟨**d**, **Fineness**⟩ and the guises **g** and **h** are obviously such that ¬(**g**=**h**)). On the other hand, Fregean believers—who allegedly may also find themselves in a situation in which they believe contradictory thoughts (at different times)—do not seem to be credited with any sort of psychological device by means of which given thoughts could be concealed from them (so as to speak); modes of presentation will not do for on a Fregean view they are taken as constituent parts of thoughts. Indeed, unlike neo-Russellian propositions, Fregean thoughts are seemingly supposed to be completely transparent to their thinkers. Hence, the Millian critic might claim that Fregean believers have the disadvantage of being prevented from not recognizing that it is one and the same proposition which is believed and disbelieved by them on different occasions (Jones's putative situation when he assents to "Today is fine" and later on dissents from "Yesterday was fine").

However, such a move is doubtful. For its supporting premise, viz. the claim that neo-Fregeanism entails the problematic result under consideration, can be shown to be false. Indeed, given that Fregean indexical thoughts of the envisaged types are (partially) individuated in terms of abilities to track the objects thought about, a belief-report such as (13)—taken as made on **d+1**—would have to be counted as **false** (with respect to Jones's story); whereas, as assumed, ascription (5)—taken as uttered on **d**—holds of Jones. Thus, supposing that this sort of result may be extended to spatial indexicals and perceptual demonstratives, whose senses in given contexts are individuated along the same lines, the consequence is not in general derivable from the neo-Fregean account that it is possible for a subject (unknowingly) to believe, on different occasions but without having meanwhile changed her mind, mutually inconsistent thoughts.

If ascription (13) is false under neo-Fregeanism, then the following report

(14) **Jones lacks the belief that yesterday was not fine**,

taken as made on **d+1** at 00:01 a.m., will come out as true under such an account (with respect to Jones's story). Thus, using

(14)' $\neg B_{00:01\ a.m., d+1}$ [Jones,⟨MP_d, ¬**Fineness**⟩]

(i.e. the negation of (13)') as the Fregean regimentation for (14), and (5)'' as the Fregean regimentation for (5), one might describe in general Jones's doxastic states by saying that on a certain occasion he believes a certain thought and on a later occasion he fails to believe the negation of that thought. Hence, the sort of case under discussion is definitely not a case in which someone holds at different times antagonistic attitudes, e.g. belief and disbelief, towards the same thought.

A different way of establishing with respect to Jones's case the falsity of **(13)**, and hence the truth of **(14)**, might be given as follows. I take it that, according to neo-Fregeanism, the following result is true of Jones at 00:01 a.m. on **d+1**:

(14)" $\neg(\exists\alpha)[T_{00:01am,d+1}(Jones,d,\alpha)\ \&\ B_{00:01am,d+1}[Jones,\langle\alpha,\neg\textbf{Fineness}\rangle]]$;

here 'α' ranges over temporal modes of presentation based on abilities to track days over time, and '**T**' stands for that relation which holds, at a given time, between a subject **x**, an object **o**, and a singular mode of presentation β if and only if, at that time, **x** thinks of **o** under β. On the other hand, I also take it that such a theory rates as being in general valid inferences from given *de dicto* belief-ascriptions to the corresponding *de re* ones; that is, propositions of the general form

(*) $B_t\ [x,\ \langle MP_o,\ \Phi\rangle]$,

where 'MP_o' and 'Φ' stand for a mode of presentation of an object **o** and a property (respectively), entail propositions of the form

(**) $(\exists\beta)[T_t(x,o,\beta)\ \&\ B_t[x,\ \langle\beta,\ \Phi\rangle]]$.

(For convenience, I have only considered the case of beliefs in thoughts of the simplest predicative form.) Now the proposition which is the scope of the negation symbol in **(14)"** clearly displays the general form **(**)**. Hence, since *ex hypothesi* **(14)"** is true in Jones's story, it follows that **that** proposition is false in his story. Therefore, the corresponding proposition of the form **(*)**, which turns out to be **(13)'**, is necessarily false in Jones's story. But **(13)'** is the Fregean regimentation for belief-ascription **(13)**. Therefore, **(13)** comes out as false with respect to Jones's story.

It is instructive to compare the above neo-Fregean results with the results a Millian theorist dealing with the same sort of case would usually obtain. Thus, under Salmon's account, a belief-report such as **(13)** would turn out to be **true** with respect to Jones's story; for its Millian regimentation is given in **(13)"** and this proposition holds with respect to Jones's case (just let '**g**' in **(13)"** be 'Yesterday was fine'). Yet, Salmon's theory would rule out a belief-report such as **(14)** as false with respect to Jones's story. Indeed, the Millian regimentation for **(14)** would be given in

(14)''' $\neg(\exists h)[G_{00:01am,d+1}(Jones,w,h) \,\&\, BEL_{00:01am,d+1}(Jones,\neg w,h)]$

(where 'w' is to be read as before); and **(14)'''** is simply the negation of **(13)"**. It should be apparent by now that the verdicts standardly given by the Millian theorist on the truth-values of attitude-ascriptions such as **(13)** and **(14)** taken in the envisaged contexts, respectively **true** and **false**, are strictly inconsistent with the verdicts given on them by the Fregean theorist, respectively **false** and **true**. Again, this would be enough to rebut the Millian Notational Variance Claim as applied to temporal and spatial indexicals and perceptual demonstratives. On the other hand, such a claim would be unsound even if the Millian theorist were to be credited with a pre-semantic notion of tracking in the way sketched earlier on: presumably, one would have the same assignments of truth-values; but one would not have the same assignments of truth-conditions.

3 The Transparency of Indexical Sense

Concerning the principle that Fregean thoughts are necessarily transparent to their thinkers,[14] it is clear that such a principle is not threatened by the Millian arguments and that it is consistent with the preceding sort of Fregean results (though one might perhaps have independent reasons for rejecting it, even from a Fregean standpoint). Indeed, the relevant form of the Transparency principle might be given as follows:

> **(T) If a rational subject x believes that p at t and disbelieves that q at t' and the thought that p = the thought that q, then x knows at t' that the thought that p = the thought that q (with t' > or = t).**

Now if $t' = t$ then a Fregean theorist would take **(T)** as being a trivially true claim for its antecedent should have to be counted as false: it is simply inconsistent with the Intuitive Criterion of Difference for thoughts. On the other hand, if $t' > t$—and this is the interesting assumption—then cases of the sort discussed before would not constitute any counter-example to claim **(T)**. In effect, an instantiation of the principle to Jones's case would turn out to be (again) trivially true since its antecedent would turn out to be false; for the second conjunct

---

[14] On the transparency of sense in general, see [4].

in the antecedent of **(T)** would not hold: it is not the case that Jones disbelieves at **00:01 a.m.** on **d+1** a thought to the effect that the preceding day was fine. Moreover, the envisaged version of the Transparency principle—strengthened in a certain way, viz. as in **(T)*** below—might even be argued to be **in general** (trivially) true from a Fregean viewpoint. Thus, consider the following claim:

> **(T)*** If a rational subject x believes that p at t and disbelieves that q at t' and the thought that p = the thought that q and x retains his belief that p from t to t', then x knows at t' that the thought that p = the thought that q (with t' different from t).

One might argue that the antecedent of **(T)*** does not hold in general on the basis that it would be inconsistent with a certain **diachronic** generalization of the Intuitive Criterion of Difference for thoughts. I have tried elsewhere[15] to put forward what I take to be a plausible formulation of such an extended principle; and if my attempt is successful then **(T)*** should be seen as a trivial truth. (It is interesting to notice that **(T)**, as well as **(T)***, would presumably hold under a Millian account of thoughts and attitude-attributions).

Of course, it does not follow that the neo-Fregean theorist should be seen as subscribing in general to the idea that thoughts are transparent to their thinkers. On the contrary, there are several senses in which Fregean thoughts are opaque to their thinkers. Indeed, there are several versions of the Transparency principle that would be regarded as false under a neo-Fregean account of indexicality. Thus, take the following sort of claim:

> **(I)** If x entertains at t the thought that p and x entertains at t' the thought that q and the thought that p is different from the thought that q, then x knows at t' that the thought that p is different from the thought that q.

This claim would be unacceptable in the light of neo-Fregeanism. Suppose that at **t** Ralph, looking at a certain object **o**, judges 'That is nice'. Meanwhile someone replaces **o** with a distinct (but rather similar) object **o'** without Ralph noticing it. Later on, at **t'**, looking at what is in fact **o'**, Ralph comes to wonder whether the object perceptually presented to him is nice. Assume that Ralph is a self-reflective Fregean thinker who is agnostic at **t'** about whether the thought he is then entertaining is the

---
[15] See [2].

same as the thought he entertained at **t**. Such thoughts are surely different from one another for they are about distinct objects. But, since the thinker is unsure whether he has successfully tracked the object thought about at **t** from **t** to **t'**, he cannot be in a position to know at **t'** that the thoughts in question are different.

It is interesting to ascertain whether the following variant of claim **(I)** would be consistent with neo-Fregeanism (it would presumably be consistent with Millianism):

> **(I)* If x entertains at t the thought that p and x entertains at t' the thought that q and the thought that p = the thought that q, then x knows at t' that the thought that p = the thought that q.**

Suppose that Ralph's case is described as before except that this time nobody replaces **o** between **t** and **t'**, while Ralph thinks that a switch has taken place. Yet, he comes to judge at **t'** 'That is nice'. Thus, he wrongly thinks at **t'** that he is then entertaining a distinct thought.

This sort of cases seem to provide us with **prima facie** straightforward counter-examples to claim **(I)***. Indeed, if the subject thinks that the thoughts he entertains on different occasions are distinct, and if such thoughts are in fact one and the same, then it follows that it will not be the case that he knows that they are identical. The problem is that, on the neo-Fregean view, the second premise of such an inference cannot be taken for granted (with respect to cases like the one above). In effect, perceptual singular modes of presentation are supposed to be based on abilities to keep track of objects over time and/or space, as well as from sensory modality to sensory modality. Hence, it is at least arguable that the Fregean thoughts entertained by Ralph at **t** and **t'** are not to be counted as being identical; indeed, one might claim that it would not make much sense to say, with respect to the above sort of circumstances, that Ralph has **in fact** tracked the object **o** from **t** to **t'**. It sounds in a sense weird to say that someone has unknowingly kept track of an object, though it surely makes sense to say that someone has unknowingly mistracked an object. Thus, one might reason as follows with a view to showing that claim **(I)*** is, in general, not inconsistent with the brand of neo-Fregeanism under consideration. Restricting our attention to indexical thoughts of the envisaged types, it seems that on such a view the only way in which particular thoughts **p** and **q**—both about a given object **o**—could be taken as identical is that the thinker who entertains them (on different occasions and/or at different places, etc.) keeps track of **o**. But if the above suggestion is correct then, in general, it does not make sense to say of a thinker that he unknowingly has kept track of an object. Therefore, the thinker could not be in a position to think that **p** and **q** are distinct thoughts; and hence the conclusion would apparently be blocked that she does not know that **p** and **q** are one and the same thought. Having our present concerns in mind, I shall not try to assess such an argument and settle the issue here; I prefer to leave it open. But one might at least conclude that, on the neo-Fregean account, a claim such as **(I)*** cannot be conclusively shown to be false on the basis of the sort of cases discussed.

References

[1] J. Branquinho. "Are Salmon's 'Guises' Disguised Fregean Senses?" In: *Analysis* 50.1 (1990), pp. 19–24.

[2] J. Branquinho. "On the Individuation of Fregean Propositions". In: *Analytic Philosophy and Logic. The Proceedings of the Twentieth World Congress of Philosophy.* Ed. by Akihiro Kanamori. Vol. 6. Bowling Green, Ohio, 2000, pp. 17–28.

[3] J. Branquinho. "On the persistence and re-expression of indexical belief". In: *Manuscrito-International Journal of Philosophy* 31 (2 2008), pp. 573–600.

[4] J. Campbell. "Is sense transparent?" In: *Proceedings of the Aristotelian Society.* Vol. 88. Aristotelian Society. 1987, pp. 273–292.

[5] G. Evans. *"Understanding Demonstratives".* Ed. by Herman Parret and Jacques Bouveresse. Berlin, 1981.

[6] G. Evans. *The varieties of reference.* Ed. by J. McDowell. Oxford: Claredon Press and New York: Oxford University Press, 1982.

[7] G. Forbes. "Indexicals and intensionality: A Fregean perspective". In: *The Philosophical Review* 96.1 (1987), pp. 3–31.

[8] G. Forbes. "Indexicals". In: *Handbook of philosophical logic.* Ed. by D.Gabbay and F.Guenthner. Vol. IV. Dordrecht: Springer, 1989, pp. 463–490.

[9] G. Frege. "The thought: A logical inquiry". In: *Philosophical Logic.* Ed. by Peter F. Strawson. Trans. by Anthony and Marcelle Quinton. Oxford: Oxford University Press, 1967, pp. 17–38.

[10] J. McDowell. "De re senses". In: *The Philosophical Quarterly* 34.136 (1984), pp. 283–294.

[11] K. C. M. Mertel. "Re-Thinking Gareth Evans' Approach to Indexical Sense and the Problem of Tracking Thoughts". In: *Grazer Philosophische Studien* 94 (1–2 2017), pp. 173–193.

[12] C. A. B. Peacocke. *Sense and Content: Experience, Thought and Their Relations*. Oxford: Claredon Press, 1983.

[13] C. A. B. Peacocke. *Thoughts: An Essay on Content*. Oxford: Basil Blackwell, 1986.

[14] F. Recanati. *Mental files in flux*. Oxford: Oxford University Press, 2016.

[15] N. Salmon. *Frege's Puzzle*. Cambridge, Mass. and London, England: The MIT Press, 1986.

[16] N. Salmon. "Illogical belief". In: *Philosophical perspectives: Philosophy of Mind and Action Theory* 3 (1989). Ed. by James E. Tomberlin, pp. 243–285.

[17] N. Salmon. "A Millian heir rejects the wages of Sinn. the Role of Content in Logic, Language and Mind". In: *Propositional Attitudes*. Ed. by C. A. Anderson and J. Owens. Stanford: CSLI, 1990, pp. 215–248.

[18] S. Schiffer. "The Mode-of-Presentation Problem. the Role of Content in Logic, Language and Mind". In: *Propositional Attitudes*. Ed. by C. A. Anderson and J. Owens. Stanford: CSLI, 1990, pp. 249–268.

[19] S. Soames. "Substitutivity". In: *On Being and Saying: Essays for Richard Cartwright*. Ed. by J.J. Thomson. Cambridge: Cambridge University Press, pp. 99–132.

[20] S. Soames. "Direct reference, propositional attitudes, and semantic content". In: *Philosophical Topics* 15.1 (1987), pp. 47–87.

[21] S. Soames. "Review of Gareth Evans: Collected Papers". In: *The Journal of Philosophy* 86.3 (1989), pp. 141–156.

A Hoard of Hidden Assumptions
DAVID MILLER

1 Introduction

Such is our fondness for deductive validity that we are at times tempted, when confronted with an invalid argument, to postulate some missing premise, or hidden assumption, that, if adopted, would render the argument valid after all. A familiar philosophical example is Hume's contention that 'all our experimental conclusions proceed upon the supposition that the future will be conformable to the past' [3, § IV, Part II, p. 35]. Much can be said, and has been said, about the usefulness of a metaphysical principle of uniformity that is anything like as general as the supposition that Hume flirted with, but what interests me in this paper is the implicit logical thesis that, in most instances of deductive failure, it is possible to identify, more or less uniquely, a statement to plug the gap in the defective argument. In this vein, some who call themselves deductivists, such as Musgrave [8], recommend that all allegedly inductive inferences (but not necessarily all invalid inferences) are best treated as enthymemes calling for systematic deductive rehabilitation.

The usual rules governing the material conditional imply that if the argument from A to C is invalid then $A \to C$ is the logically weakest additional premise that is strong enough to make it valid. That consideration would not have been congenial to Hume, even if it had occurred to him, since he evidently thought that there is some general statement underwriting all 'reasonings from experience' [2, p. 651]. Yet it can be proved ([6, Chapter 8, §§ 1f.]) that there does exist a logically weakest strictly universal statement that restores validity to the kinds of argument that Hume denounced, and it is perhaps possible that something like that statement was skulking, unformed, at the back of his mind.

2 Critical thinking

In the *critical thinking movement*, an approach to the teaching of argumentation (and sometimes also of logic) that has become startlingly

---

Copyright © 2019 by David Miller. All rights reserved.

popular in the last thirty years or so, the question of how invalid arguments are to be reinvented as valid is handled somewhat differently. It is accepted that, in most cases in which the argument from A to C is invalid, there are ever so many further premises B that, although sufficient to make the argument from A to C into a valid argument, cannot be identified with the premise that was missing. Indeed, it is generally acknowledged that B cannot qualify as the missing premise if it alone validly implies the conclusion C (in particular, B cannot be identified with C). These uncontroversial observations generate conditions **a** and **b** in §4 below. What is distinctive in the treatment in some critical thinking texts of the essential 'thinking skill' called *assumption spotting* is the method deployed to identify a statement B not only as sufficient for the argument from A and B to C be valid, but also as necessary.

According to the authors of [1, §IV, p. 563], who are highly critical, the so-called *negative test*, or *reverse test*, requires additionally that 'the negation of the original conclusion is derivable from the addition of the negation of [the missing] assumption to the original [premises]'. In short, the statement B is the assumption missing from the invalid argument from A to C only if the argument from A and ¬B to the conclusion ¬C is valid. This generates condition **c** in §4 below. It is noted further that 'those who advance the negative test appear to abide by ...[a] fourth requirement, even though it is not stated' ([1, §V, p. 567]), to the effect that 'the negation of the conclusion is not derivable from the negation of the [missing] assumption alone', in short that the argument from ¬B to the conclusion ¬C be not valid. This generates condition **d** in §4 below.

These authors conclude their rather relaxed discussion of the two forms of the negative test by wondering whether 'there are invalid arguments for which no assumption satisfies all four requirements and other invalid arguments where multiple alternatives satisfy those requirements', and 'what the underlying rationale of the negative test actually is' (*ibidem*). In the spirit of critical rationalism ([4, 6]), which advocates criticism rather than attempted justification as the only proper way to evaluate contested hypotheses, I discharge myself from pursuing the second inquiry, but I do wish to pursue the first. It will be shown in the rest of the paper how far the negative test (in both variants) is from enabling us to identify the assumptions that are missing from invalid arguments.

For exposure, exposition, and criticism, of some of the epistemological shortcomings of the critical thinking movement, see [5] and [7, §5].

3 A few technicalities

To begin, let us be explicit about terminology and notation. Given the class of meaningful sentences of some unspecified language that incorporates at least elementary sentential logic, sentences that are logically equivalent will be identified and called *statements*. We shall (non-standardly) use ⊩ and ⊢, in analogy with ≤ and <; that is, A ⊩ B (rather than the usual A ⊢ B) signifies that B *is derivable from* A, while A ⊢ B means that B *is properly (or unilaterally) derivable from* A; that is, A ⊩ B but A ≠ B. As is customary, ⊤ is the *tautological* or *logically true* statement, and ⊥ is the *inconsistent* or *logically false* statement.

The following not unexpected results concerning interpolation will be useful. Suppose that Z ⊢ X, so that X ∧ ¬Z is consistent. Provided that the theory X ∧ ¬Z is not maximal (that is, negation complete), there exists a *proper interpolant* between Z and X, that is to say, a statement Y such that Z ⊢ Y ⊢ X. Such an interpolant Y may be constructed as follows. Since X ∧ ¬Z is consistent and not maximal, there exists at least one statement U that is undecided by X ∧ ¬Z; that is, neither X ∧ ¬Z ⊩ U nor X ∧ ¬Z ⊩ ¬U. For any such undecided U, let Y = Z ∨ (X ∧ U). Since Z ⊢ X, we have Z ⊩ Y ⊩ X. If Y = Z, then X ∧ U ⊩ Z, and hence X ∧ ¬Z ⊩ ¬U; and if X = Y then X ∧ ¬Z ⊩ U. Since U (and ¬U) are undecided by X ∧ ¬Z, we have shown that X ≠ Y ≠ Z. In other words, Z ⊢ Y ⊢ X; that is, Y is a proper interpolant between Z and X. So too, of course, is Z ∨ (X ∧ ¬U). It is easy to see that, by the distributive law, these two interpolants can be expressed as X ∧ (Z ∨ U) and X ∧ (Z ∨ ¬U).

The converse holds too: if Z ⊢ Y ⊢ X for some Y, then X ∧ ¬Z is consistent and not maximal, and indeed, Y is undecided by X ∧ ¬Z. For if X ∧ ¬Z ⊩ Y then, since Z ⊢ Y, we may conclude that X ⊩ Y, contrary to assumption. Likewise, if X ∧ ¬Z ⊩ ¬Y, then X ∧ Y ⊩ Z; and since Y ⊢ X, we may conclude that Y ⊩ Z, again contrary to assumption. Since Y = Z ∨ (X ∧ Y), it follows that every proper interpolant between Z and X has the form Z ∨ (X ∧ U) where U is a statement undecided by X ∧ ¬Z.

4 Statement of the problem

Given statements A, C such that the conclusion C cannot be validly derived from the premise or assumption (sometimes called the reason) A,

♠ $$A \nVdash C,$$

the problem is to characterize all those *missing premises* or *hidden assumptions* B that provide solutions to the following set of conditions:

a	A, B ⊩ C	c	A, ¬B ⊩ ¬C
b	B ⊮ C	d	¬B ⊮ ¬C.

By contraposition, these conditions may be more succinctly rewritten:

a	A, B ⊩ C		c	A, C ⊩ B
b	B ⊮ C		d	C ⊮ B.

Interest will be restricted to the conditions **a–c** and the conditions **a–d**.

5 Deducibility relations between A and C

It is easy to see that only some relations of deducibility can hold between A and C when **a–d**, or just **a–c**, hold. A ⊩ C is at once ruled out by ♠. If ¬A ⊩ C, then ¬A ⊩ B → C, and by **a**, A ⊩ B → C, whence ⊩ B → C, which contradicts **b**. If C ⊩ A then by **c**, C ⊩ B, which contradicts **d**. If A and C are mutual contraries (that is to say, A ⊩ ¬C or, equivalently, C ⊩ ¬A, or ⊩ ¬A ∨ ¬C), then by **a**, A and B are also mutual contraries.

In brief, if all of the conditions **a–d** are to hold, then C ⊩ ¬A is the only possible deducibility relation between A and C, while if only **a–c** are to hold, then C ⊩ A and C ⊩ ¬A are both possible. They are indeed possible simultaneously, since C may be the inconsistent statement ⊥.

6 Truth-functional solutions

The conditions **a** and **c** together imply that A∧C ⊩ B ⊩ A → C. The only truth functions B of A and C that are in accordance with this restriction are (tritely) the conjunction A ∧ C and the material conditional A → C, and (hardly less tritely) the biconditional A ↔ C and the statement C. But neither A ∧ C nor C satisfies **b**, and A → C does not satisfy **d**.

In brief, A → C and A ↔ C are the only solutions of **a–c** that are truth functions of A and C. These solutions coincide if & only if C ⊩ A. The biconditional A ↔ C is the only truth-functional solution of **a–d**.

7 Necessary & sufficient conditions for solutions

As noted in §5, there are no solutions of **a–c** if ¬A ⊩ C, and there are no solutions to **a–d** if either ¬A ⊩ C or C ⊩ A. Moreover, if ¬A ⊮ C, then neither A → C ⊩ C nor A ↔ C ⊩ C, and so both A → C and A ↔ C satisfy **b**. But both A → C and A ↔ C satisfy **a** and **c** for all A, C.

In brief, ¬A ⊮ C is a necessary & sufficient condition (i) for **a–c** to have any solutions; (ii) for A → C to be a solution of **a–c**; and (iii) for A ↔ C to be a solution of **a–c**. These solutions are different if & only if C ⊮ A. It is a necessary & sufficient condition for (i) **a–d** to have any solutions, and for (ii) A ↔ C to be a solution of **a–d**, that ¬A ⊮ C ⊮ A.

8 Non-truth-functional solutions

It is not implied in §6 that A → C and A ↔ C are the only possible solutions of **a–c**, or that A ↔ C is the only possible solution of **a–d**. What is implied is that no other solution is a truth function of A and C.

It is also implied in §6 that any further solution B of **a–d** has to satisfy $A \wedge C \vdash B \vdash A \to C$. There are therefore three disjoint ranges in which such an assumption B could be located: (α) $A \wedge C \vdash B \vdash A \leftrightarrow C$; ($\beta$) $A \leftrightarrow C \vdash B \vdash A \to C$; and ($\gamma$) $A \wedge C \vdash B \vdash A \to C$, where B and $A \leftrightarrow C$ are logically incomparable (that is to say, $A \leftrightarrow C \nvdash B \nvdash A \to C$). We shall establish that, provided that A and C are neither too strong nor too weak, the ranges (α) and (β) yield ample opportunities for new solutions to **a–c** and **a–d**. No similar result is known for the range (γ).

It was shown in §3 that, for there to exist an interpolant B between $A \wedge C$ and $A \to C$, it is necessary & sufficient that $(A \to C) \wedge \neg(A \wedge C)$, which is identical with $\neg A$, be consistent and not maximal. Now $A = \bot$ is ruled out by ♠, and hence $\neg A$ is consistent. That is, there exist solutions of **a–d** that are distinct from $A \leftrightarrow C$ if & only if $\neg A$ is not maximal (that is, if & only if A is not irreducible in the sense of [9, §4]).

9 (α) Solutions properly between $A \wedge C$ and $A \leftrightarrow C$

It follows from (α) $A \wedge C \vdash B \vdash A \leftrightarrow C$ that $A, B \Vdash C$ and $A, C \Vdash B$; that is, **a** and **c** are satisfied. It is true also that if $B \Vdash C$ then $B \Vdash A$, and hence $B \Vdash A \wedge C$, contradicting (α); whence $B \nVdash C$, and **b** is satisfied. By §3, for there to be any B satisfying (α), it is necessary and sufficient that $(A \leftrightarrow C) \wedge \neg(A \wedge C)$, which is identical with $\neg A \wedge \neg C$, be consistent and not maximal. But the consistency of $\neg A \wedge \neg C$ is equivalent to $\neg A \nVdash C$, which by §7, is necessary and sufficient for **a–c** to have solutions.

By §7 again, if **a–c** are satisfied, then **d** is satisfied if & only if $C \nVdash A$.

In brief, provided that $\neg A \wedge \neg C$ not a maximal theory, and that the conditions **a–c** have solutions, then there exists a statement B such that $A \wedge C \vdash B \vdash A \leftrightarrow C$, and every such statement B is a solution of **a–c**. For B to be a solution of **a–d**, it is necessary and sufficient that $C \nVdash A$.

10 (β) Solutions properly between $A \leftrightarrow C$ and $A \to C$

It is clear that if the biconditional $A \leftrightarrow C$ and the conditional $A \to C$ are logically equivalent (that is, identical) the ranges (β) and (γ) identified in §8 are empty, and there is nothing more to be said. Since the equivalence holds if & only if $C \Vdash A$, we now add to the assumption ♠ that $\neg A \nVdash C$ the assumption that $C \nVdash A$. The combined assumptions may be written

♣ $\qquad\qquad A \wedge C \vdash C \vdash A \to C.$

As shown in §7, ♣ is necessary & sufficient for **a–d** to have solutions.

It follows from §7 that if **a–c** has any solutions then every B that satisfies (β) $A \leftrightarrow C \vdash B \vdash A \to C$ is also a solution of **a–c**. For $A \to C$ satisfies **a**, and therefore any stronger B does; while $A \leftrightarrow C$ satisfies **b** and **c**, and therefore any weaker B does. By §3, for there to be any B satisfying (β), it is necessary and sufficient that $(A \to C) \land \neg(A \leftrightarrow C)$, which is identical with $\neg A \land C$, be consistent and not maximal. But for $\neg A \land C$ to be consistent it is necessary & sufficient that $C \nvdash A$. Since $C = A$ is ruled out by ♣, this condition can be strengthened to $C \nVdash A$.

Moreover, if $C \Vdash B$ then $C \lor (A \leftrightarrow C) \Vdash B$; that is, $A \to C \Vdash B$, which (β) declares impossible. In other words, **d** is satisfied by any B that lies in the open interval between $A \leftrightarrow C$ and $A \to C$ under consideration.

In brief, provided that $\neg A \land C$ is not a maximal theory, and that the conditions **a–c** have solutions, then there exists a statement B such that $A \leftrightarrow C \vdash B \vdash A \to C$, and every such statement B is a solution of **a–d**.

11 Conclusion

It is not been possible to determine whether there are solutions to **a–c** or **a–d** that lie within the range (γ), but enough has been said to make it evident that, except in extreme circumstances, even the stronger set of conditions does not allow identification of the assumptions missing from invalid arguments. I doubt that those in the critical thinking movement will be much impressed by this technical result. Nevertheless, the status and significance of the negative test remain as obscure as they ever were.

References

[1] P. Gardner and S. Johnson. "Teaching the Pursuit of Assumptions". In: *Journal of Philosophy of Education* 49.4 (2015), pp. 558–570.

[2] D. Hume. *An Abstract of a Book Lately Published, Entitled A Treatise of Human Nature, Etc.* 1740. (Ed. by L. Selby-Bigge. Oxford: Clarendon Press 1888, 1978).

[3] D. Hume. *An Enquiry Concerning Human Understanding.* 1748. (Ed. by L. Selby-Bigge. Oxford: Clarendon Press 1893, 1975).

[4] D. W. Miller. *Critical Rationalism. A Restatement and Defence.* Chicago and La Salle IL: Open Court Publishing Company, 1994.

[5] D. W. Miller. "Do We Reason When We Think We Reason, or Do We Think?" In: *Learning for Democracy* 1.3 (2005), pp. 57–71.

[6] D. W. Miller. *Out of Error. Further Essays on Critical Rationalism.* Aldershot and Burlington VT: Ashgate Publishing Company, 2006. (now published by Routledge, Abingdon and New York).

[7] D. W. Miller. "Mashinnoye oogadivaniye, chast' II". In: *Voprosy Filosofii* 8 (2012), pp. 117–126. (Spanish translation, "Adivinación automática", in G. Guerrero P. & L. M. Duque M. (eds.), *Filosofía de la ciencia: Problemas contemporáneos*, pp. 19–44, Programa Editorial, Universidad del Valle, Santiago de Cali, 2015. English version, http://www.warwick.ac.uk/go/dwmiller/jsm.pdf).

[8] A. E. Musgrave. "Popper and Hypothetico-Deductivism". In: *Handbook of the History of Logic. Inductive Logic*. Ed. by D. M. Gabbay, S. Hartmann, and J. Woods. Vol. 10. Handbook of the History of Logic. Amsterdam: Elsevier B. V., 2011, pp. 205–234.

[9] A. Tarski. "Grundzüge des Systemenkalkül". In: *Fundamenta Mathematicae* 25.4 (1935), pp. 342–383, 26.2 (1936) pp. 283–301. (English translation, "Foundations of the Calculus of Systems", Chapter XII, of A. Tarski, *Logic, Semantics, Metamathematics*, Oxford: Clarendon Press, 1956, 2nd edition, Indianapolis: Hackett Publishing Company, 1983).

Nous Sommes des Prisonniers du Discours: Theorie- Langage- Processus Cognitif dans L'Epistémologie de Quine

Vera Vidal

1 Le Projet Épistémologique de Quine

Notre lecture de Quine nous a conduit à penser que l'ensemble de sa philosophie est une réponse à ce que nous appelons son projet épistémologique, lequel apparaît clairement formulé dès la première page de l'ouvrage **The Roots of Reference:**

Given only the evidence of our senses, how do we arrive at our theory of the world? (N'étant donné rien d'autre que l'évidence sensorielle, comment aboutissons-nous à nos théories du monde?) [3, p. 1]

Par "théorie du monde" il entend la totalité de nos visions du monde, y compris celle du sens commun, ainsi que les visions scientifique et philosophique, lesquelles ne se distinguent des visions du sens commun que par leurs degrés de complexité linguistique et analytique.

Selon Quine, aucune vision du monde n'est naïve, ni capable de refléter la réalité tel un miroir. Elle est toujours le produit de conjectures plus ou moins sophistiquées que les hommes ont "inventées" à travers l'histoire pour leur permettre de s'expliquer eux-mêmes et tout ce qui les entoure. Ces conjectures dépassent infiniment les informations fournies par les données de notre sensibilité- le seul critère d'évidence vraiment fiable selon lui.

En dépit du fait que ce projet épistémologique n'a été clairement formulé que dans un ouvrage publié en 1974, on se rend compte qu'il a toujours été le principe moteur de ses investigations philosophiques.

Cette préoccupation pour les rapports entre notre discours et le monde réapparaît à plusieurs reprises. On peut détacher déjà dans [1], ouvrage publié en 1960, le passage suivant :

Nous nous servons d'une structure verbale très étendue d'énoncés qui sont liés les uns aux autres de diverses façons. Mais cette structure constitue une unité- et c'est cet ensemble qui comprend tout ce que nous sommes capables de connaître du monde. [1, p. 2]

Dans son livre [6], la même conviction réapparaît au premier paragraphe du premier chapitre consacré à l'analyse de la notion d'évidence:

A partir des impacts sur nos surfaces sensorielles, nous avons, au cours des générations, à l'intérieur de notre créativité collective et cumulative, projeté notre théorie systématique du monde extérieur. Notre système réussit à prédire les stimulations sensorielles subséquentes. Comment avons-nous fait cela? [6, p. 1]

Au vu de ces affirmations et aussi de bien d'autres que l'analyse de ses textes nous révèlent, nous croyons légitime l'inférence selon laquelle la question fondamentale de Quine, celle que nous appelons son projet épistémologique, repose sur trois présupposés fondamentaux de sa philosophie, à savoir:

- la conviction que le seul critère d'évidence qu'on ne peut mettre en cause est l'évidence empirique;
- la constatation que nous avons tous une théorie du monde;
- la croyance que notre théorie du monde ne se phénoménalise qu'à travers un réseau d'énoncés qui peut être organisé de différentes manières tout en gardant toujours une unité et qui résume tout ce que nous sommes capables de connaître du monde intérieur et extérieur.

Etant donné que, si nous ne déduisions toutes nos croyances qu'à partir des informations fournies par l'évidence sensible, nous n'arriverions jamais à justifier la totalité de notre savoir puisqu'il dépasse énormément ces seules données, le problème qui se pose aux épistémologues est celui de la justification des rapports entre notre discours et la réalité, entre les mots et les objets.

Ce projet épistémologique se révèle par conséquent comme un effort pour répondre à la question: Comment lier les excitations de nos surfaces sensibles par les stimuli physiques venus du monde extérieur à la production d'une théorie du monde?

Selon Quine, une théorie est un ensemble d'énoncés ordonnés par des règles très strictes, faisant partie d'un système linguistique qui leur accorde signification et degré de vérité. Pourvu que tout ce que nous connaissons du monde extérieur et intérieur soit compris dans cette vaste et complexe structure verbale, il conclut que notre connaissance se révèle à travers le discours, qu'elle est imbriquée dans la structure du langage, ce qui fait de nous des "prisonniers du discours".

On comprend ainsi pourquoi Il a choisi d'aborder la question épistémologique qui l'intéresse par le biais de l'analyse du langage: Il croit au parallélisme langage-théorie et il considère toutes nos théories comme des sous-ensembles du langage.

Le choix quinéen de suivre la méthode d'analyse linguistique n'exclut pas le monde extra-linguistique que le discours envisage de décrire. Il considère l'existence des objets physiques comme incontestable puisqu'elle se manifeste par la stimulation de ces objets sur nos surfaces sensibles. Ainsi, une telle croyance s'appuie non seulement sur le seul critère d'évidence qu'il trouve fiable – l'évidence empirique- mais aussi sur "les croyances de notre race".

Il affirme qu'avec le lait maternel nous ingérons "une naïve philosophie naturelle" qui nous amène à accepter le postulat de l'existence du monde physique composé d'objets de taille moyenne, d'une durée limitée, indépendants et différents de nous-mêmes. Cette croyance se révèle utile et efficace tout au long de notre existence. Pour la mettre en cause, il faut trouver une raison très forte qui montrerait plus d'efficacité pour la vie pratique et les besoins théoriques. Sinon, toujours fidèle à son principe de conservatisme, Quine soutient qu'il vaut mieux garder ce puissant postulat des "croyances de la race humaine".

Les présupposés physicalistes, réalistes et empiriques de Quine sont pourtant parfaitement cohérents avec sa méthode analytique d'investigation philosophique. Certes, le monde physique existe et il est indépendant de nous mais, pour exprimer nos théories du monde, nous n'avons pas d'autre moyen que l'utilisation du discours.

L'analyse du discours revèle que nous faisons des affirmations qui dépassent énormément les inférences que l'évidence sensible nous autorise. On "invente" des lois générales, des principes causaux qui, même s'ils dérivent logiquement d'autres énoncés fortement appuyés sur les données d'observation, les dépassent de beaucoup. Il faut alors que les épistémologues puissent rétablir les rapports discours-réalité pour essayer d'y distinguer, le plus clairement possible, ce qui est objectif de ce qui est "invention". Pour cela, il trouve que la meilleure méthode de travail est la démarche de la philosophie analytique.

Toujours fidèle à ses postulats empiriques, Quine est convaincu que toute analyse du langage doit suivre une méthode rigoureuse d'observation du comportement des locuteurs ainsi que de la relation du discours avec les stimuli physiques dont il dérive par des processus plus ou moins évidents.

Le rapport que l'épistémologue établit entre le discours et les données d'observation- rapport qui peut être scientifique ou issu du sens commun- doit être accessible à l'ensemble de la communauté linguistique. En dépit de la diversité de nos perceptions individuelles, les analyses doivent aboutir à des résultats qui atteignent un consensus, le plus homogène possible, des membres de la communauté. Les énnoncés d'observation sont ceux qui obtiendront le plus grand degré de consensus, mais il faut que les autres l'obtiennent également pour que notre théorie puisse être reconnue et validée.

Quine soutient qu'on doit suivre en épistémologie et en linguistique la méthode de travail des sciences de la nature. Il répudie toutes les analyses qui font appel à des entités mentales ou non observables:

Il est préférable de comprendre comment la science évolue que de fabriquer une structure fictive pour aboutir à des résultats semblables. [2, p. 168]

Il suggère que les épistémologues, les linguistes, les psychologues, les neurologues et autres chercheurs réunissent leurs analyses pour mieux comprendre, selon une démarche d'investigation rigoureuse, scientifique, le processus qui amène les hommes à répondre par un discours aux stimulations physiques de leurs terminaisons nerveuses et à expliquer comment ils arrivent à des énoncés irréductibles à d'autres qui dérivent directement des données d'observation.

Ce travail interdisciplinaire est la seule option qui reste aux épistémologues pour attaquer la question du rapport discours-réalité après que "le rêve réductiviste des empiristes" s'est avéré irréalisable, dont le dernier grand et remarquable effort de construction fut celui de Carnap, qui a échoué.

Si l'épistémologie accepte sa suggestion de se "naturaliser", elle deviendra un chapitre des sciences de la nature, puisqu'elle étudie un phénomène naturel: l'homme, un objet physique parmi d'autres qui stimulent ses terminaisons nerveuses et l'amènent à répondre avec un discours qui est lui aussi un objet physique. Tels sont les présupposés du physicalisme quinéen.

D'après ce que nous venons de dire, nous croyons légitime d'inférer que c'est son projet épistémologique qui l'a amené à essayer de comprendre et de justifier la transcendance du discours par rapport à l'évidence des données sensorielles à travers l'analyse de deux situations:

- celle de l'enfant qui apprend sa langue maternelle;

-celle d'un linguiste qui cherche à comprendre le langage d'un peuple dont il ne connait ni la langue ni même la culture. Il appelle ce proccès *traduction radicale*.

2 La Théorie du Langage de Quine

Quand on essaie de compprendre l'ensemble du système philosophique quinéen, on s'aperçoit de l'importance qu'y occupe sa théorie du langage. Elle est au centre, voire au fondement, de son système.

Il est important de signaler quelques composants fondamentaux de sa théorie du langage:

- Le **naturalisme**- qui se révèle dans la conviction que les théories du langage doivent être soumises à la vérification empirique puisqu'elles s'occupent de phénomènes naturels et observables: le discours, les actes de langage, les réponses que ces actes provoquent chez ceux qui les comprennent, sans négliger le processsus d'acquisition du langage;

- Le **behaviorisme**- selon lequel toute analyse linguistique doit s'appuyer sur l'observation du comportement des locuteurs et éviter le recours à des interprétations mentalistes.

Le mécanisme behavioriste stimulus-réponse joue un rôle décisif dans sa théorie de l'apprentissage du langage. On apprend à parler par un processus qui commence comme une réponse à des stimuli physiques et verbaux. De plus, il croit que la totalité de notre discours est une réponse élaborée à partir des stimulations physiques du monde extérieur sur nos terminaisons nerveuses.

Le rôle du mécanisme behavioriste est important non seulement pour l'apprentissage mais aussi pour la pratique du langage: c'est grâce à lui que le langage a le statut d'acte public observable dont l'analyse peut conduire à la formulation d'hypothèses capables d'être testées empiriquement.

A noter que Quine reconnaît les insuffisances de la méthode behavioriste pour rendre compte de la complexité du phénomène linguistique, mais il s'insurge contre l'idée de la remplacer par des analyses faisant appel à des entités mentales ou abstraites.

Son projet de naturalisation de l'épistémologie l'amène à chercher des explications rigoureuses pour l'analyse du processus du langage et à les atteindre du côté des sciences de la nature plutôt que des théories mentalistes. Nous croyons que beaucoup de ce que Quine envisageait- à la fois à titre de méthode et de résultats- se rapproche du travail développé aujourd'hui dans le champ des sciences cognitives.

- **Le pragmatisme** qui se révèle dans sa procédure méthodologique.

Comme les théoriciens du pragmatisme linguistique, Quine critique toute attitude qui envisag la réification de la signification. Il affirme à plusieurs reprises les risques qu'on court si l'on accepte le mythe de la signification, titre de l'un de ses articles. Mais, contrairement au pragmatisme, sa théorie sémantique aboutit à l'impossibilité radicale de trouver une définition quelconque de la signification linguistique à partir de l'observation du comportement des locuteurs qui puisse prétendre être rigoureuse et univoque, selon l'idéal scientifique qui inspire ses analyses du langage. Il croit que le même ensemble de données observables justifiera des relations de synonymie et conséquemment des traductions conflictuelles selon les hypothèses analytiques qu'on assume.

En plus de cette coïncidence partielle de fondements avec le mouvement du pragmatisme linguistique, Quine utilise une procédure très semblable à celle des théoriciens de ce courant de philosophie du langage. Cela est dû à l'un des présupposés fondamentaux de sa théorie: on apprend à parler par un processus de dialogue avec notre précepteur et/ou les autres membres de notre communauté linguistique, ce qui suppose:

- leur approbation ou incompréhension de nos énoncés dans certaines situations, qui est la mesure de notre réussite ou de notre échec dans ce processus;

- l'observation du comportement des autres locuteurs qui amènera l'apprenant, peu à peu, à la compréhension de ses actes de langage.

On peut inférer ces présupposés de son affirmation:

Le langage est un art social qu'on acquiert par l'observation du comportement des locuteurs et dans des circonstances publiquement observables [2, p. 168]

La lecture de Quine révèle que le dialogue intersubjectif, l'interaction entre les locuteurs, l'observation de ce qu'ils font avec le langage, jouent un rôle décisif dans le processus d'acquisition de la compétence linguistique et aussi dans la performance, puisque autrui est la mesure de la réussite ou de l'échec pendant le développement de ce processus. Et parce que nous croyons que c'est son analyse de l'acquisition du langage qui est à la base de son système philosophique, on comprend l'importance que les principes du pragmatisme occupent dans l'ensemble de sa pensée.

D'habitude on rapproche le behaviorisme quinéen des thèses de Skinner, mais nous soutenons qu'il est beaucoup plus proche de Wittgenstein. Le mécanisme stimulus-réponse, fortement présent dans ses analyses, ne fonctionne pas dans le cadre étroit du behaviorisme: il suppose une sémantique qui incorpore la notion du suivi des règles d'usage, lesquelles sont dépendantes des normes établies par la communauté linguistique qu'on intègre et qui sont assimilées par l'apprenant selon un processus de dialogue et d'observation. Cela l'amène à la conclusion que la signification et la référence des mots émergent dans le contexte des actes de langage. Les racines de la référence se trouvent dans la façon dont on apprend à parler des objets de toute sorte: physiques, mathématiques, mentaux. Et cet apprentissage exige qu'on aboutisse à une uniformité sociale qui s'accorde aux règles d'usage linguistique de notre communauté.

Bref, on apprend à parler selon les príncipes logiques, ontologiques, enfin culturels de cette communauté. Les règles linguistiques sont rattachées au contexte culturel, aux "formes de vie" si l'on veut utiliser le langage wittgensteinien.

3 Théorie de l'apprentissage du Langage

Cette théorie occupe une place importante et joue un rôle clé pour la bonne compréhension des thèses de Quine. Elle a été construite à partir de l'analyse de deux situations: l'apprentissage du langage maternel par un enfant et ce qu'il a appelé "la traduction radicale".

3.1 L'apprentissage du langage maternel par un enfant

De l'observation de ce processus, Quine constate que l'enfant commence par établir une étroite liaison du langage avec les données sensibles: les premiers mots appris sont ceux qui se réfèrent aux objets physiques. C'est dans une étape plus complexe qu'il apprendra les mots qui s'appliquent aux sensations, aux entités abstraites, aux ensembles, aux relations, aux entités mathématiques. De plus, les énoncés d'observation sont les premiers qu'un enfant prononce quand il apprend à parler.

Selon Quine, l'apprentissage du langage traverse les étapes suivantes:

1ère: Réponse aux stimuli physiques et verbaux

Au début de ce processus les enfants reçoivent des stimuli d'objets qui touchent leurs terminaisons nerveuses, en même temps qu'une stimulation verbale. La répétition de cet ensemble (stimulation physique et verbale dans le cadre de quelques situations semblables) leur permet de commencer à employer correctement quelques tournures de la langue.

Quine reconnaît que ce mécanisme behavioriste serait insuffisant pour justifier la totalité du processus. Il n'arrive pas à justifier le fait qu'un enfant, d'après quelques situations concrètes d'apprentissage, soit capable de généraliser et d'utiliser telles expressions dans toutes les situations possibles.

C'est pourquoi il introduit la notion *espace qualitatif inné*. Il s'agit d'une norme de similarité qui nous permet de généraliser et d'acquérir des habitudes sans lesquelles tous les stimuli seraient semblables. Si cette norme est fondamentale pour tous les processus d'apprentissage, elle ne pourrait pas être apprise et serait innée.

Pour se défendre de l'accusation de développer un argument antiempiriste, il affirme qu'il s'agit d'une propriété assez connue dans les études de psychologie du comportement et qu'on peut l'explorer et l'observer en laboratoire. Les animaux sont aussi régis par cette norme. Elle est au fondement de tous les processus inductifs, de nos généralisations et expectatives face à l'avenir. Puisqu'il s'agit d'une capacité commune à tous les locuteurs, elle garantit une acceptabilité assez générale des inductions individuelles.

Un critère de décision qui permet à l'enfant de savoir s'il a utilisé l'expression adéquate à la situation dans laquelle il l'a employée est la compréhension ou l'incompréhension de son locuteur. C'est le dialogue qui permet de constater si l'apprenti a réussi ou échoué. L'approbation l'amènera à répéter ce comportement dans des situations semblables; la negation l'amènera à l'éviter. C'est par un processus behavioriste de réussite ou d'échec dans son dialogue que l'apprenant progresse.

Par cette analyse, Quine élabore la notion de signification-stimulus: ce sont des paires ordonnées de classes de stimulation capables de provoquer une réponse positive ou négative du sujet devant l'émission de cette expression dans des situations concrètes. La signification-stimulus n'est assimilable ni à la stimulation verbale, ni à la stimulation physique en tant que telles, mais à leur conjugaison avec une certaine situation, ce qui provoque une réponse positive ou négative du sujet qui se trouve dans cette situation.

2ème: La substitution par analogie

A cette phase du processus, l'enfant devient capable d'effectuer des substitutions dans des structures déjà connues. A partir de "Voilà mon bras" et "ma main", il peut dire: "Voilà ma main". Les substitutions sont faites d'abord dans les énoncés d'observation et, plus tard, dans les énoncés abstraits.

Quine est convaincu que ces deux étapes du processus d'acquisition du langage couvrent une partie très restreinte de la totalité de notre performance linguistique.

"D'autres associations inter-verbales sont nécessairess; elles rendent possible l'usage de nouveaux énoncés sans qu'ils soient attachés même secondairement à un quelconque domaine fixe de stimulations verbales". [2, p. 6]

"Apprendre par ostension c'est apprendre par simple induction et le mécanisme d'un tel apprentissage est le conditionnement. Mais cette méthode est incapable de nous conduire très loin dans le langage" [3, p. 37]

3ème: L'interanimation des énoncés

Même si l'origine de notre discours a été une réponse aux stimulations physiques, et même si le processus d'apprentissage linguistique a, lui-aussi, commencé comme réponse à de telles stimulations, ce n'est que quand nous dépassons la phase behavioriste que nous progressons effectivement. Dans cette troisième étape, le discours s'éloigne de la stimulation directe venue des objets physiques.

A cette phase du processus, les énoncés interagissent. Cela veut dire que même les énoncés d'observation peuvent ne plus être la réponse immédiate à une stimulation physique. Leur émission dépend alors d'un réseau d'énoncés, quelquefois très complexe, réseau qui leur apporte à la fois signification, référence et valeur de vérité. Par exemple, quand un chercheur, dans un laboratoire de chimie, réalise une série de mélanges de produits et observe que la substance résultante est bleue, il prononce l'énoncé d'observation: "Il y a du cuivre". Il s'agit d'un énoncé d'observation, prononcé en réponse à une stimulation physique, mais sa signification et sa valeur de vérité proviennent de l'ensemble de la théorie chimique dont l'énoncé fait partie. Quelqu'un qui ne connaîtrait pas la théorie ne serait pas en mesure de prononcer un tel énoncé en présence de cette stimulation physique.

On se rend compte que la plus grande partie du réseau linguistique est constituée d'énoncés qui sont des réponses à d'autres énoncés, et que leur cohérence dérive de leurs rapports réciproques à l'intérieur de la théorie à laquelle ils appartiennent. Ils dépendent des règles logiques, des lois causales, des présupposés et principes internes des théories, enfin, des composants culturels.

La force du réseau théorique est telle qu'on ne peut plus retrouver dans le complexe le simple dont il est issu et qu'on n'est plus capable d'identifier l'énoncé d'observation qui a été à l'origine du processus ayant abouti à un certain énoncé théorique.

La question de savoir dans quel cas un énoncé a été appris en réponse directe aux stimulations physiques, et dans quel cas dans le cadre d'un processus d'inter-animation, appartient à l'histoire déjà oubliée (et irrécupérable) de chaque locuteur. Selon Quine, on est ainsi en mesure de localiser, dans le processus même d'inter-animation, le moment où le discours dépasse la phase behavioriste initiale pour entrer dans un processus de production de nouveaux énoncés à partir de l'ensemble qui constitue la théorie du monde de chaque individu à chaque époque.

La priorité qu'acquiert le linguistique sur le non linguistique entraîne l'affaiblissement progressif du contenu sensoriel du discours encore prépondérant en début de processus. La connaissance indirecte prédomine. En effet, le locuteur se trouve entouré d'instruments linguistiques, d'un ensemble de règles qui favorisent l'homogénéisation de ses perceptions individuelles en vue d'un possible accord qui permettra le dialogue. Le locuteur parlera selon le modèle linguistique et culturel de sa communauté.

Le discours devient incapable d'être le miroir de la réalité et d'atteindre une neutralité ou une objectivité véritables. Le processus d'apprentissage du langage exige qu'on utilise des structures linguistiques, référentielles, logiques, enfin, culturelles, ce qui dépasse largement le cadre des réponses aux stimulations physiques qui ont déclenché le processus. Des conventions, des "inventions" complètent les lacunes que créent les données d'observation. Impossible de distinguer dans le discours ce qui est objectif de ce qui est "production humaine", ce qui est "substantiel" de ce qui est "style".

Par cette analyse, Quine accomplit son projet épistémologique. On est désormais en mesure de comprendre comment les stimulations de nos terminaisons nerveuses se lient à notre théorie du monde. De comprendre aussi pourquoi cette théorie se compose d'énoncés qui affirment beaucoup plus que ce qu'on aurait pu affirmer si on s'était limité à produire des énoncés qui ne seraient que des réponses aux stimulations physiques.

On est aussi en mesure de comprendre le rôle que les principes du pragmatisme jouent dans cette analyse. Apprendre à parler, c'est apprendre à suivre des règles de comportement linguistique, processus qui dépasse énormément le mécanisme stimuli-réponse du behaviorisme linguistique et qui est très proche des propositions du pragmatisme linguistique.

L'apprenant du langage progresse quand il dépasse le mécanisme behavioriste et domine l'étape de l'interanimation des énoncés. A ce moment, il s'éloigne du rapport direct de l'observation et il produit des énoncés qui sont des réponses à d'autres énoncés. Un réseau linguistique très complexe s'établit où on n'arrive plus à trouver la relation directe de chaque énoncé avec le stimulus physique qui a déclenché le processus. Ainsi, la signification et la référence des expressions linguistiques et le degré de vérité des énoncés dépendront de l'ensemble théorique qu'ils intègrent.

3.2 La traduction radicale

Cette situation hypothétique est celle d'un linguiste qui envisage d'apprendre le langage et de construire un manuel de traduction des expressions linguistiques d'un peuple dont il méconnait la langue et la culture. Ce peuple vit dans un état très primitif sans aucune similitude de comportements avec ceux de la culture dont est issu le linguiste. Cela rend encore plus difficile ses analyses. Le linguiste ne dispose ni d'interprètes, ni de dictionnaire, n'ayant comme point de repère que l'observation du comportement des locuteurs.

Ce linguiste sera obligé d'établir des hypothèses analytiques qui lui permettront d'aboutir à un parallélisme entre la signification des expressions du langage de ce peuple et celles de son langage maternel. Selon Quine, il sera obligé d'établir certaines hypothèses, par exemple:

- *empathie*- le linguiste doit éviter des résultats qui impliquent que ce peuple puisse avoir des croyances très aberrantes par rapport aux siennes;

- *la continuité des croyances*- le manuel de traduction qu'il construit ne peut pas faire penser que ce peuple change de croyances tout le temps;

- *simplicité*- le manuel doit révéler une structure linguistique qui ne soit pas trop compliquée.

Le degré d'efficacité du manuel est fonction de la réussite ou de l'échec du dialogue du linguiste avec son interlocuteur. De même que dans la situation de l'apprentissage du langage maternel par un enfant, c'est dans un cadre pragmatique qu'on pourra valider l'efficacité des hypothèses construites.

La situation qui est prise à titre d'exemple est celle où le linguiste entend l'expression "gavagai" et s'efforce de la traduire. En connaissant déjà les expressions qui expriment l'accord ou le désaccord dans ce langage, il reproduit une série de situations expérimentales de dialogue avec le peuple qu'il observe, destinées à l'aider à appréhender la signification et la référence de "gavagai". Dans ces situations, ses locuteurs doivent dire l'équivalent de "oui" ou "non" face aux questions qu'il pose. Ces questions permettront au linguiste de se rendre compte des situations correctes ou incorrectes d'emploi de telle expression.

Il constate que cette expression s'applique à des situations qui concernent les lapins, mais aucune observation expérimentale ne lui permettra de décider avec certitude si sa traduction doit être l'expression "lapin", "lapinité", ou "parties non-détachées de lapin". Toutes les expériences possibles justifieraient également bien le choix de l'une ou de l'autre de ces expressions comme étant la traduction correcte.

Quand le linguiste décide de traduire "gavagai" par "lapin", il se base sur des hypothèses analytiques qu'il a construites, qui dépendent fondamentalement de sa théorie du monde. Le choix entre les expressions "lapin", "lapinité", "parties non- détachées de lapin" comme la traduction correcte de "gavagai" suppose une option de théories du monde logiquement incompatibles mais empiriquement équivalentes. Pour effectuer ce choix, le linguiste doit dépasser les données de ses observations, puisqu'elles justifieraient aussi bien l'option par chacune de ces théories. Ces trois expressions diffèrent non seulement en ce qui concerne leur signification, mais aussi leur référence étant donné qu'elles se réfèrent à des choses différentes. Et aucun critère de décision ne sera capable de justifier l'affirmation qu'il y a une meilleure option parmi les trois.

Quine est convaincu que si deux linguistes entreprennent cette tâche et qu'ils n'entrent pas en contact l'un avec l'autre, ils produiront des manuels de traduction différents, incompatibles du point de vue logique. Néanmoins, leurs deux manuels seront également compatibles avec le comportement du peuple observé.

De l'analyse de cette situation, Quine renforce sa position de la transcendance du discours par rapport à toute observation possible et sa théorie de l'apprentissage du langage lui fournit des arguments et des fondements pour l'ensemble de son système. Puisqu'il ne s'agit pas de traiter dans cet article le rapport de sa théorie du langage avec toutes les thèses de son système, je suggère à ceux qui s'intéressent à ce sujet la lecture de mes écrits cités dans la bibliographie [8, 9, 10, 11, 12, 13, 15, 14, 16, 17, 18].

4   Evidence, observation et prévision chez Quine

Poussé par son projet épistémologique de bien comprendre comment les stimulations du monde extérieur sur nos terminaisons nerveuses nous ont amené à construire une théorie du monde qui transcende énormément le niveau des dites stimulations et en plus nous permettent de faire des prévisions, Quine est obligé d'essayer de comprendre la nature et les limites de l'évidence pour notre processus cognitif. Il soutient que les épistémologues peuvent s'aider de plusieurs sciences dans son analyse des mécanismes de la perception et de ses rapports avec notre discours: la neurologie, la génétique évolutive, la psychologie, la psycholinguistique, l'Histoire des Sciences, l'analyse logique. Cette recherche interdisciplinaire pourrait aboutir à comprendre les bases évidentielles de notre savoir et de notre discours qui est la phénoménalisation du résultat du processus cognitif.

Il s'intéresse à l'analyse de la prédiction parce qu'il s'agit du rapport direct entre le discours et la réalité physique. En d'autres termes, cela constitue le test empirique des théories, question fondamentale pour les investigations des épistémologues.

La prédiction suppose l'observation et des critères d'évidence. Les épistémologues, Quine compris, sont conscients des problèmes techniques qu'on trouve pour établir le rapport entre les objets et les stimulations qui touchent les terminaisons nerveuses des sujets, lesquels essaient de les analyser en faisant appel à leurs processus cognitifs. Il choisit d'analyser la question de l'évidence en ayant recours à la notion d'énoncés d'observation et aussi à un composant pragmatique:

Une exigence de plus est l'intersubjectivité; à la différence d'un rapport de sentiment, l'énoncé doit entraîner le même verdict de tous les témoins, linguistiquement compétents, de cet événement. [4, p. 10]

Les rapports du discours avec le monde dépendent des énoncés d'observation et de l'intersubjectivité et Quine affirme que c'est l'évidence qui rend la science objective. Il s'abstient de définir "évidence". Il choisit d'affirmer que les énoncés d'observation sont les véhicules de l'évidence scientifique. Ils sont la pierre angulaire du processus cognitif et de l'apprentissage du langage, le point de rencontre du discours et du monde et le seul point de repère de la notion d'évidence.

Même si pour qu'un énoncé soit considéré comme un énoncé d'observation il faut qu'il soit étroitement lié aux stimulations physiques, cela ne veut pas dire que tous ont été appris comme une réponse directe aux dites stimulations. La plupart l'ont été par des constructions très complexes, selon des procédures basées sur l'analogie et l'interanimation des énoncés, aidées par des rapports intersubjectifs, comme nous l'avons décrit dans notre exposition sur le processus d'apprentissage du langage.

Il est important de comprendre que le critère d'évidence en faveur d'une hypothèse scientifique n'est pas nécessairement reconnu comme tel par tous les locuteurs. Cela veut dire que la notion ou le critère d'évidence sont relatifs à une certaine théorie du monde. Un énoncé d'observation reconnu comme tel pour un groupe de scientifiques peut ne pas l'être pour tout le monde. Alors, un énoncé d'observation peut provoquer des réponses positives ou négatives des membres de la communauté scientifique, sans avoir aucun effet stimulant sur les autres membres de la même communauté linguistique à laquelle ceux-là appartiennent.

L'importance des critères pragmatiques est claire dans ces affirmations de Quine. Les énoncés d'observation sont les plus proches des récepteurs sensoriels, ils possèdent le plus fort degré d'évidence et aussi d'accord intersubjectif. Quine est convaincu que dans ces énoncés l'information objective est tellement mélangée aux "inventions" qu'on doit reconnaître que tous les énoncés sont théoriques et qu'on n'est plus en mesure de distinguer ce qui est objectif de ce qui est une "création de la race humaine".

Les critères d'évidence et d'objectivité devront faire appel à la complicité intersubjective. Il faut supposer qu'autrui aperçoit le monde de manière semblable à la mienne, qu'il soit d'accord avec les rapports que j'établis entre les théories et les données d'observation, qu'il croit que nous partageons nos perceptionss et que nous arrivons à communiquer sans équivoque. Ce sont des suppositions qui ne trouvent aucune justification rigoureuse dans l'observation. Elles sont postulées par une sorte d'empathie qui est nécessaire à tout processus d'apprentissage et de communication intersubjective.

Cette position de Quine ne le conduit pas au scepticisme ni au relativisme du discours scientifique. Il soutient que la prédiction est un bon point de repère de la certitude en science. Il ne définit pas la prédiction comme une norme mais comme un certain "jeu de langage" selon la terminologie wittgensteinienne:

Le jeu de la science, en contraste avec d'autres jeux de langage comme la fiction et la poésie" [6, p. 3]

Quine soutient que l'aspect un peu flou du critère d'évidence n'empêche pas qu'on puisse parler de vérité en science. Il croit qu'on peut trouver des points de repère pour choisir, avec un certain degré de certitude, entre les énoncés vrais et faux. Ce choix dépendra de la théorie du monde qu'on accepte. Il est possible de démontrer la fausseté d'une théorie, mais on pourra toujours construire des théories logiquement incompatibles et empiriquement équivalentes. Cette conviction lui fait soutenir la thèse de l'indétermination de toute théorie par l'expérience.

5 Théorie- Langage- Processus Cognitif

Après notre exposition, nous sommes en mesure de comprendre pourquoi Quine affirme que nous sommes des "prisonniers du discours".

Notre processus cognitif se développe parallèlement à l'apprentissage du langage. Le discours écrit, oral ou gestuel, est la phénoménalisation de nos connaissances.

L'ensemble des résultats du processus cognitif constitue les théories du monde qui se phénoménalisent par le langage et deviennent des objets physiques qui peuvent être analysés par les méthodes des sciences de la nature, d'où sa suggestion de la naturalisation de l'épistémologie.

Poussé par son projet épistémologique de recherche de liens entre les stimulations physiques de nos organes de la sensibilité et la production de la théorie du monde qui en dérive, Quine a cherché les rapports empiriques entre les objets qui sont les mots et ceux qui ne sont pas des mots. Tel est le thème de son livre [1]. Insatisfait de l'analyse qu'il avait faite du processus d'apprentissage du langage dans cet ouvrage pour arriver à expliquer comment nous apprenons à parler des objets, il publie quatorze ans plus tard [3]. Il reprend cette thématique dans [5], [6] et [7]. Ces ouvrages et d'autres articles reflètent le même présupposé philosophique: tout ce qu'on peut connaître sur nous-mêmes et le monde est exprimé à travers un vaste réseau d'énoncés. Pour mieux comprendre la signification et la valeur de vérité du discours que les hommes ont "inventé" et qui est un objet physique, il faut trouver les liens avec les autres objets dont le discours s'occupe. Pour cela il trouve que le meilleur chemin est l'analyse du processus par lequel nous apprenons à parler des objets, parce que c'est dans cette situation que nous arrivons à mieux dévoiler les rapports entre les mots et les choses, le discours et la réalité.

Cette analyse lui revèle l'indétermination des théories et du langage. Même si, à la périphérie, l'ensemble du réseau théorique touche nécessairement l'expérience, il est indéterminé par toute expérience possible. Des théories logiquement incompatibles peuvent affronter avec succès le tribunal des tests de la vérification expérimentale.

On constate ainsi que Quine diffère du projet réductiviste de l'empirisme classique qui suppose que tous les énoncés sont capables de se soumettre plus ou moins directement à la confirmation expérimentale indépendamment de leur contexte.

Sa conviction de l'indétermination des théories et du discours l'amène à soutenir que tous les énoncés sont susceptibles d'une révision, y compris les énoncés éternels- ceux qui sont vrais ou faux indépendamment des circonstances de leur utilisation-, les énoncés mathématiques et ceux de la logique. Il soutient une position anti-dogmatique mais en croyant qu'on peut trouver des critères plus au moins rigoureux pour justifier le choix de certaines théories par rapport à d'autres en fonction de leur degré de certitude, de démonstration empirique et logique. Dans le cadre d'une communauté linguistique, il sera toujours possible d'obtenir un certain consensus en faveur d'une thèse par rapport à d'autres qui passent aussi bien les tests empiriques.

References

[1] W. V. O. Quine. *Word and Object*. Mass.: M.I.T. Press, 1960.

[2] W. V. O. Quine. *Ontological Relativity and other essays*. Columbia U. P., 1969.

[3] W. V. O. Quine. *The Roots of Reference*. La Salle III: Open Court, 1974.

[4] W. V. O. Quine. *Mind and language*. Ed. by S.Guttenplan. Oxford: Clarendon Press, 1975.

[5] W. V. O. Quine. "Theories and Things". Cambridge, Mass., 1981.

[6] W. V. O. Quine. *Pursuit of Truth*. Cambridge, Mass.: Harvard U.P., 1990.

[7] W. V. O. Quine. *From Stimulus to Science*. Cambridge, Mass.: Harvard U.P., 1995.

[8] V. Vidal. "O Aspecto Pragmático da Filosofia da Linguagem de Quine". In: *Reflexão X (32), p.49 e Revista Filosófica Brasileira (2), p. 116* (1985).

[9] V. Vidal. "Contribuições do Sistema Filosófica de Quine para as Investigações da Filosofia Analítica". In: *Paradigmas Filosóficos da Atualidade*. Ed. by M.C Carvalho. Papirus, 1990, pp. 39–81.

[10] V. Vidal. *Sur la Thèse Quinéenne de l'Indétermination de la Traduction*. Thèse de Doctorat em Philosophie- Université de Paris I, 1991.

[11] V. Vidal. "Le Pragmatisme dans la Philosophie de Quine". In: *Verbum* 13 (4 1994), pp. 273–288.

[12] V. Vidal. "Sur la thèse Quinéenne de l'Indétermination de la Traduction". In: *TTR (Traduction, Terminologie,Rédaction)* 3 (1 1994).

[13] V. Vidal. "Empatia e Transcendência: reflexões sobre o sistema filosófico de Quine". In: Florianópolis: Principia, 2003, pp. 205–228.

[14] V. Vidal. "O papel da Empatia na Teoria da Verdade de Quine". In: *A Questão da Verdade: da metafísica moderna ao pragmatismo*. Ed. by V. Vidal and S. Castro. Rio de Janeiro: 7 Letras, 2006, pp. 90–100.

[15] V. Vidal. "Quine e Wittgenstein: um diálogo impossível?" In: *Colóquio Wittgenstein*. Ed. by G. Imaguire, M.A. Montenegro, and T. Pequeno. Edições U.F.C, 2006, pp. 90–100.

[16] V. Vidal. "Ontologia analítica de Quine". In: *Metafísica contemporânea*. Ed. by G. Imaguire, C.L.S. Almeida, and M.A. Oliveira. Petrópolis, RJ: Vozes, 2007, pp. 98–120.

[17] V. Vidal. "Quine e a Crítica às Entidades Mentais". In: *Mente, Cognição, Linguagem*. Ed. by C. Candioto. Champagnat, 2008, pp. 99–119.

[18] V. Vidal. "Características do Pragmatismo Quineano". In: Univ. Gama Filho, 2009, pp. 51–67.

Dice: a hazardous symbol for chance?
JEAN-YVES BÉZIAU

Le hasard de la vie
Nous conduit tout droit au fond de l'oubli
Pour renaître à jamais dans l'amour de la nuit
D'un souffle qui nous tire du coeur de l'ennui
Baron de Chambourcy

Contents
1. Give Philosophy a Chance
2. The Chance of being Lucky
3. Dicing on the Beach of Infinity
4. The Secret Mechanism of the Key Code
5. Falling in Love and Miracle
6. Warum the Rose?
7. Squaring Chance
8. Dedication and Personal Recollections
9. References and Further Readings

Copyright © 2019 by Jean-Yves Béziau. All rights reserved.

1 Give Philosophy a Chance

What is chance? This is what we are examining in the present paper. We do so by discussing if *dice*, or *dice throwing* to be more explicit, is a good representation of chance. We are at the same time investigating a notion and developing a methodology about how we can do that. For chance or other notions. Right now chance is the lucky girl but we keep an eye on other nice animals: Siberian tigers, Teddy bears, Guinea pigs.

The heart of the methodology here is *symbolization*, in a very simple sense, symbolized by the balance. The balance is a renowned symbolization of justice. Something very general, untouchable, not to say intangible, is presented very concretely in front of us. Justice certainly does not reduce to a balance how beautiful it can be, but the balance is a starting point, a runway for our thought to take off in the direction to the sky of ideas.

We have to be careful: if we choose the wrong springboard, we will not go very far, or go in the wrong direction, reaching Columbia instead of India, Hesperus instead of Phosphorus, as it happened once upon a time.

This symbolic methodology goes here hand to hand with *imagination* and *structure*. Images may be limited and illusory but they can mirror reality to help us to go through the looking glass. Structure, not to say structuralism, means establishing relation with other notions, considering that nobody lives in isolation and that relation, one of the four basic aspects of the Logos, is a key for understanding. Beside, outside, upside, inside chance there are many other notions that make sense of it.

By using this threefold strategy, we will by chance experimenting philosophy. Let's throw the dice: *Alea iacta est! Rien ne va plus...*

2 The Chance of being Lucky

Everybody knows the word "chance". But what does it mean exactly? And how can we better know what chance itself is?

Most of the words we are using have a meaning which is: fuzzy, incomplete and confused, not to say inconsistent. And these three features often come together. This can be seen as a problem... or not! They allow flexibility, fundamental for the development and creativity of thought.

This contrasts with the dream of a perfect and rigorous language promoted at the end of the 19th century by people like Frege. A radical solution like wanting a very clean house and body without any star dust, a perfect beautiful society as it was promoted by the Nazis, without Rats and Jews. No chance...

One may reply: I don't want sprawling in the mud like a pig. Of course, but it is important to go beyond dichotomy, to see that the rejection of one tendency is not the acceptance or promotion of a diametrically opposed tendency, as ugly and ridiculous as the *prima facie* one is. These are two sides of the same coin and it is not just a matter of flipping the coin. We can look for a third option. Not necessarily to go beyond, in the sense of Hegel's *Aufhebung*, leading to the synthetic queen of his trilogical dialectic, crowning thesis and antithesis. Maybe more like the *middle way* of Buddhism. Precision without preciosity.

The meaning of chance, like the one of many other notions, cannot be definitely fixed, arrested by any thought police or locked into some language boxes. There should always be a second chance.

When examining the meaning of a word, one may want to go "inside" the word, to decompose it, to look for its origin and etymology. This makes sense up to a certain point and it depends how it is done. This analytic method may straightforwardly lead to nonsense, a vivisection driving to death. If we decompose the word "chance" in its six letters and look for the meaning of the word in these letters or in their combination, we need to be very lucky to reach any understanding.

A more intelligent decomposition is to look for morphemes, but the meaning of a word does not reduce to morphology. This is striking if we are aware of the multiplicity of languages: the same common notion can be expressed in so many different ways. For example the word "cause" used in English to express the notion of causality has two pretty different expressions in Latin and Greek, "cause" and "$αἴτιος$". And this is true also of its classical opposite, "chance", in Greek: "$ευκαρια$".

We need to keep this in our mind when looking at the etymology of a word. Etymology is only one key and it can open the wrong door. But to be too cautious can block any chance to access to paradise. With an open mind, let's have a look at the etymology of chance (*Online Etymology Dictionary*):

> from Old French *cheance* "accident, chance, fortune, luck, situation, the falling of dice" (12c., Modern French *chance*), from Vulgar Latin *cadentia* "that which falls out," a term used in dice, from neuter plural of Latin *cadens*, present participle of *cadere* "to fall," from PIE root **kad-** "to fall."

Fall is here the key. But before falling in love with the root of "chance", let's have a look at its semantic web. Here is an interesting pseudo-Socratic dialogue:

Thesaurus.plus

What can we infer from that? Better not to bet all eggs in one basket. There are different words for the same thing and different things for the same word. Meaning is circulating and fluctuating through words. This was magisterially pointed out by Michel Bréal in his book *Essai de Sémantique* (1897), coining the word "semantics". The English word "chance" has not the same meaning as the French word "chance", which curiously has exactly the same spelling (but a more beautiful sound line).

Are they true enemy brothers? Not really, because one of the meanings of the French "chance" is one of the submeanings of its English cousin, the Lucky one. And vice versa, the aleatory English aspect of the word "chance" is not outside of the semantic field of its French neighbor. On the one side of the so-called English Channel, one meaning prevails, on the other side another meaning prevails. At the end we have a symmetric inequality.

The English word "chance" is generally translated in French by a word of Arabic origin "hazard". It is related to a mysterious castle in Syria where people in the middle age were throwing dice and nowadays even more dangerous things.

Hazardous means in English dangerous or bad luck, contrarily to the French meaning of "hasard", mainly equivalent to the English chance, more neutral, not to say contingent. To complete the picture of our semantical navigation and in memory of the great sailor Ferdinand Magellan, we will also consider the Portuguese case.

We have then the following table:

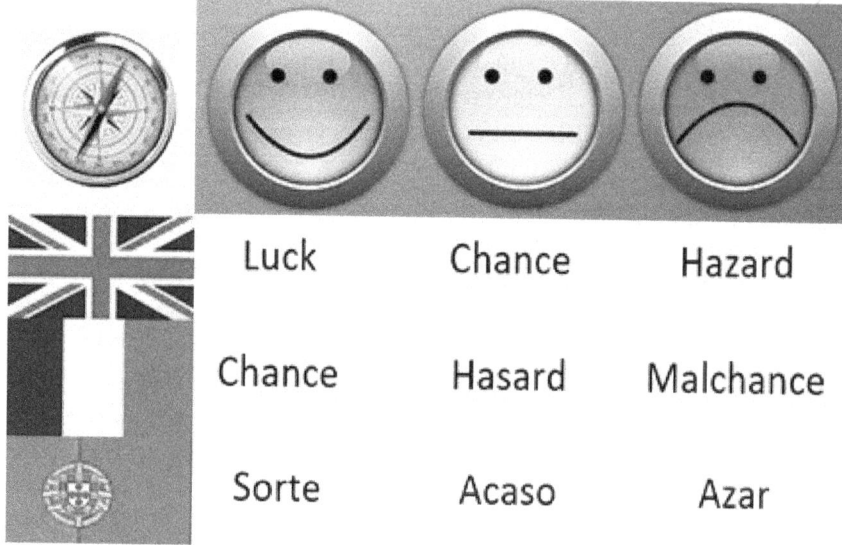

We have three notions, or better, three variations of the same notion, expressed by different words in these three languages. It can be useful, for the good or for the bad, to have in mind these semantical fluctuations. Especially if we are on the road to the end of the word, and we want to escape any *accident*, in the sense of Aristotle or Ayrton Senna.

3 Dicing on the Beach of Infinity

When we throw dice, the result is physically very difficult to determine in particular due to the homogeneity of the dice cubes.

Can we equate this very high indetermination with an absolute *Indetermination*, if any. If we do so, are we not in the same situation as when equating a huge quantity with infinity? And can we say that grains of sand on a beach is the right symbol for infinity? We have to be careful not to confuse quantity with quality. Very hot is not the same as boiling. To live 1.000 years is not the same as being eternal.

It is true that we don't know if physical reality is in itself completely determined or not. If it is, if God does not play dice, the indetermination is human not ontological. If it is not, if the universe is a Satanic roulette, throwing dice is anyway mainly a physical phenomenon, even if the dice are thrown by human hands. Chance is therefore physically symbolized. This vision of chance seems too restricted, not good enough for an open perspective free of physicalism. The general notion of chance is not limited to physical phenomena, it is related to anything that can happen, in particular psychological, emotional, sociological, historical, biological events.

And a physical phenomenon is not necessarily a very good example of epistemological indetermination because physics is the easiest way for describing and predicting what will happen, with the so-called laws of physics. We can predict a solar eclipse, the trajectory of a missile, etc. And laws implicitly mean here that these phenomena are ontologically determined. *La Mécanique Céleste* or *Cosmic Karma*.

Of course things have changed in modern physics, in particular in quantum physics. But dice are not a very good example of quantum phenomenon. Dice appear more as particles, we don't see much wave here, unless considering handwaving, which is also part of the game. In this case dice throwing can be seen as an expression of the mysterious wave/particle duality.

But without developing a new age fantasy or a remake of Little Red Riding Hood in Las Vegas, we can however still defend dice. Dice used for gam(bl)ing are not big rough cubes on which you can sit, or that you can use as tables for the last supper. They are small and polish, with rounded corners and perfectly identical - generally pairwise used. There is an almost perfect physical identity between two dice. And each dice is identical to itself, not like anybody else is, but its six faces and eight round corners are identical. We are in the kingdom of regularity, uniformity and symmetry. Notions which are far beyond the daily physical world made of incongruities, like the paving stone on which you stumble when going to go the bakery to buy a baguette with an irregular shape. Gaming dice are therefore from another word, a mathematical world.

But the divine symmetry of a gaming dice is broken, not to say soiled, by some marks, not to say stains. Devil prints? If you closely look at one of these dice, you will see that there are different signs on each of its faces, or better, different arrangements of similar signs, dots precisely. And that makes all the difference, according to which the game makes sense. What would be the interest of playing dice if all the faces were exactly the same?

The result of throwing dice is meaningful not because of the positions of the dice, but because of the "numbers" which appear at the top of each dice. And there are no connections between these inscriptions and the physical moves of the dice. This lack of connection is a good symbol for chance.

Each of the six faces is different from the other faces, but this is a rather symbolic difference. Symbolic here in the distorted loose sense of formal logic. Symbols without meaning that need to be interpreted. They need a top model!

Depending on model theoretical numerology, you lose or win, or can use the "result" as a sign for action, as a revelation for anything you want. The chance is on your side...

The hands throwing the dice are also a good symbol for chance because they express the opposite of manipulation and/or control. The hand is considered as the symbol of human being. Hands are tools and have been used for developing artificial tools, like fork and knife, not to speak about cranes and bulldozers, that enable human being to transform and shape reality, like a god. When playing dice, he stops to be a god, maybe he just let God plays for him. Blowing dice can also be an interesting symbolization, a more feminine and meteorological version. *Gone with the wind...*

Dicing is therefore in many ways a good symbol for chance, balancing between hazard and luck, something rather absurd. Now even if we agree that dice are good chance representatives, we have to go through the looking glass, to the other side. A symbol is good but it is just a sign. The sign in itself has no value. Its cash value, or to speak in a more platonic way, its true value, is what is behind/beyond it, what it is pointing at. But we will not go straight to the point, if any. We will first make a detour, or worse, we will go in the opposite direction, trying to understand what determinism is. And we will do that also using symbolism.

4   The Secret Mechanism of the Key Code

Face to face to chance, we have determinism. A good symbol for determinism is something mechanical. We will not take here a big machine like a locomotive or the universe. Let's choose a more familiar and smaller machine: a watch. If you have a plastic watch you may wonder what we are talking about. But we are talking about real watches, a Patek Philippe Calibre 89, or the 2015 Vacheron Constantin with 57 complications.

Another example of the same kind, domestic and of human dimension, is a key. Both cases are interesting because the relation between the mechanism and its use and/or meaning is quite mysterious. A relation not as absurd as between a dice face and the number on it, but quite strange.

Can we say that the mechanism of a watch captures or expresses time? It is even difficult to say that it does this analogically, unless we believe in Ptolemaic astrology. We can think that the solar system and the universe are big rotating machinery but the mechanism of a watch does not work in the same way. Rotation is the only common ground. The solar system is only metaphorically a gearing mechanism, its teeth, if any, can be seen only with the binocular eye of our reason. The exact time given by a watch and what is going on earth and elsewhere belong to two different determinisms, causal chains, that mysteriously coincide.

Regarding the key, we have a mechanism which has as an effect and use completely different from its nature. With this mechanism you can protect your house and/or open a door to a safe with gold bars. The key to open the safe can be in gold but need not. Using a Saussurean language, we can say that there is no relation between the signifier and signified. The key is therefore, funny enough for such symbolic queen, the opposite of a symbol. It can be considered as the symbol of the anti-symbol, the arbitrary sign: there is no relation between the word "consciousness" and the reality of consciousness, but the word opens the door to this reality, makes you conscious of it.

The essential nature of the key(lock) is that it is a complex mechanism that cannot so easily be replicated or shaped (from the keylock it is difficult to create the key). It has a complicated physical specification. This sophisticated physical determination tends to be replaced nowadays by the high mathematical indetermination of a secret code, similar to dice throwing. Instead of keys and gold bars we now have plastic money and its secret codes.

With good material technology you can replicate the key. For the secret code there is no key because there is no key to indetermination. You just have to fix a high improbability by setting the data. That's not difficult: to generate such an improbability, you don't need to count to 100.

It is like playing dice. But the good thing is that you always win: knowing the pin number of your credit card or your safe, you can have access to it. On the other side the thief will nearly always lose. Her "chances" to enter your house or to have access to your bank account are so remote that you can sleep in a very relax way, voluptuously dreaming of all you will buy if you win at the lottery...

But let's come back to the key question. Determinism is absurd: the mechanic trail takes you straight to death. If everything is played in advance, what is the meaning of the game? We are in the reign of fatality, as absurd as pure chance. Is there no middle way to zigzag between Hazardous Scylla and Fatal Charybdis without trespassing? Hopefully leading us to Ithaca or - why not? - Heaven...

5 Falling in Love and Miracles

There is something in human life which is against or upside all kinds of mechanisms, physical, emotional, sociological. This is love. It is interesting to note that the expression "falling in love" is rooted in the same idea as "chance": to fall - cf. the etymology of chance and its symbolization through dice throwing.

Falling is something out of control, unless we give the credit to gravity, symbolically manifested by an apple, which for some reason, or by pure coincidence, became also the symbol of sin. The gravity of love, which in Portuguese leads to "gravidez", i.e. pregnancy. This is not automatic, but you have better chance than at the lottery. You can also pray for God. This is what Fançoise de Sionnaz did and the result was the birth of Saint François de Sales, August 21, 1567, source of miracles.

What is the difference between love and miracle, if any? It seems that love is higher than miracle, because a miracle makes sense within some determined circumstances. A paralytic starts to walk: without paralysis there would be no miracle.

Love is from scratch. It is in some sense completely absurd, but this absurdity is not like pointless determinism or the aleatoric nonsense of chance. Love makes sense. By itself. Without a goal to achieve. Life may look really absurd and uninteresting and love opens or awakes the meaning. It is the key to life.

Is love dangerous, hazardous? Love is not back luck, but illusion can always show up. Mirage, smoke screen, phantasm. On the one side there is the *femme fatale*, on the other side there is wedding, both anchored in material determinism, not to say dialectical materialism.

Is falling in love like dice throwing? No, because dice throwing in itself has no sense. You throw two dice and then get a "result", say 45. What is the problem? There is no problem! And also this is not a solution. You may "interpret" this as a solution but in itself it has no meaning. When you fall in love everything just starts to be full of meaning. Out of nothing. Love is not the result of throwing dice or another mechanism. It is something completely undetermined. And it is not a game, children don't fall in love.

6 Warum the Rose?

The rose is customizely associated with love. But can we really consider that this flower is a good symbol for love?

Die Rose ist ohne Warum. The rose is without 'why';
Sie blühet, weil sie blühet. it blooms simply because it blooms.
Sie achtet nicht ihrer selbst, It pays no attention to itself,
fragt nicht, ob man sie siehet. nor does it ask whether anyone sees it.

The above poem is due to Angelus Silesius. Heidegger quotes it in his book *The principle of reason* opposing it to the favorite axiom of Leibniz, *Nihil est sine ratione*.

However a rose is a flower, which emerges, grows and dies. The rose has been tragically staged in a song by Cécile Caulier, entitled *Mon amie la rose*, originally interpreted by Françoise Hardy (1964, both in French and English), having a second life with Natacha Atlas' interpretation in 1999.

On est bien peu de chose	A lifetime comes and goes
Et mon amie la rose	And as my friend the rose
Me l'a dit ce matin	said only yesterday
À l'aurore je suis née	This morning I was born
Baptisée de rosée	and baptized in the dawn
Je me suis épanouie	I flowered in the dew
Heureuse et amoureuse	and life was fresh and new
Aux rayons du soleil	The sun shone through the cold
Me suis fermée la nuit	And through the day I grew,
Me suis réveillée vieille	by night-time I was old
Pourtant j'étais très belle	At least there's never been
Oui j'étais la plus belle	No, you have never seen,
Des fleurs de ton jardin	a rose more bright and gay
On est bien peu de chose	A lifetime comes and goes
Et mon amie la rose	And as my friend the rose
Me l'a dit ce matin	said only yesterday
Vois le dieu qui m'a faite	The good lord smiled on me,
Me fait courber la tête	so why then should it be
Et je sens que je tombe	I feel I'm falling now,
Et je sens que je tombe	oh yes, I'm falling now
Mon cœur est presque nu	My heart no-one can save
J'ai le pied dans la tombe	My head begins to bow,
Déjà je ne suis plus	my feet are in the grave
Tu m'admirais hier	The rose God smiled upon
Et je serai poussière	Tomorrow will be gone
Pour toujours demain	forever gone away
On est bien peu de chose	A lifetime comes and goes
Et mon amie la rose	And so my friend the rose
Est morte ce matin	was dead at break of day
La lune cette nuit	The moon is shining bright
A veillé mon amie	and in my dreams tonight
Moi en rêve j'ai vu	Beneath the starlit sky,
Éblouissante et nue	my friend the rose goes by
Son âme qui dansait	He has seen my dreams I see
Bien au-delà des nues	A soul that wouldn't die,
Et qui me souriait	still watching over me
Crois celui qui peut croire	Whatever fortune brings
Moi, j'ai besoin d'espoir	I'll hope for better things
Sinon je ne suis rien	or life will just be grey
Ou bien si peu de chose	A lifetime comes and goes
C'est mon amie la rose	That's what my friend the rose
Qui l'a dit hier matin	said only yesterday.

Between the extreme of Silesius' No Reason Rose and the tragic destiny of Caulier's Friendly Rose there is something at the middle that the rose better symbolizes. It is something to which we can give a paradoxical name: *free determinism*. A caricature of it are games, not gambling games, but less hazardous games, like soccer or tennis, on the physical side, or chess and poker, on the intellectual side. These games are not completely undetermined. There are some rules and these rules make sense of the game, determine the game, what you can do or not. But upon these rules you can freely act, exercising and showing your ability.

This does not restrict to organized games. Riding a horse, surfing the wave, or proving a theorem is also something like that as many things in human life. And also in nature, like the rose. Rules permit to create and new rules can also be created. Evolution is a middle term between absolute determinism and pure creation. This is where the rose stands. *Cada macaco no seu galho.*

7 Squaring Chance

Let's now figure the whole of our inquiry. By investigating the symbolization of chance as dice throwing we have delineated three other notions which are opposed or/and different.

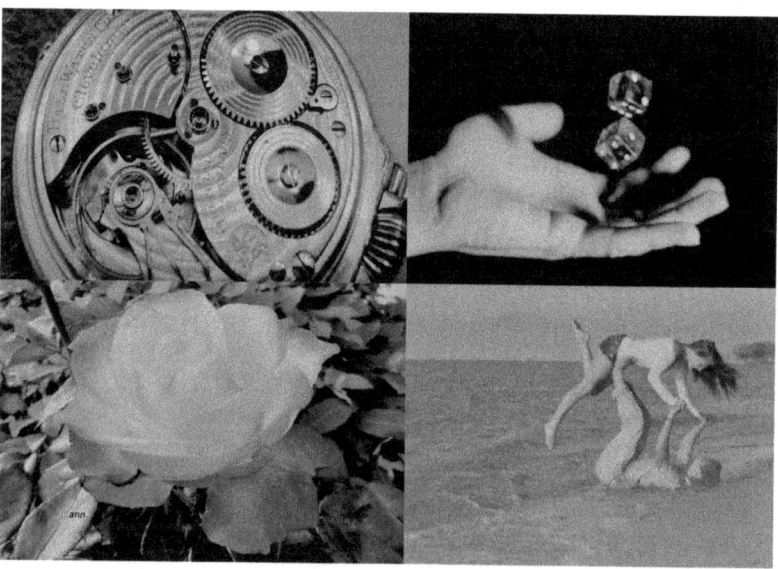

Dice throwing is a good symbol for chance for inner reasons, as we have explained in section 3, but also for outer reasons because it characterizes the similarities and differences with three siblings. At the upper level we see the contrast between an inexorable mechanism and aleatory rambling, both are absurd because they are precisely defined but they don't make sense by themselves. On the lower slice we have side by side, the rose as the expression of the beautiful creativity of nature, and falling in love, not the expression of free will, but nor the manifestation of an absurd fatality.

These four aspects of reality most of the time don't clearly show up because it's all mix up:

What we have done in this paper can be called *conceptual clarification* and is connected with the methodology preached by Alfred Tarski, when he was titillating truth:

> I hope nothing which is said here will be interpreted as a claim that the semantic conception of truth is the "right" or indeed the "only possible" one. I do not have the slightest intention to contribute in any way to those endless, often violent discussions on the subject: "What is the right conception of truth?" Disputes of this type are by no means restricted to the notion of truth. They occur in all domains where – instead of an exact, scientific terminology – common language with its vagueness and ambiguity is used; and they are always meaningless, and therefore in vain. It seems to me obvious that the only rational approach to such problems would be the following: We should reconcile ourselves with the fact that we are confronted, not with one concept, but with several different concepts which are denoted by one word; we should try to make these concepts as clear as possible (by means of definition, or of an axiomatic procedure, or in some other way); to avoid further confusions, we should agree to use different terms for different concepts; and then we may proceed to a quiet and systematic study of all concepts involved, which will exhibit their main properties and mutual relations.

We could say something similar about what we done here with chance. But we are not a blind follower of the Polish King of Logic. We think that besides, or better, upstream *definition* and *axiomatic procedure*, we can clarify our thinking with symbol, imagination and structure (SIS-mic methodology). And we don't necessarily want to fix things with words or/and fix the meaning of the words. We prefer fixing ideas. We can maybe define philosophy in this way, a philosopher as an idea fixer. And we hope we succeeded to fix chance pretty well.

8 Dedication and Personal Recollections

I met Tarcísio for the first time in Rio de Janeiro in 1996 for the defense of the PhD of Arthur Buchsbaum. Then Tarcísio invited me to develop projects with his Artificial Intelligence Logic (LIA) team in Fortaleza, Ceará.

My first visit to Fortaleza was in 1997 and the latest one in 2015. In between I have been there numerous times for short or long visits, being in particular a visiting professor/researcher for 2 years (2008/2010) of FUNCAP/CNPq at the Federal University of Ceará (UFC). I have extensively visited Ceará from North to South, East to West, Mountain to Sea. Thanks to Tarcísio I discovered the amazing land of Ceará and its creatures. Tarcísio introduced me to several princesses: Iracema, Ypioca, Guaraminga... and a most beautiful one I will not reveal the name here.

Tarcísio and Jean-Yves close to Guaraminga, 2004

I had many discussions with Tarcísio about a great variety of topics. ranging from Asclepius to Astrologius through Autobus. I don't remember all of them. And anyway, more important than the Topics themselves is the logico-philosophical way to deal with them. Tarcísio has a real philosophical spirit. And also a philosophical way of behaving.

The topic of the present paper is directly related to another topic I much discussed with Tarcísio, the notion of game, leading to our joint paper "The rules of the games". Under the title "Dice: a hazardous symbol for chance?" I gave a talk, first draft of the present paper, at the *III Latin American Analytic Philosophy Conference* that took place, May 27-30, 2014, in Fortaleza, Brazil. For this Festschrift I decided therefore to go on sculpting this topic, to offer a beautiful logico-philosophical piece to Tarcísio.[1]

Tarcísio and Jean-Yves, Praia de Iracema, 2004

9 References and Further Readings

Here you will find (papers, books, a movie) mentioned, directly or not, in my contribution and some useful further readings.

J.-Y.Beziau, "Modeling causality", in J.-Y.Beziau, D.Krause and J.R. Becker Arenhart (eds), *Conceptual Clarifications - Tributes to Patrick Suppes (1922-2014)*, College Publication, London, 2015, pp.187-205.

J.-Y.Beziau, "The Contingency of Possibility", *Principia*, **20** (2016), pp.99-115.

---

[1]Thanks to Catherine Chantilly and Jeremy Narby for discussion and comments.

J.-Y.Beziau, "Possibility, Contingency and the Hexagon of Modalities", *South American Journal of Logic*, **3** (2017).

J.-Y.Beziau, "An analogical hexagon", *International Journal of Approximate Reasoning*, **94**, (2018), pp.1–17

J.-Y.Beziau, "The Pyramid of Meaning", in J.Ceuppens et al. (eds), *A coat of many colours - Dany Jaspers Festschrift*, Brussels, 2018.

J.-Y.Beziau (ed), *La pointure du symbole*, Petra, Paris, 2014.

J.-Y.Beziau (ed), *The arbitrariness of the sign in question*, College Publication, London, 2018.

D.Bohm, *Causality and chance in modern physics*, Harper, New York, 1961.

M.Bréal, *Essai de sémantique, science des significations*, Paris, Hachette, 1897.

P.Cohen, *Undergangens arkitektur (Architecture of doom)*, movie, Stokholm, 1989.

G. Frege, "Über die wissenschaftliche Berechtigung einer Begriffsschrif", *Zeitschrift für Philosophie und philosophische Kritik*, **81** (1882), pp.48–56.

M.Günther, *The Zurich Axioms: The rules of risk and reward used by generations of Swiss bankers*, New American Library, New York, 1995.

M.Heidegger, *Der Satz vom Grund (The principle of reason)*, Günther Neske, Pfullingen,1957.

D.Parrochia, *Coïncidences Philosophie et épistémologie du hasard, Le corridor bleu*, Saint Pierre, 2015.

T.Pequeno and J.-Y.Beziau, "Rules of the game", in J.-Y.Beziau and M.E.Coniglio (eds), *Logic without frontiers*, College Publication, London, 2011, pp.131-144.

C.Rosset, *La logique du pire: éléments pour une philosophie tragique*, Presses Universitaires de France, Paris, 1971.

F. de Saussure, *Cours de linguistique générale*, Payot, Lausanne et Paris, 1916.

J.-F-Suscillon, *Pouvoir de Saint François de Sales, ou miracles et guérisons opérés par le saint évéque*, Burdet, Annecy, 1865,

A.Tarski, "The Semantic Conception of Truth and the Foundations of Semantics", *Philosophy and Phenomenological Research*, **4** (1944), 341–376.

F.Trochu, *La maman de Saint François de Sales*, Apostolat des éditions, Paris 1963.

S.Zweig, *Magellan. Der Mann und seine Tat*, Reichner, Vienna, 1938.

Vichy, February 15, 2018

Jean-Yves Beziau
Universidade do Brasil, Rio de Janiero
Ecole Normale Supérieure, Paris

www.ingramcontent.com/pod-product-compliance
Lightning Source LLC
Chambersburg PA
CBHW051032160426
43193CB00010B/913